T0142877

Advances in Intelligent Systems and Computing

Volume 518

Series editor

Janusz Kacprzyk, Polish Academy of Sciences, Warsaw, Poland
e-mail: kacprzyk@ibspan.waw.pl

About this Series

The series "Advances in Intelligent Systems and Computing" contains publications on theory, applications, and design methods of Intelligent Systems and Intelligent Computing. Virtually all disciplines such as engineering, natural sciences, computer and information science, ICT, economics, business, e-commerce, environment, healthcare, life science are covered. The list of topics spans all the areas of modern intelligent systems and computing.

The publications within "Advances in Intelligent Systems and Computing" are primarily textbooks and proceedings of important conferences, symposia and congresses. They cover significant recent developments in the field, both of a foundational and applicable character. An important characteristic feature of the series is the short publication time and world-wide distribution. This permits a rapid and broad dissemination of research results.

Advisory Board

Chairman

Nikhil R. Pal, Indian Statistical Institute, Kolkata, India
e-mail: nikhil@isical.ac.in

Members

Rafael Bello Perez, Universidad Central "Marta Abreu" de Las Villas, Santa Clara, Cuba
e-mail: rbellop@uclv.edu.cu

Emilio S. Corchado, University of Salamanca, Salamanca, Spain
e-mail: escorchado@usal.es

Hani Hagras, University of Essex, Colchester, UK
e-mail: hani@essex.ac.uk

László T. Kóczy, Széchenyi István University, Győr, Hungary
e-mail: koczy@sze.hu

Vladik Kreinovich, University of Texas at El Paso, El Paso, USA
e-mail: vladik@utep.edu

Chin-Teng Lin, National Chiao Tung University, Hsinchu, Taiwan
e-mail: ctlin@mail.nctu.edu.tw

Jie Lu, University of Technology, Sydney, Australia
e-mail: Jie.Lu@uts.edu.au

Patricia Melin, Tijuana Institute of Technology, Tijuana, Mexico
e-mail: epmelin@hafsamx.org

Nadia Nedjah, State University of Rio de Janeiro, Rio de Janeiro, Brazil
e-mail: nadia@eng.uerj.br

Ngoc Thanh Nguyen, Wroclaw University of Technology, Wroclaw, Poland
e-mail: Ngoc-Thanh.Nguyen@pwr.edu.pl

Jun Wang, The Chinese University of Hong Kong, Shatin, Hong Kong
e-mail: jwang@mae.cuhk.edu.hk

More information about this series at http://www.springer.com/series/11156

Pankaj Kumar Sa · Manmath Narayan Sahoo
M. Murugappan · Yulei Wu
Banshidhar Majhi
Editors

Progress in Intelligent Computing Techniques: Theory, Practice, and Applications

Proceedings of ICACNI 2016, Volume 1

 Springer

Editors
Pankaj Kumar Sa
Department of Computer Science
  and Engineering
National Institute of Technology
Rourkela, Odisha
India

Manmath Narayan Sahoo
Department of Computer Science
  and Engineering
National Institute of Technology
Rourkela, Odisha
India

M. Murugappan
School of Mechatronic Engineering
Universiti Malaysia Perlis (UniMAP)
Arau, Perlis
Malaysia

Yulei Wu
The University of Exeter
Exeter, Devon
UK

Banshidhar Majhi
Department of Computer Science
  and Engineering
National Institute of Technology
Rourkela, Odisha
India

ISSN 2194-5357   ISSN 2194-5365   (electronic)
Advances in Intelligent Systems and Computing
ISBN 978-981-10-3372-8   ISBN 978-981-10-3373-5   (eBook)
DOI 10.1007/978-981-10-3373-5

Library of Congress Control Number: 2017933561

Printed on acid-free paper

This Springer imprint is published by Springer Nature
The registered company is Springer Nature Singapore Pte Ltd.
The registered company address is: 152 Beach Road, #21-01/04 Gateway East, Singapore 189721, Singapore

Foreword

Message from the Honorary General Chair Prof. Mike Hinchey

Welcome to the 4th International Conference on Advanced Computing, Networking and Informatics. The conference is hosted at the Centre for Computer Vision and Pattern Recognition, NIT Rourkela, Odisha, India. For this fourth event, held September 22–24, 2016, the theme is Computer Vision and Pattern Recognition.

Following the great success of the last three years, we are very glad to recognize the co-organization of the Center for Computer Vision and Pattern Recognition, at National Institute of Technology Rourkela, India; the Faculty of Engineering and Technology, Liverpool John Moores University, UK; the College of Engineering, Mathematics and Physical Sciences, University of Exeter, UK; and the Faculty of Science, Liverpool Hope University, UI.

Having selected 114 articles from more than 500 submissions, we are glad to have the proceedings of the conference published in the *Advances in Intelligent Systems and Computing* series of Springer.

I am very pleased to have published the special issues of papers from ICACNI in *Innovations in Systems and Software Engineering*: *A NASA Journal*, published by Springer and of which I am editor-in-chief, in each of the preceding years, and all of which were truly excellent and well received by our subscribers.

The accepted papers this year will be considered again for this journal and for several other special issues.

I would like to acknowledge the special contribution of Prof. Sunil Kumar Sarangi, Former Director of NIT Rourkela, as the chief patron for this conference.

The conference is technically co-sponsored by the following professional organizations/laboratories:

1. Joint Control Systems Society and Instrumentation and Measurement Society Chapter, IEEE Kolkata Section
2. IEEE Communications Society Calcutta Chapter
3. Aerospace Electronics and Systems Division, CSIR National Aerospace Laboratories, Govt. of India

4. Dependable Computing and Networking Laboratory, Iowa State University, USA
5. Multimedia Content Security Innovative Research Group, Henan University of Science and Technology, China
6. Poznan University of Technology Vision Laboratory, Poland

We are grateful to all of them for their co-sponsorship and support. The diversity of countries involved indicates the broad support that ICACNI 2016 has received. A number of important awards will be distributed at this year's event, including Best Paper Awards from ACM Kolkata Chapter, a Best Student Paper Award from IEEE ComSoc Koltaka Chapter, a Student Travel Award from INNS, and a Distinguished Women Researcher Award.

I would like to thank all of the authors, contributors, reviewers, and PC members for their hard work. I would especially like to thank our esteemed keynote speakers and tutorial presenters. They are all highly accomplished researchers and practitioners, and we are very grateful for their time and participation.

But the success of this event is truly down to the local organizers, local supporters, and various chairs who have done so much work to make this a great event. We hope you will gain much from ICACNI 2016 and will plan to submit to and participate in ICACNI 2017.

Best wishes,

Professor Mike Hinchey

ICACNI 2016 Honorary General Chair
President, International Federation for Information Processing (www.ifip.org)
Director, Lero-the Irish Software Engineering Research Centre (www.lero.ie)
Vice Chair and Chair Elect, IEEE UK and Ireland Section
mike.hinchey@lero.ie

Preface

It is indeed a pleasure to receive overwhelming response from academicians and researchers of premier institutes and organizations of the country and abroad for participating in the 4th International Conference on Advanced Computing, Networking, and Informatics (ICACNI 2016), which makes our endeavor successful. The conference organized by Centre for Computer Vision and Pattern Recognition, National Institute of Technology Rourkela, India, during September 22–24, 2016, certainly marks a success toward bringing researchers, academicians, and practitioners in the same platform. We have received more than 550 articles and very stringently have selected through peer review 114 best articles for presentation and publication. We could not accommodate many promising works as we tried to ensure the highest quality. We are thankful to have the advice of dedicated academicians and experts from industry and the eminent academicians involved in providing technical co-sponsorship to organize the conference in good shape. We thank all people participating and submitting their works and having continued interest in our conference for the fourth year. The articles presented in the two volumes of the proceedings discuss the cutting-edge technologies and recent advances in the domain of the conference. The extended versions of selected works would be re-reviewed for publication in reputed journals.

We conclude with our heartiest thanks to everyone associated with the conference and seeking their support to organize the 5th ICACNI 2017 at National Institute of Technology Goa during June 01–03, 2017.

Rourkela, India   Pankaj Kumar Sa
Rourkela, India   Manmath Narayan Sahoo
Arau, Malaysia   M. Murugappan
Exeter, UK   Yulei Wu
Rourkela, India   Banshidhar Majhi

Organizing Committee

Advisory Board Members

Arun Somani, FIEEE, Iowa State University, USA
Costas StasoPoulos, IEEE Region 8 Director 2015–2016, Electricity Autjority of Cyprus, Cyprus
Dacheng Tao, FIEEE, FIAPR, FIET, University of Technology, Sydney, Australia
Friedhelm Schwenker, University of Ulm, Germany
Ishwar K. Sethi, FIEEE, Oakland University, USA
Kenji Suzuki, SMIEEE, Illinois Institute of Technology, USA
Mohammad S. Obaidat, FIEEE, Fordham University, USA
Nikhil R. Pal, FIEEE, Vice President for Publications IEEE Computational Intelligence Society (2015–16), Indian Statistical Institute Kolkata, India
Rajkumar Buyya, FIEEE, The University of Melbourne, Australia
San Murugesan, SMIEEE, Western Sydney University Sydney, Australia
Sanjeevikumar Padmanaban, SMIEEE, Ohm Technologies, India
Subhas Mukhopadhyay, FIET, FIEEE, Massey Unniversity, New Zealand
Subhash Saini, The National Aeronautics and Space Administration (NASA), USA
Vincenzo Piuri, FIEEE, Vice President for Technical Activities IEEE 2015, University of Milano, Italy

Chief Patron

Sunil Ku. Sarangi, FNAE
Former Director
National Institute of Technology Rourkela, India

Patron

Banshidhar Majhi
Dean (Academic)
Professor, Department of Computer Science and Engineering
National Institute of Technology Rourkela, India

Honorary General Chair

Mike Hinchey, FIET, SMIEEE
President, International Federation for Information Processing
Director, Lero-the Irish Software Engineering Research Centre
Vice Chair and Chair Elect, IEEE UK and Ireland Section
Former Director and Expert, Software Engineering Laboratory,
NASA Goddard Space Flight Centre
Professor, University of Limerick, Ireland

General Chairs

Durga Prasad Mohapatra, National Institute of Technology, Rourkela, India
Manmath Narayan Sahoo, National Institute of Technology, Rourkela, India

Organizing Co-chairs

Pankaj Kumar Sa, National Institute of Technology, Rourkela, India
Sambit Bakshi, National Institute of Technology, Rourkela, India

Programme Co-chairs

Atulya K. Nagar, Liverpool Hope University, UK
Dhiya Al-Jumeily, Liverpool John Moores University, UK
Yulei Wu, University of Exeter, UK

Technical Programme Committee

Abir Hussain, Liverpool John Moores University, UK
Adam Schmidt, Poznan University of Technology, Poland
Akbar Sheikh Akbari, Leeds Beckett University, UK
Al-Sakib Khan Pathan, SMIEEE, UAP and SEU, Bangladesh/ Islamic University in Madinah, KSA
Andrey V. Savchenko, National Research University Higher School of Economics, Russia
Annappa B., SMIEEE, National Institute of Technology Karnataka, Surathkal, India
Asutosh Kar, Aalborg University, Denmark
Biju Issac, SMIEEE, FHEA, Teesside University, UK
C.M. Ananda, National Aerospace Laboratories, India
Ediz Saykol, Beykent University, Turkey
Enrico Grisan, University of Padova, Italy
Erich Neuhold, FIEEE, University of Vienna, Austria
Igor Grebennik, Kharkiv National University of Radio Electronics, Ukraine
Iti Saha Misra, Jadavpur University, India
Jerzy Pejas, Technical University of Szczecin, Poland
Laszlo T. Koczy, Szechenyi Istvan University, Hungary
Palaniappan Ramaswamy, SMIEEE, University of Kent, UK
Patrick Siarry, SMIEEE, Université de Paris, France
Prasanta K. Jana, SMIEEE, Indian School of Mines, Dhanbad, India
Robert Bestak, Czech Technical University, Czech Republic
Shyamosree Pal, National Institute of Technology Silchar, India
Sohail S. Chaudhry, Villanova University, USA
Symeon Papadopoulos, Centre for Research and Technology Hellas, Greece
Valentina E. Balas, SMIEEE, Aurel Vlaicu University of Arad, Romania
Xiaolong Wu, California State University, USA
Yogesh H. Dandawate, SMIEEE, Vishwakarma Institute of Information Technology, India
Zhiyong Zhang, SMIEEE, SMACM, Henan University of Science and Technology, China

Organizing Committee

Banshidhar Majhi, National Institute of Technology, Rourkela, India
Bidyut Kumar Patra, National Institute of Technology, Rourkela, India
Dipti Patra, National Institute of Technology, Rourkela, India
Gopal Krishna Panda, National Institute of Technology, Rourkela, India
Lakshi Prosad Roy, National Institute of Technology, Rourkela, India

Manish Okade, National Institute of Technology, Rourkela, India
Pankaj Kumar Sa, National Institute of Technology, Rourkela, India
Ramesh Kumar Mohapatra, National Institute of Technology, Rourkela, India
Ratnakar Dash, National Institute of Technology, Rourkela, India
Sambit Bakshi, National Institute of Technology, Rourkela, India
Samit Ari, National Institute of Technology, Rourkela, India
Sukadev Meher, National Institute of Technology, Rourkela, India
Supratim Gupta, National Institute of Technology, Rourkela, India
Umesh Chandra Pati, National Institute of Technology, Rourkela, India

Contents

About the Editors

Dr. Pankaj Kumar Sa received the Ph.D. degree in Computer Science in 2010. He is currently working as assistant professor in the Department of Computer Science and Engineering, National Institute of Technology Rourkela, India. His research interests include computer vision, biometrics, visual surveillance, and robotic perception. He has co-authored a number of research articles in various journals, conferences, and books. He has co-investigated some research and development projects that are funded by SERB, DRDOPXE, DeitY, and ISRO. He has received several prestigious awards and honors for his excellence in academics and research. Apart from research and teaching, he conceptualizes and engineers the process of institutional automation.

Dr. Manmath Narayan Sahoo is an assistant professor in Computer Science and Engineering Department at National Institute of Technology Rourkela, Rourkela, India. His research interest areas are fault tolerant systems, operating systems, distributed computing, and networking. He is the member of IEEE, Computer Society of India and The Institutions of Engineers, India. He has published several papers in national and international journals.

Dr. M. Murugappan is a senior lecturer at School of Mechahtronics Engineering at Universiti Malaysia Perlis (UniMAP), Perlis, Malaysia. He received his Ph.D. degree in Mechatronic Engineering from Universiti Malaysia Perlis (UniMAP), Malaysia in 2010, Master of Engineering degree in Applied Electronics from Government College of Technology, Anna University, Tamilnadu, India, in 2006, and Bachelor of Electrical and Electronics Engineering from Adiparasakthi Engineering College, Melmaruvathur, Tamilnadu in 2002. His research interest areas are signal processing (EEG, ECG, HRV, ECG), affective computing (emotion, stress, emotional stress), pattern recognition, brain computer interface (BCI), human–machine interaction (HMI), digital image processing, statistical analysis, neuromarketing and neurobehavioral analysis. He has published over 45 research papers in refereed journals and over 50 papers in national and international conferences.

Dr. Yulei Wu is a lecturer in Computer Science at the University of Exeter. He received his Ph.D. degree in Computing and Mathematics and B.Sc. degree in Computer Science from the University of Bradford, UK, in 2010 and 2006, respectively. His recent research focuses on future network architecture and protocols, wireless networks and mobile computing, cloud computing, and performance modelling and analysis. He has published over 30 research papers on these areas in prestigious international journals, including IEEE Transactions on Parallel and Distributed Systems, IEEE Transactions on Communications, IEEE Transactions on Wireless Communications, IEEE Transactions on Vehicular Technology, and ACM Transactions on Embedded Computing Systems, and reputable international conferences. He was the recipient of the Best Paper Awards from IEEE CSE 2009 and ICAC 2008 conferences. He has served as the guest editor for many international journals, including Elsevier Computer Networks and ACM/Springer Mobile Networks and Applications (MONET). He has been the chair and vice chair of 20 international conferences/workshops and has served as the PC member of more than 60 professional conferences/workshops. He was awarded the Outstanding Leadership Awards from IEEE ISPA 2013 and TrustCom 2012, and the Outstanding Service Awards from IEEE HPCC 2012, CIT 2010, and ScalCom 2010. His research interest areas are Future Internet Architecture: software-defined networking, network functions virtualization, clean-slate post-IP network technologies (e.g., information-centric networking), cloud computing technologies, mobility; Wireless Networks and Mobile Computing; Cloud Computing; Performance Modelling and Analysis.

Dr. Banshidhar Majhi is a professor in Computer Science and Engineering Department at National Institute of Technology Rourkela, Rourkela, India. Dr. Majhi has 24 years of teaching and 3 years of industry experience. He has supervised eight Ph.D. students, 40 M.Tech, and 70 B.Tech students, and several others are pursuing their courses under his guidance. He has over 50 publications in journals and 70 publications in conference proceedings of national and international repute. He was awarded Gold Medal for Best Engineering Paper from IETE in 2001, and from Orissa Engineering Congress in 2000. He visited Department of Computer Engineering, King Khalid University, Abha, Kingdom of Saudi Arabia, as a professor from October 2010 to February 2011 and Department of Computer Engineering and Information Technology, Al-Hussein Bin Talal University, Ma'an, Jordan, as assistant professor from October 2004 to June 2005. His research interests are image processing, data compression, security protocols, parallel computing, soft computing, and biometrics.

Part I
Invited Papers

How Meta-heuristic Algorithms Contribute to Deep Learning in the Hype of Big Data Analytics

Simon Fong, Suash Deb and Xin-she Yang

Abstract Deep learning (DL) is one of the most emerging types of contemporary machine learning techniques that mimic the cognitive patterns of animal visual cortex to learn the new abstract features automatically by deep and hierarchical layers. DL is believed to be a suitable tool so far for extracting insights from very huge volume of so-called big data. Nevertheless, one of the three "V" or big data is velocity that implies the learning has to be incremental as data are accumulating up rapidly. DL must be fast and accurate. By the technical design of DL, it is extended from feed-forward artificial neural network with many multi-hidden layers of neurons called deep neural network (DNN). In the training process of DNN, it has certain inefficiency due to very long training time required. Obtaining the most accurate DNN within a reasonable run-time is a challenge, given there are potentially many parameters in the DNN model configuration and high dimensionality of the feature space in the training dataset. Meta-heuristic has a history of optimizing machine learning models successfully. How well meta-heuristic could be used to optimize DL in the context of big data analytics is a thematic topic which we pondered on in this paper. As a position paper, we review the recent advances of applying meta-heuristics on DL, discuss about their pros and cons and point out some feasible research directions for bridging the gaps between meta-heuristics and DL.

Keywords Deep learning · Meta-heuristic algorithm · Neural network training · Nature-inspired computing algorithms · Algorithm design

S. Fong (✉)
Department of Computer Information Science, University of Macau, Macau SAR, China
e-mail: ccfong@umac.mo

S. Deb
INNS-India Regional Chapter, Ashadeep, 7th Floor Jorar, Namkum, Ranchi 834010, Jharkhand, India
e-mail: suashdeb@gmail.com

X. Yang
School of Science and Technology, Middlesex University, London NW4 4BT, UK
e-mail: x.yang@mdx.ac.uk

© Springer Nature Singapore Pte Ltd. 2018
P.K. Sa et al. (eds.), *Progress in Intelligent Computing Techniques: Theory, Practice, and Applications*, Advances in Intelligent Systems and Computing 518, DOI 10.1007/978-981-10-3373-5_1

1   Introduction

Deep learning (DL) is a new branch of machine learning mainly in the aspects of supervised learning. Given some suitable neural network architecture, logics on neuron weight updates and activation function, deep learning models and extracts high-level abstractions from voluminous data. It is usually done by using a series of interconnected multiple processing layers setup in hierarchical structure. Since its inception in 2006 by Hinton [1], DL now is becoming one of the hottest research areas in the machine learning research community. DL has various versions which centered on collectively concept of a series of algorithms and models including but not limited to convolutional neural networks (CNNs), deep Boltzmann machines (DBMs), deep belief networks (DBNs), deep representation, recursive auto-encoders and restricted boltzmann machines (RBMs), just to name a few. While their potential capabilities are to be exploited, some of the most popular applications are computer vision and image classification, by RBM and DBN.

Deep learning is regarded to be "deep" as the name coined in comparison with the well-known "shallow learning" algorithms such as support vector machine (SVM), boosting and maximum entropy method and other discriminative learning methods. Those shallow learning recognizes data features mostly by artificial sampling or empirical sampling from the data, so the induced model or knowledge network learns the mappings between the features and prediction targets in a nonlayer memory structure. In contrast, deep learning learns the relations between the raw data which are characterized by feature values and the targets, layer by layer, through transforming the data from raw feature space to transformed feature space. Additionally, deep structure can learn and approach nonlinear function. All these advantages are beneficial to classification and feature visualization [2].

With the objective of useful deriving insights from the bottom of big data, deep learning has been formulated and tested in different research areas with respect to various performance indicators such as processing speed, accuracy and capabilities to adapt to continuous data collection environment. Case studies of DL applied on various industrial areas including image classification, pattern recognition and natural language processing (NLP) and more seem to come. For computer visions, there are proven successful examples as demonstrated in the large (CNN)-scale visual recognition challenge (ILSVRC) by ImageNet [3]. Convolutional neural network is the first implemented DL tool in image classification and it showed effectiveness. The error rate drops from 25 to 15% when CNN was used over conventional neural network. After the success, the combination of techniques, namely deep learning for learning and prediction, big data or data warehousing and GPU for parallel processing, is integrated into large-scale image classification applications. Companies of search engine giants such as Baidu and Google have upgraded their image searching capability using DL technology [4] in this big data analytics era.

Although DL outperformed most of the shallow learning methods and it has been tested in industrial applications, its design still carries some shortcomings.

A large CNN typically is configured with millions of parameters and it is mostly trained by contrastive divergence (CD) learning algorithm which is iterative and known to be time-consuming [5]. The most significant problem is that when facing very large-scale data the DNN will take several days or even months to learn, even though the greedy search strategy is in place. Regarding this, many companies who are seriously considering deploying CNN would try to alleviate the speed limitation by investing heavily into hardware capabilities. Using high-power processing equipment such as multiprocessors, large capacity of fast memories and parallel computing environment are common approach. Some researchers alternatively try to use other training algorithms than CD to marginally speed up the training process. Apart from the hardware requirement and learning algorithm, the shortcomings lie in the fundamental structure of CNN where many parameters must be tuned properly. Some examples are training the weights for effective learning, controlling the "attractors" which are related to stabilizing the system dynamics of the neural network states. The history of the inputs may need to be stored in a large set of attractors. All these could be possibly solved by some kind of optimization algorithms. This belongs to hyperparameter search problem; just like any machine learning algorithm, CNN is able to induce a representative model that can capture some useful insights from a large data, given the model parameters are fine-tuned to its optimal state. A tuned machine learning model replies on balancing the learning at appropriate level of model complexity. Overfitting occurs if the model is trained with too much examples, making the model too complex. Consequently it overly fits the data into constructing the model on almost every instance was considered, but it lacks of generalization power to unseen data. On the other extreme, when the complexity of the model is too low, it will not be able to all the essential information in the data. This phenomenon is called underfitting. In the case of CNN, a set of hyperparameters should be determined before training big data commences. The choice of hyperparameters can remarkably affect the final model's performance in action. However, determining appropriate values of parameters for optimal performance is a complex process. Claesen and Moor in 2015 [6] have argued that it is an open challenge inherent to hyperparameter such as optimizing the architecture of neural networks [7], whereas the count of hidden layers of a neural network is one such hyperparameter, and the amount of neurons that associate with each layer gives rise to another set of additional hyperparameters. The search space gets increasingly complex when they depend conditionally upon the number of layers in the case of CNN. Tuning all these parameters is quite difficult in real-time computing environment. Instead of finding the perfect balance, as suggested by most hyperparameter optimization strategies, meta-heuristic could be used [8], allowing the best model in terms of optimization solution emerges by itself by stochastic and heuristic search over certain iterations in lieu of brute force or Monte Carlo that tries through all the alternatives.

Meta-heuristic algorithms are designed to find global or near-optimal solutions within acceptable search time, at reasonable computational cost. In case of CNN, the whole model of neural network could technically be represented by a solution vector which could be optimized to produce the best fitness in terms of prediction

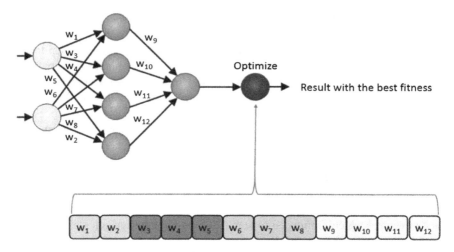

Fig. 1 Encoding the weights of neural network into a solution vector

accuracy. This can be easily done by encoding a vector of weights from the neural network, with the value in each vector cell representing the weight of a linkage between a pair of neuron in the neural network. Figure 1 illustrates this simple encoding concept that represents a neural network to be optimized as a solution vector.

Once the encoding is in place, we can train a neural network using a meta-heuristic search algorithm, for finding a solution vector that represents a combination of weights that gives the highest fitness. In this position paper, the relevant works of applying meta-heuristic algorithms for artificial neural network's training are reviewed. In addition, we survey the possibility of implementing meta-heuristic algorithms on restricted Boltzmann machine's (RBM) parameter training process.

The reminder of this paper is organized as follows: meta-heuristic algorithms and their fundamental design constructs are introduced in Sect. 2. Section 3 describes meta-heuristic algorithms are implemented on optimizing neural network training. Section 4 shows a case of deep learning and restricted Boltzmann machine which could be empowered by meta-heuristics. Discussion about the prospects of apply meta-heuristics on DL and conclusion is given in Sect. 5.

2 Meta-heuristic Algorithm

Meta-heuristic is another emerging trend of research, mostly found its application in optimization including combinatorial optimization, constraint-based optimization, fixed-integer, continuous numeric and mixed-type search space optimization [9]. Meta stands for some high-level control logics that controls some rules underneath or embraced in an iterative operation which try to improve the current solution generation after generation. It is a collective concept of a series of algorithms including evolutionary algorithm; the most famous one is genetic algorithm (GA) [10]. Recently a branch of population-based meta-heuristics has gained popularity and showed effectiveness in stochastic optimization. Optimal or near-optimal answers are almost always guaranteed, by going through the search repeatedly through certain number of iteration. The movements of the search agents are mostly nature-inspired or biological algorithm, foraging the food hunting patterns and/or social behavior of insects/animals toward global best situations which sets as the objective of the search operation [11]. These search-oriented algorithms found success recently in many optimization applications, from scientific engineering to enhancing data mining methods. Lately, with the hype of big data analytics, and the rise of neural network in the form of CNN being shown useful in DL within big data, meta-heuristics may again show its edge in probably complementing the shortcomings of CNN, improving its ultimate performance such as fitting a hand into glove.

Some of the most prevalent population-based meta-heuristics is particle swarm optimization (PSO) [12], trajectory algorithm, such as tabu search [13], and so on. In this review paper, we focus mainly on the underlying logics of GA and PSO, partly because these two are the most popular meta-heuristics that have demonstrated their merits. Another reason is that GA and PSO represent two very fundamental concepts in terms of the movement logics: how they converge and how the solutions emerge through trying out heuristically alternative solutions in the search process. GA, on the one hand, represents a tightly coupled evolutionary mechanism, having a population of chromosomes that represent the solutions, being mutate, crossover and the fittest ones pass onto the future generations till only the few fittest solutions stay.

PSO, on the other hand, works in similar approach but with an explicit concept of separating global velocity and local velocity among the swarm of moving particles (search agents). While the global velocity is controlling the search momentum toward the best possible solution (to be found), the local velocity associated with each particle enables some randomness and alternatives in the hope of finding better solutions just than the current ones. In a way, PSO unifies two subtle objectives in the logic design, namely local intensification (LI) and global exploration (GE) [14]. These two very underlying forces which are usually embedded in the search mechanism empower the overall search, often yielding good results. Since PSO was launched, and the two subtle searching forces that complement each other toward the final goal were discovered, a number of variants of meta-heuristics were created in the meta-heuristic research community. Many of the new variants are either extensions of PSO embracing the two concepts of LI and GE or hybrids of the existing meta-heuristic algorithms, interchanging some implementations of LI and GE.

The latter type of new variants which are hybrid is founded on the shortcomings of the original prototype where the algorithm is inspired by a certain nature phenomenon. The original prototype normal would faithfully follow the salient features

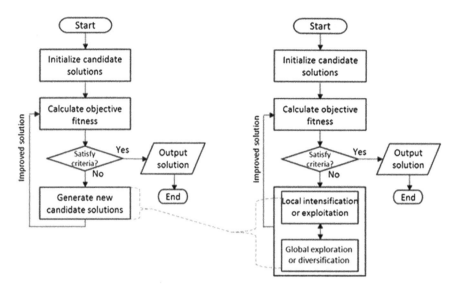

Fig. 2 LI and GE in a general meta-heuristic algorithmic logic

of an animal or natural manifestation, thereby limiting its algorithmic efficacy for mimicking the animal as closely as possible. As a result, stand-alone and original prototype may work as efficiently as it is wished to be, leaving some room for improvement by modifications.

Under this observation that mods are better than the original (they have to be better in order to get the papers published), researchers have been trying to combine multiple meta-heuristics in the hope of yielding some better results since it is already known that no meta-heuristic alone is able to offer the best. In the process of thinking of a new hybrid, the original meta-heuristic algorithm is dissected into different parts, checking of its unique function(s) and how they were built suitable for some particular types of complex problem. Blum and Roli [15] explained that the power of the meta-heuristics search is somehow due to the dual efforts and cooperation between the local exploitation and global exploration strategies. Figure 2 shows the dual steps are integral part of the original meta-heuristic algorithm, in general sense.

By understanding these two underlying forces in the meta-heuristic design, it helps in finding suitable optimization constructs for improving any machine learning algorithms, including CNN of course. GI is designed to continuously find a global optimum from some afar positions of the search space. Therefore, in some meta-heuristics the GI components enable the search agents to widespread the search agents from their current positions through some random mechanism. So it is in the hope that by venturing far away, the swarm is able to escape from being stuck at local optimum as well as finding a better terrain undiscovered previously. On the other hand, LI is designed to guide the search agents to scout intensively at the current proximity for refining the locally best solution they have found so far.

Referring to the other trend of meta-heuristics than hybrids, variants of PSO in the names of some animals have subsequently been arisen. They can be considered as enhanced versions of PSO, with unique features of keeping LI and GE explicitly defined. Typical examples are wolf search algorithm (WSA) [16] and elephant search algorithm (ESA) [17] that have been recently proposed, and they are considered as "semi-swarm" meta-heuristics. The search agents by the semi-swarm design have certain autonomous capacity focusing in LI, yet they cooperatively follow the guidelines of GE to outreach for global exploration. Their pseudo-codes are shown in Figs. 3, 4 and 5, respectively. In particular, the LI and GE parts are highlighted in the pseudo-codes showing their similarities, yet loosely coupled movements by the search agents enforcing these two crucial search forces.

In Fig. 3, it can be seen that PSO the global velocity and the individual particles' velocities are tightly coupled in the rule. By the design of WSA as shown in Fig. 4, WSA relaxes this coupling by allowing every wolf search agent to roam over the search space individually by taking their local paths in various dimensions. They sometimes merge when they are bound within certain visual ranges. At a random chance, the wolf search agents jump out of their proximity for the sake of GE. At the end, the search agents would unite into their nuclear family packs which eventually would have migrated to the highest achievable solution (that has the maximum fitness found so far by both GE and LI). In Fig. 5, ESA is comprised of search agents of two genders. Each gender group of elephant search agents will explicitly do GE and LI. The two search forces are separately enforced by the two

---

Input: P: particle population size, D: dimensions, *epoch*: max #iterations, w,C_1,C_2: step sizes for evolution, local and global vel.; Output: b^{global}

  Initialize(); // initialize all particles and parameters
  While ($t<epoch$ && b^{global} not satisfactory) do
    For each particle p in P do
      // update the local best position for each particle   Local effort
      If *fitness*$(x_p) \geq$ *fitness*(b_p^{local}) then
        $b_p^{local} = x_p$
      End-if
      // update the global best position from all particles   Global effort
      If *fitness*$(b_p^{local}) \geq$ *fitness*(b^{global}) then
        $b^{global} = b_p^{local}$
      End-if
    End-for
    // update particle's velocity and position
    For each particle p in P do
      For each dimension d in D do
        $r_1=Rand()$; $r_2=Rand()$;
        $v_{p,d}=w \times v_{p,d}+C_1 \times r_1 \times [b_{p,d}^{local}-x_{p,d}]+C_2 \times r_2 \times [b^{global,d}-x_{p,d}]$
        $x_{p,d} = x_{p,d} + v_{p,d}$
      End-for
    End-for
    t++
  End-while

Fig. 3 Simplified version of PSO with local and global search efforts *highlighted*

```
Input: W: wolf population size, D: dimensions, epoch: max. #iterations, r: radius of
the visual range, s: step size by which a wolf moves at a time, α: velocity factor of
wolf, e: escape threshold; Output: b^global
   Initialize(); // initialize all wolves and parameters
   While (t<epoch && b^global not satisfactory) do
      For each wolf w in W do
            w=Prey_food_proactively(w_{t-1}); // local search              Local effort
            Up_new_location(w);
            If fitness(w) ≥fitness(w:∀peers) // check if w should merge
               b_w^local =w←w_best-peer
            Else-if
               w=Prey_food_passively(w_{t-1}); // random walk
            End-if
            rs=Rand();
            If rs>e then // trigger a possible global search               Global effort
               w=w+Escape(w_{t-1});
            End-if
            // update the global best position from all wolves
            If fitness(b_w^local) ≥ fitness(b^global) then
               b^global = b_w^local
            End-if
      End-for
      For each wolf w in W do // update wolves' positions for next swarm
         For each dimension d in D do
               If Prey_food_proactively(w_{t-1}) is activated then
               w(t,d) ← w(t,d)+ α ×r×Rand()
               Else if Prey_food_passively(w_{t-1}) is activated then
               w(t,d) ← w(t,d)+ α ×Rand()
               Else if Escape(w_{t-1}) is activated then
               w(t,d) ← w(t,d)+ α ×s×Escape()
         End-for
      End-for
      t++
   End-while
```

Fig. 4 Simplified version of WSA with local and global search efforts *highlighted*

genders in ESA. The leader of the female herd which has the local best fitness guides her peers in doing local search intensively around the proximity. The male elephants venture far away to search for terrains that may yield better fitness. When positions of higher fitness are found by the male elephants, the female elephant herds will migrate over there. Unlike PSO and WSA, the two underlying forces are being fulfilled diligently by two genders of elephant groups, totally separately and autonomously. In addition to separation of GE and LI, the unique design of ESA is all elephants follow a limited life span. Aged elephants will expire, be relinquished from the search process, and new elephants will be born in locations which are inferred from a mix of best positions from their parents, female group leaders and some randomness. This extra force, evolution, is shown in Fig. 4. To certain perspective, ESA carries the goodness of semi-swarm meta-heuristics with respect to GE and LI, and combine this virtue into some evolutionary mechanism, ensuring the future generations progress into better solutions than the existing ones. Such

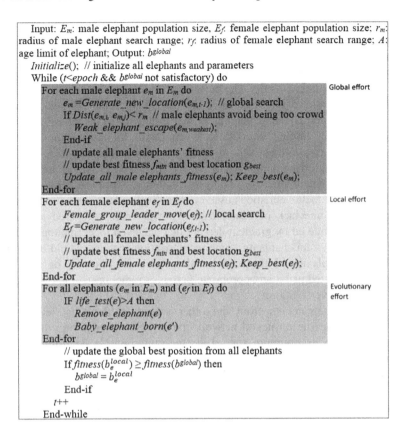

Input: E_m: male elephant population size, E_f: female elephant population size; r_m: radius of male elephant search range; r_f: radius of female elephant search range; A: age limit of elephant; Output: b^{global}

Initialize(); // initialize all elephants and parameters

While (t<*epoch* && b^{global} not satisfactory) do

    For each male elephant e_m in E_m do   Global effort

        e_m=*Generate_new_location*($e_{m,t-1}$); // global search

        If *Dist*($e_{m,i}$, $e_{m,j}$)< r_m // male elephants avoid being too crowd

            Weak_elephant_escape($e_{m,weakest}$);

        End-if

        // update all male elephants' fitness

        // update best fitness f_{min} and best location g_{best}

        Update_all_male_elephants_fitness(e_m); *Keep_best*(e_m);

    End-for

    For each female elephant e_f in E_f do   Local effort

        Female_group_leader_move(e_f); // local search

        E_f=*Generate_new_location*($e_{f,t-1}$);

        // update all female elephants' fitness

        // update best fitness f_{min} and best location g_{best}

        Update_all_female_elephants_fitness(e_f); *Keep_best*(e_f);

    End-for

    For all elephants (e_m in E_m) and (e_f in E_f) do   Evolutionary effort

        IF *life_test*(e)>A then

            Remove_elephant(e)

            Baby_elephant_born(e')

    End-for

        // update the global best position from all elephants

        If *fitness*(b_e^{local}) ≥ *fitness*(b^{global}) then

            $b^{global} = b_e^{local}$

        End-if

        t++

    End-while

Fig. 5 Simplified version of ESA with local and global search efforts *highlighted*

designs will hopefully shed some light into applying meta-heuristic into optimizing DL tools such as CNN.

Depending on the complexity of the hyperparameter challenges in CNN optimization and the presences of multimodal (local optima) meta-heuristics ranging from fully swarm, to semi-swarm and loosely coupled with evolutionary abilities are available.

3 Applying Meta-heuristic Algorithm on Neural Network Training

There have been some debates among researchers in the computer science research community, on the choice of optimization methods for optimizing shallow and deep neural networks such as CNN. Some argued that meta-heuristic algorithms should be used in lieu of classical optimization methods such as gradient descent, Nesterov and Newton–Raphson because meta-heuristics were designed to avoid falling stuck at local minima.

Researchers who are skeptical about the use of meta-heuristics usually share the concern that local minima are not of a serious problem that needs to be heavily optimized at the neural networks. The presences of local minima which are believed to come in mild intensity and quantity are caused by some permutation of the neurons at the hidden layers, depending on the symmetry of the neural network. It was supposed that finding a good local minimum by minimizing the errors straightforwardly is good enough. It is an overkill using extensive efforts in searching for the global minima to the very end. Moreover, some are wary that overly optimizing a neural network limits its flexibility, hence leading to overfitting the training data if meta-heuristic is used or excessively used. The overfitted neural network may become lack of generalization power when compared to a neural network that was trained by gradient descent that achieved local minima which are good enough. When this happens, some regularization function would be required to keep the complexity of the model under check. To the end of this, some researchers suggested using an appropriate kernel or radial basis function provides simple and effective solution.

Nevertheless, researchers from the other school of thoughts believed that applying meta-heuristic on neural network training has its edge on providing the weight training to its optimum state. Like the epitome of a doctrine, the name meta-heuristic consists of the terms "meta" and "heuristic" which are Greek where "meta" is "higher level" or "beyond" and heuristics implies "to find," "to know," "to guide an investigation" or "to discover." Heuristics are simple logics to find best (or near-best) optimal solutions which are on par with the absolute best (which is extremely difficult to find) at a reasonable computational cost. In a nutshell, meta-heuristics are a collection of simple but intelligent strategies which could fit into a wide range of application scenarios for enhancing the efficiency of some heuristic procedures [18]. Optimizing the configuration of neural network is one of such heuristic procedures.

By tapping the searching ability of global optimum by meta-heuristic algorithms, researchers aim to train a neural network to execute faster than traditional gradient descent algorithm. In the following part, I reviewed four researchers' work on implementing meta-heuristic on neural network training including GA on NN, PSO on NN and hybrid GA and PSO on NN.

Though in the above-mentioned section, the basic constructs of most of the meta-heuristics algorithms are GE and LI (in addition to initialization, stopping criteria checking and other supporting functions), meta-heuristic operate by implementing different forms of agents such as chromosome (GA), particles (PSO), fireflies (firefly algorithm). These agents collectively keep moving close to the global optimum or near global optimum through iterative search. Many strategies such as evolutionary strategy, social behavior and information exchange are implemented; therefore, many versions were made possible. Readers who want to probe into details of the variety of meta-heuristics are referred to a latest review [19].

Artificial neural network (ANN) is traditionally constructed in layout of multi-layered feed-forward neural network; some used back-propagation (BP) as error feedback to modify the weights for training up the cognitive power of the network. The weight training strategy has been traditionally gradient descent (GD). However, in the literature, there have been numerous cases of applying meta-heuristic on optimizing such traditional neural network for speeding up the training process. This is primarily achieved by replacing the GD strategy with iterative evolutionary strategy or swarm intelligence strategy by meta-heuristics [20–24].

Gudise and Venayagamoorthy [20] compared the performance of feed-forward neural network optimized by PSO and feed-forward neural network with BP; the experiment result shows that feed-forward network with PSO is better than that with BP in terms of nonlinear function.

Leung F.H. et al. [24] showed their work on the efficacy of tuning up the structure and parameters of a neural network using an improved genetic algorithm (GA). The results indicate that the improved GA performs better than the standard GA when the neural networks are being tested under some benchmarking functions.

Juang C.F. [22] proposed a new evolutionary learning algorithm based on a hybrid of genetic algorithm (GA) and particle swarm optimization (PSO), called HGAPSO. It takes the best of the both types of meta-heuristics: swarming capability and evolutionary capability. Defining the upper half of the GA population as elites and enhancing them by PSO, while the rest of the population are processed by GA, the hybrid method outperforms PSO and GA individually in training a recurrent or fuzzy neural network.

Meissner M. et al. [23] used optimized particle swarm optimization (OPSO) to accelerate the training process of neural network. The main idea of OPSO is to optimize the free parameters of the PSO by generating swarms within a swarm. Applying the OPSO to optimize neural network training, it aims to build a quantitative model. OPSO approach produces a suitable parameter combination which is able to improve the overall optimization performance.

Zhang J.R. Zhang [21] proposed a hybrid algorithm of PSO coupled with BP for neural network training. By leveraging the advantage of PSO's global searching ability as well as BP's deep search capability, the hybrid algorithm showed very good performance with respect to convergent speed and convergent accuracy.

Optimizing shallow learning by traditional neural network approaches has been shown successfully possible using the meta-heuristic methods as mentioned above. In contrast, DL by CNN is a relatively unexplored field. Very few papers have been published except for one by [25] which used simulated annealing (SA) algorithm to optimize the performance of CNN and showed improved results. Nonetheless, there are still good prospects in trying out different meta-heuristics for optimizing CNN for DL, because the fundamental problems and solutions are about the same: you have a number of unknown variables on hand, and meta-heuristics attempt to offer the best possible solution.

Structures of deep learning model is similar to the traditional artificial neural network, except for some modifications are implemented for better learning ability. For instance, the CNN is a traditional ANN modified with pooling procession and the structure of RBM is an undirected graph or a bidirectional neural network. DL model shares similar models with neural network; more importantly, different training algorithm may be called upon instead of gradient descent strategy. This warrants further exploration into this research arena. Several important contributions which have been mentioned about are elaborated as follows.

3.1 Genetic Algorithm on Neural Network

Leung et al. [24] first tried implementing genetic algorithm (GA) on neural network training in 2003. Though Leung et al. may not be the pioneer in apply GA on neural network, an improved version of GA that is made suitable for ANN is put forward. Crossover operations, mutation operations and fitness function of GA are all redefined, custom-made. Firstly, when it comes to encoding the chromosome and perform the crossover operation, four possible offspring candidates will be generated and the one with the largest fitness value will be chosen as offspring. The four possible crossover offspring are generated as regulations listed below:

$$os_c^1 = \left[os_1^1, os_2^1, \ldots os_n^1\right] = \frac{p_1 + p_2}{2}$$
$$os_c^2 = \left[os_1^2, os_2^2, \ldots os_n^2\right] = p_{max}(1 - w) + max(p_1, p_2)w$$
$$os_c^3 = \left[os_1^3, os_2^3, \ldots os_n^3\right] = p_{min}(1 - w) + min(p_1, p_2)w$$
$$os_c^4 = \left[os_1^4, os_2^4, \ldots os_n^4\right] = \frac{(p_{min} + p_{max})(1 - w) + (p_1 + p_2)w}{2}$$

where the p_{max} and p_{min} are calculated, respectively, according to the parameters of the neural network by maximizing and minimizing them such as $p_{max} = \left[para_{max}^1, para_{max}^2, \ldots, para_{max}^n\right], p_{min} = \left[para_{min}^1, para_{min}^2, \ldots, para_{min}^n\right]$. For example, Max([1, −1, 4], [−3, 3, 2]) = [1, 3, 4] and Min([1, −1, 4], [−3, 3, 2]) = [−3, −1, 2]. Secondly, the mutation operations are redefined. The regulations are given below:

$$os = os + \left[b_1 \Delta nos_1, b_2 \Delta nos_2, \ldots, b_n \Delta nos_n\right]$$

os is the chromosome with biggest fitness value in all four possible offspring. b_1 random equals to 0 or 1 and Δnos_i is random number making sure $para_{min}^i \leq os_i + b_i \Delta nos_i \leq para_{max}^i . os'$ is the final generation after crossover operation and mutation operation.

Thirdly, the fitness value is defined. By adding parameters in the neural network mathematical expression, the actual output of GA-optimized neural network y_k equals to:

$$y_k = \sum_{j=1}^{n_h} \delta\left(s_{jk}^2\right) w_{jk} logsig\left[\sum_{i=1}^{n_i} \delta\left(s_{ij}^1\right) w_{ij} x_i - \delta\left(s_j^1 b_j^1\right)\right] - \delta(s_k^2) logsig\left(b_k^2\right)$$

in which $k = 1, 2,..., n_{out}$, s_{ij} denotes link from ith neuron in input layer to jth neuron in hidden layer, s_{jk} denotes link from jth neuron in hidden layer to kth neuron in output layer, w_{jk} denotes weight between each neuron, b_k^1 and b_k^2 denote bias in input layer and hidden layer, respectively, n_{in}, n_h and n_{out} denote the number of neurons of input layer, hidden layer and output layer, respectively. The error of the whole network is defined as mean of all chromosomes:

$$error = \sum_{k=1}^{n_{out}} \frac{\left|y_k - y_k^d\right|}{n_d}$$

in which n_d denotes the number of chromosome used in the experiment, y_k^d denotes the desired output of output neuron k. Given the error of the network, GA is implemented to optimize the network thus minimize the error. The fitness function is defined as

$$fitness = \frac{1}{1 + error}$$

where the smaller the error and the bigger the fitness value. GA is implemented to find the global optimum of the fitness function; thus, the parameter combinations of weight w are the trained weight for the network.

3.2 Particle Swarm Optimization on Neural Network

Gudise and Venayagamoorthy [5] implemented PSO on neural network training in 2003. The fitness value of each particle (member) of the swarm is the value of the error function evaluated at the current position of the particle and position vector of the article corresponds to the weight matrix of the network.

Zhang et al. [21] developed a hybrid algorithm of BP and PSO that could balance training speed and accuracy. The particle swarm optimization algorithm was showed to converge rapidly during the initial stages of a global search, but around global optimum, the search process will become very slow. On the contrary, the gradient descending method can achieve faster convergent speed around global optimum, and at the same time, the convergent accuracy can be higher.

When the iteration process is approaching end and current best solution is near global optimum, if the change in the weight in PSO is big, the result will vibrate severely. Under this condition, Zhang supposed with the increase in iteration time that the weight in PSO should decline with the iteration time's increasing to narrow the search range and thus pay more attention to local search for global best. He

Fig. 6 Change in inertia weight through linear and nonlinear curve

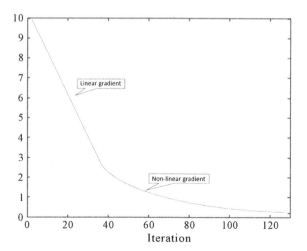

suggests the weight decline linearly first and then decline nonlinearly as shown in Fig. 6.

The concrete working process is summarized as follows: For all the particles p_i, they have a global best location p^{global_best}. If the p^{global_best} keeps unchanged for over 10 generations that may infer the PSO pays too much time on global search, BP is implemented for p^{global_best} to deep search for a better solution.

Similar to GA's implementation in neural network, the fitness function defined is also based on whole network's error and to minimize the error as the optimization of PSO. The learning rate μ of neural network is also controlled in the algorithm, as

$$\mu = k \times e^{-\mu_0 \cdot epoch}$$

where μ is learning rate, k and μ_0 are constants, *epoch* is a variable that represents iterative times, through adjusting k and μ_0, the acceleration and deceleration of learning rate can be controlled.

By implementing the strategy that BP focusing on deep searching and PSO focusing on global searching, the hybrid algorithm has a very good performance.

3.3 Hybrid GA and PSO on Neural Network

Juang [22] proposed, in 2004, hybrid genetic algorithm and particle swarm optimization for recurrent network's training.

The hybrid algorithm called HGAPSO is put forward because the learning performance of GA may be unsatisfactory for complex problems. In addition, for the learning of recurrent network weights, many possible solutions exist. Two individuals with high fitness values are likely to have dissimilar set of weights, and the recombination may result in offspring with poor performance.

Juang put forward a conception of "elite" of the first half to enhance the next generation's performance. In each generation, after the fitness values of all the individuals in the same population are calculated, the top-half best-performing ones are marked. These individuals are regarded as elites.

In every epoch, the worse half of the chromosome is discarded. The better half is chosen for reproduction through PSO's enhancement. All elite chromosomes are regarded as particles in PSO. By performing PSO on the elites, we may avoid the premature convergence in elite GAs and increase the search ability. Half of the population in the next generation are occupied by the enhanced individuals and the others by crossover operation. The working flow of algorithm has mechanism of evolving the offsprings as well as enhanced elites through crossover and mutation.

The crossover operation of HGAPSO is similar to normal GA, random selecting site on chromosome and exchange the sited piece of chromosome to finish the crossover operation. In HGAPSO, uniform mutation is adopted; that is, the mutated gene is drawn randomly, uniformly from the corresponding search interval.

3.4 Simulated Annealing on CNN

Rere et al. [25] claimed to be the first which applied a meta-heuristic called simulate annealing (SA) on optimizing CNN for improved performance. Their results showed that SA is effective on optimizing CNN with better results than the original CNN although it comes with an increase in computation time cost. The SA works by progressively improving the solution; the logics are outlined as follows: Step 1. Encode the initial solution vector at random, compute the objective function; Step 2. Initialize the temperature which is a crucial parameter relating to the convergence speed and hitting onto the global optimum; Step 3. Pick a new solution in the neighborhood of the current position. New solutions are generated depending on T—this is similar to the local intensification; Step 4. Evaluate the goodness of the new solution candidates. Update the global best if a better candidate is found; Step 5. Periodically lower the temperature during the search process so that the chance of receiving deteriorating moves drops, so the search is converging; Step 6. Repeat the above steps until the stopping criterion is satisfied.

In the meta-heuristic search design, SA is trying to minimize the standard error on fitness function of the vector solution and the training set. The fitness function is formulated as follows:

$$fitness = 0.5 \times \sqrt{\frac{\sum_{i=1}^{N}(|y-y'|)^2}{N}}$$

where y' is the expected output, y is the actual output, N is the amount of training samples. SA should converge and hence stop when the minimum neural network complexity has attained the most optimal (lowest) state and the approximate error accuracy indicates a very low value. In the optimization process, the CNN computes the weights and the bias of the neural network, the results are past from the last layers to evaluate the lost function. This is being proceeded as SA tries to find the best solution vector, such as a wrapper process. SA attempts at scouting for a better solution in each iteration by randomly picking a new solution candidate from the proximity of the current solution by adding some random Δx. This optimization is relatively simple, because only the last layer of the CNN is used to instill the solution vector, where the convolution part is almost left untouched. The SA seems doing only local search, and it does not require a large number of search agents, nor doing much of global exploration. The results show that the larger the neighborhood sizes (10, 20 and 50), the better the accuracy is achieved. However, the gain in accuracy was most obvious in small numbers of epoch, from 1 to 9. From epoch 10 onwards, the marginal difference in performance between the original CNN and SA applied on CNN diminish. The time cost, however, becomes more than double at epoch 10 between the CNN and SA + CNN with 50 neighbor size. This is doubtful where SA is a suitable meta-heuristic to be applied on CNN especially if the epoch reaches a large number in magnitudes greater than 10 for some complex big data analytics. This work nevertheless promotes the use of meta-heuristics; researchers would be encouraged to try other meta-heuristics.

4 Deep Learning and Restricted Boltzmann Machine

When dealing with image classification or other problems, traditional method is using pre-processing transforming data as input values for neural network learning; while using deep learning method for classification, raw data (pixel values) are used as input values. This will keep to the maximum extent protecting all information regardless of useful or not from being destroyed by extraction methods. The most advantage lies that all the extraction methods are based on expert knowledge and expert choice thus are not extensible to other problems, while deep learning algorithm can overcome these limitations by using all data with its powerful processing ability. A convolutional neural network (CNN) is shown in Fig. 7.

The Boltzmann machine only has two layers of neurons, namely input and output layers, respectively. The first layer is the input layer and the second layer is the output layer, although the structure is very simple only contains two layers, but it mathematical function in it is not simple. Here, we need to introduce the following probability equation to know the RBM.

From the equation, we know that there are three items in it; they are the energy of the visible layer neural, the energy of the hidden layer neural and the energy consisted of the two layers. From the energy function of the RBM, we can see there

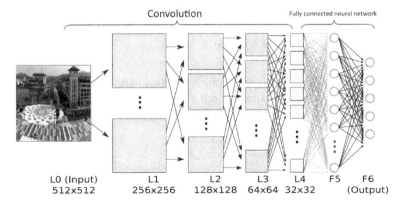

Fig. 7 Typical architecture of a convolutional neural network

are three kinds of parameters lie in the RBM; different from the neural network, each Neuron has a parameter too, but for the neural network, only the connection of two layers has the parameters. Also, the neural network does not carry any energy function. They use the exponential function to express the potential function. There are also the probabilities existing in the RBM. They work exactly in the same way as the two kind of the probabilities such as $p(v|h)$ and $p(h|v)$. It has a similarity with probability graph model in detail, and examples are Bayesian network and Markov network. It looks like a Bayesian network because of the conditional probability. On the other hand, it does not look like a Bayesian because of the two directions probabilities. These probabilities are lying on the two variables that have only one direction.

Compared with the Markov network, the RBM seems to be having a little bit relation with it, because the RBM has the energy function just as the Markov network. But it is not so much alike because the RBM's variable has parameter, and the Markov network does not have any parameter. Furthermore, the Markov network does not have conditional probability because it has no direction but just the interaction. From the graph's perspective, variables in the Markov network use cliques or clusters to represent the relations of close and communicated variables. It uses the production of the potentials of the clique to express the joint probability instead of conditional probability just like the RBM. Its input data are the kind of the Boolean data, within the range between 0 and 1.

The training way of the RBM is to maximize the probability of the visible layer and to generate the distribution of the input data. RBM is a kind of special Markov random function and a special kind of Boltzmann machine. Its graphical model is corresponding to the factor product analysis. Different from the probability graphical model, the RBM's joint distribution directly uses the energy function of both visible layer v and hidden layer h to define instead of potential of it given as

$$E(v, h) = -a^T v - b^T h - v^T W h$$

$$P(v, h) = \frac{1}{Z} e^{-E(v, h)}$$

Later we will know that Z is the partition function defined as the sum of $e^{-E(v,h)}$ over all the possible configurations. In other words, it is just a constant normalizing the sum over all the possible hidden layer configurations.

$$P(v) = \frac{1}{Z} \sum_h e^{-E(v, h)}$$

The hidden unit activations are mutually independent given the activations. That is, for m visible and n hidden units, the conditional probability of a configuration of the visible unit v, given a configuration, is

$$P(v|h) = \prod_{i=1}^{m} P(v_i|h)$$

Conversely, the conditional probability of h given v is $P(h|v)$. Our goal is to infer the weights that maximize the marginal of the visible; in detail, we can step through the following equation to infer and learn the RBM.

$$\arg \frac{Max}{w} E\left[\sum_{v \in V} log P(v) \right]$$

As for the training algorithm, the main idea is also applied gradient descent idea into RBM. Hinton put forward contrastive divergence [26] as a faster learning algorithm. Firstly, the derivative of the log probability of a training vector with respect to a weight is computed as

$$\frac{\partial log P(v)}{\partial w_{ij}} = \langle v_i h_j \rangle_{data} - \langle v_i h_j \rangle_{model}$$

where the angle brackets are used to denote expectations under the distribution specified by the subscript that follows. This leads to a very simple learning rule for performing stochastic steepest ascent in the log probability of the training data:

$$\Delta w_{ij} = \varepsilon \left(\langle v_i h_j \rangle_{data} - \langle v_i h_j \rangle_{model} \right)$$

where ε is a learning rate.

Because there are no direct connections between hidden units in an RBM, it is very easy to get an unbiased sample of $\langle v_i h_j \rangle_{data}$. Given a randomly selected training image, v, the binary state, h_j, of each hidden unit, j, is set to 1 with probability

$$p\left(h_j = 1 | v\right) = \sigma\left(b_j + \sum_i v_i w_{ij}\right)$$

where b_j is the current state of hidden neuron j, $\sigma(x)$ is the logistic sigmoid function $\sigma(x) = \frac{1}{1+exp(-x)}$. $v_j h_j$ is then an unbiased sample. The contrastive divergence (CD) is used to calculate the latter part $\langle v_i h_j \rangle_{model}$. Details can be found in their respective publications.

Considering the complicated computation of implementing CD, the training process of RBM is not easy. Under this condition, implementing meta-heuristic on RBM training to substitute CD is of high possibility.

5 Discussion and Conclusion

Meta-heuristic has successfully implemented in neural network training in the past years. The algorithm used includes GA, PSO, their hybrids as well as many other meta-heuristic algorithms. Moreover, feed-forward back-propagation neural network and spiking neural network [27] are all trained with tests on famous classification problems, where benchmarking their performance is easy. Given the rises of these two emerging trends of meta-heuristics and deep learning which are gaining momentums in both academic research communities and industries, these two trends do cross-road. At this junction, this paper contributes as a timely review on how meta-heuristics contribute to optimizing deep learning tools such as convolutional neural network (CNN).

The basic structure of deep learning network is similar to traditional neural network. CNN is a special neural network with different weight computation regulations, and restricted Boltzmann machine (RBM) is a weighted bio-dimensional neural network or bio-dimensional graph. Their training processes are also mainly executed through iterative formula on error which is similar to traditional neural network's training. Given their iterative natures in execution and the fact that a neural network (at least a part of it) could be coded as a solution vector, it is natural to tap onto the power of meta-heuristics to optimize their layouts, configurations, weights computations and so forth.

In the near future, it is anticipated that it is of high possibility of applying meta-heuristic in deep learning to speed up training without declining performance. However, relevant publications along this direction are still rare. Nevertheless, the interests which are expressed in terms of Google Trends show some takeoff. Fig. 8 shows that the four trends are indeed converging to about the same level of popularity in 2016 and beyond. The four trends are by the keywords of deep learning, machine learning, mathematical optimization and big data. Interestedly, optimization was very much hyped up in the early millennium, and now, it is approaching a steady state (Fig. 9).

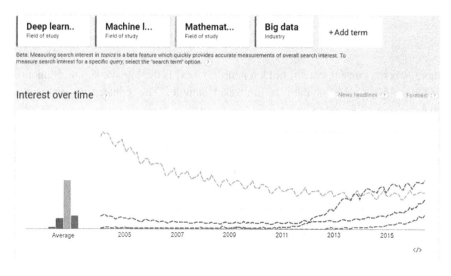

Fig. 8 Google Trends screenshot that shows the popularities of the four fields of study from 2004 till present

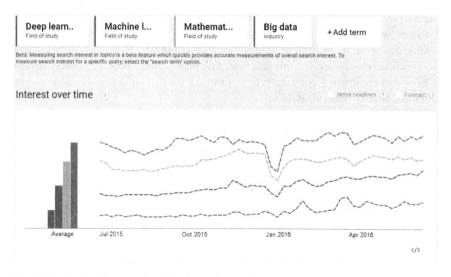

Fig. 9 Google Trends screenshot that shows the popularities of the four fields of study from the past 12 months

The other three trends, especially big data rose up from nowhere since 2011, exceeding the rest by now. The trends of machine learning and deep learning share the same curve patterns. These trends imply sufficient interest levels for these techniques to fuse and cross-apply over one another. In Fig. 7, which is the Google Trends result of these four fields of study, the timeline is of past 12 months. It is apparent to see that they do share similar patterns and fluctuation over time, though they are setting at different interest levels.

Furthermore, there still exists a question that goes under today's computation ability, especially the GPU parallel computing whose computation ability is many times stronger than CPU that has been widely used in industrial area [28]. This would further propel the use of meta-heuristic in solving DL problems. The original design of meta-heuristics especially those population-based search algorithms is founded on parallel processing. The search agents are supposed to operate in parallel, thereby maximizing their true power in GPU environment. When the time cost overhead is removed from the design consideration in GPU environment, searching for near-optimal results by meta-heuristic is still attractive as they can offer an optimal configuration of CNN. It is also anticipated that the parameters and structures of CNN will only become increasingly complex when they are applied to solve very complex real-life problems in the big data analytic era. It is foreseen that classical meta-heuristics such as PSO and GA will be first applied to solve DL problems, followed by variants and hybrids of those. While most of the papers in the first wave will concentrate on local intensification in scouting for new solution candidates by adding some randomness and Levy flight distribution, more are expected to follow which embrace both local intensification and global exploration. GE will be useful only when the search space is sufficiently huge, e.g., self-adaptive versions of meta-heuristics such as bat algorithm [29] and wolf search algorithm [30], and the search space is the summation of the dimensions of the existing possible solutions and the allowable ranges of the parameters values. This advanced meta-heuristic technique translates to parameter-free optimization algorithms which are suitable to deal with only very complex neural network configuration and performance tuning problems. Then again, considering the big challenges of big data analytics in solving big problems (e.g., analyzing in real time of road traffic optimization problems using Internet of things), corresponding powerful DL tools should be equipped with effective optimization strategies.

Acknowledgements The authors are thankful for the financial support from the Research Grant called "A Scalable Data Stream Mining Methodology: Stream-based Holistic Analytics and Reasoning in Parallel," Grant no. FDCT/126/2014/A3, offered by the University of Macau, FST, RDAO and the FDCT of Macau SAR government.

References

1. Hinton, G. E., Osindero, S., and Teh, Y. W. "A fast learning algorithm for deep belief nets", Neural Computation. 2006;18(7):1527–1554
2. Yu Kai, Jia Lei, Chen Yuqiang, and Xu Wei. "Deep learning: yesterday, today, and tomorrow", Journal of Computer Research and Development. 2013;50(9):1799–1804
3. ILSVRC2012. Large Scale Visual Recognition Challenge 2012 [Internet]. [Updated 2013-08-01]. Available from: http://www.imagenet. Org/challenges/LSVRC/2012/

4. Izadinia, Hamid, et al. "Deep classifiers from image tags in the wild". In: Proceedings of the 2015 Workshop on Community-Organized Multimodal Mining: Opportunities for Novel Solutions; ACM; 2015
5. Gudise, V. G. and Venayagamoorthy, G. K. "Comparison of particle swarm optimization and back propagation as training algorithms for neural networks". In: Proceedings of In Swarm Intelligence Symposium SIS'03; 2006. p. 110–117
6. Marc Claesen, Bart De Moor, "Hyperparameter Search in Machine Learning", MIC 2015: The XI Metaheuristics International Conference, Agadir, June 7–10, 2015, pp. 14-1 to 14-5
7. Steven R. Young, Derek C. Rose, Thomas P. Karnowski, Seung-Hwan Lim, Robert M. Patton, "Optimizing deep learning hyper-parameters through an evolutionary algorithm", Proceedings of the Workshop on Machine Learning in High-Performance Computing Environments, ACM, 2015
8. Papa, Joao P.; Rosa, Gustavo H.; Marana, Aparecido N.; Scheirer, Walter; Cox, David D. "Model selection for Discriminative Restricted Boltzmann Machines through meta-heuristic techniques". Journal of Computational Science, v.9, SI, p. 14–18, July 2015
9. Xin-She Yang, "Engineering Optimization: An Introduction with Metaheuristic Applications", Wiley, ISBN: 978-0-470-58246-6, 347 pages, June 2010
10. Goldberg, D. E. and Holland, J. H. "Genetic algorithms and machine learning". Machine Learning. 1988;3(2):95–99
11. Iztok Fister Jr., Xin-She Yang, Iztok Fister, Janez Brest, Dusan Fister, "A Brief Review of Nature-Inspired Algorithms for Optimization ", ELEKTROTEHNISKI VESTNIK, 80(3): 1–7, 2013
12. Kennedy, J. "Particle Swarm Optimization"; Springer, USA; 2010. p. 760–766
13. Glover, F. "Tabu search-part I". ORSA Journal on Computing. 1989;1(3):190–206
14. Xin-She Yang, Suash Deb, Simon Fong, "Metaheuristic Algorithms: Optimal Balance of Intensification and Diversification", Applied Mathematics & Information Sciences, 8(3), May 2014, pp. 1–7
15. C. Blum, and A. Roli, "Metaheuristics in combinatorial optimization: Overview and conceptual comparison", ACM Computing Surveys, Volume 35, Issue 3, 2003, pp. 268–308
16. Simon Fong, Suash Deb, Xin-She Yang, "A heuristic optimization method inspired by wolf preying behavior", Neural Computing and Applications 26 (7), Springer, pp. 1725–1738
17. Suash Deb, Simon Fong, Zhonghuan Tian, "Elephant Search Algorithm for optimization problems", 2015 Tenth International Conference on Digital Information Management (ICDIM), IEEE, Jeju, 21–23 Oct. 2015, pp. 249–255
18. Beheshti, Z. and Shamsuddin, S. M. H. "A review of population-based meta-heuristic algorithms". International Journal of Advances in Soft Computing & Its Applications, 2013;5 (1):1–35
19. Simon Fong, Xi Wang, Qiwen Xu, Raymond Wong, Jinan Fiaidhi, Sabah Mohammed, "Recent advances in metaheuristic algorithms: Does the Makara dragon exist?", The Journal of Supercomputing, Springer, 24 December 2015, pp. 1–23
20. Gudise, V. G. and Venayagamoorthy, G. K. "Comparison of particle swarm optimization and back propagation as training algorithms for neural networks". In: Proceedings of In Swarm Intelligence Symposium SIS'03; 2006. p. 110–117
21. Zhang, J. R., Zhang, J., Lok, T. M., and Lyu, M. R. "A hybrid particle swarm optimization–back-propagation algorithm for feed forward neural network training". Applied Mathematics and Computation. 2007;185(2):1026–1037
22. Juang, C. F. "A hybrid of genetic algorithm and particle swarm optimization for recurrent network design". Systems, Man, and Cybernetics, Part B: Cybernetics, IEEE Transactions 2004;34(2):997–1006
23. Meissner, M., Schmuker, M., and Schneider, G. "Optimized particle swarm optimization (OPSO) and its application to artificial neural network training". BMC Bioinformatics. 2006;7 (1):125

24. Leung, F. H., Lam, H. K., Ling, S. H., and Tam, P. K. "Tuning of the structure and parameters of a neural network using an improved genetic algorithm". IEEE Transactions on Neural Networks. 2003;14(1):79–88
25. L.M. Rasdi Rere, Mohamad Ivan Fanany, Aniati Murni Arymurthy, "Simulated Annealing Algorithm for Deep Learning", The Third Information Systems International Conference, Procedia Computer Science 72 (2015), pp. 137–144
26. Hinton, G. E., "Training products of experts by minimizing contrastive divergence", Neural Computing, 2002 Aug;14(8):1771–800
27. Maass, W. "Networks of spiking neurons: The third generation of neural network models". Neural Networks. 1997;10(9):1659–1671
28. Simon Fong, Ricardo Brito, Kyungeun Cho, Wei Song, Raymond Wong, Jinan Fiaidhi, Sabah Mohammed, "GPU-enabled back-propagation artificial neural network for digit recognition in parallel", The Journal of Supercomputing, Springer, 10 February 2016, pp. 1–19
29. Iztok Fister Jr., Simon Fong, Janez Brest, and Iztok Fister, "A Novel Hybrid Self-Adaptive Bat Algorithm," The Scientific World Journal, vol. 2014, Article ID 709738, 12 pages, 2014. doi:10.1155/2014/709738
30. Qun Song, Simon Fong, Rui Tang, "Self-Adaptive Wolf Search Algorithm", 5th International Congress on Advanced Applied Informatics, July 10–14, 2016, Kumamoto City International Center, Kumamoto, Japan

Using Games to Solve Challenging Multimedia Problems

Oge Marques

Abstract There are many challenging problems in multimedia research for which state-of-the-art solutions fall short of performing perfectly. The realization that many of these tasks are arduous for computers yet are relatively easy for humans has inspired many researchers to approach those problems from a human computation viewpoint, using methods that include crowdsourcing and games. This paper offers a discussion and advices how to use games to supplement traditional content analysis techniques and assist in the solution of hard multimedia problems.

Keywords Games · Multimedia · Crowdsourcing · Human computation · Computer vision · Image processing

1 Introduction

There are many contemporary problems in multimedia research for which fully automatic solutions are still not available, and the performance of state-of-the-art algorithms is far inferior to humans performing comparable tasks [8]. In the past few years, following the success of the ESP game [11] and its ability to assist in the solution to the problem of image tagging (i.e., associating keywords to an image that describe its contents), a growing number of research projects in the intersection between games and multimedia have emerged. These so-called serious games (or "games with a purpose") belong to a category of approaches that include human users in the loop and attempt to extract valuable information from the user input, which can then inform and improve the underlying algorithms designed to solve a particular problem. In addition to games, human input (and knowledge) can also be leveraged through the use of *crowdsourcing*, where inputs from large numbers of human participants are processed to serve as a basis for statistical analysis and inference [8].

O. Marques (✉)
Florida Atlantic University, Boca Raton, FL 33431, USA
e-mail: omarques@fau.edu

© Springer Nature Singapore Pte Ltd. 2018   27
P.K. Sa et al. (eds.), *Progress in Intelligent Computing Techniques: Theory,*
Practice, and Applications, Advances in Intelligent Systems and Computing 518,
DOI 10.1007/978-981-10-3373-5_2

Games and crowdsourcing belong to the broader field of *human computation*, whose main idea consists of collecting users' input and interactions and using them to progressively refine a computational model designed to solve a particular problem. In crowdsourcing efforts, such input is captured in the form of very specific tasks, for which users are often paid a modest amount. In games, the goal is to keep the user entertained and—after the game has been played a significant number of times—mine information from the game logs and use them to feed the associated computational model.

In this paper, we postulate that games—if properly designed—have the potential to assist in the solution of challenging multimedia research problems. We note, however, that there are some problems with the vast majority of "gamification" approaches to the solution of research problems in the field of multimedia proposed over the past decade. More importantly, we offer advice for researchers who want to overcome those problems and build successful games for solving relevant scientific problems.

2 Background and Motivation

The success of using games to solve scientific problems in the field of image analysis, computer vision, and visual scene understanding can be traced back to the pioneering work of Luis von Ahn and his *Extra Sensory Perception (ESP)* game [11]. This game was eventually made more popular in the form of a Google Labs project (Google Image Labeler) and has helped collect an enormous amount of useful tags that describe the contents of a large collection of images crawled from the Web at large. The ESP game involves two players, matched at random, which are shown the same image and asked to type words that they believe could be used as tags to describe the image contents. When the game finds an agreement between the words used by both players, they are rewarded with points. A label (or tag) is accepted as *correct* when different pairs of users agree about it.

The ESP game has inspired many efforts, including ARTigo, a Web-based platform (http://www.artigo.org) containing six artwork annotation games as well as an artwork search engine in English, French, and German. Funded by the German Research Foundation (DFG), the ARTigo project has—during the period between 2008 and 2013—successfully collected over 7 million tags (mostly in German) to describe the artworks in the collection and engaged more than 180,000 players (about a tenth of whom are registered). It remains active at the time of this writing.

The seminal work by von Ahn also coined the expression "games with a purpose" and its acronym, GWAP. Until 2011, the research group responsible for ESP and several other games used to maintain a Website (http://www.gwap.com), which contained links to a variety of games developed to assist in the solution of specific problems, such as *Peekaboom* [13] for object detection, *TagATune* [5] for music and sound annotation, and *Verbosity* [12], a game for collecting common sense knowledge for the Open Mind Common Sense project. More recently, von Ahn has devoted most of his efforts to create, improve, and promote Duolingo (http://www.duolingo.com), an extremely successful multi-platform app that gamifies the process of learning foreign languages.

Outside of the realm of multimedia research, prominent examples of games designed to solve scientific problems include FoldIt (http://fold.it) (a 3D puzzle used to assist in understanding protein folding and potentially leading to the discovery of cures for diseases such as AIDS and cancer) and EyeWire (http://eyewire.org), a game for finding the *connectome*—a comprehensive map of neural connections—of the retina.

Games (with a purpose) remain a valid, meaningful, and not yet fully explored avenue for assisting in the solution of challenging scientific problems. Games can be particularly useful when the research question at hand that can be mapped to *tasks* that satisfy one or more of the following criteria: (i) are easy for humans and hard for computers; (b) require intensive labor; (iii) enable noble scientific pursuits; and (iv) improve human life.

3   Our Work

In this section, we present a brief description of three of our recent research projects involving the use of games and crowdsourcing for solving challenging multimedia research problems, namely

- Ask'nSeek: A two-player game for object detection, segmentation, and labeling.
- Click'n'Cut: An interactive intelligent image segmentation tool for object segmentation.
- Guess That Face: A single-player game for face recognition under blurred conditions.

At the end of the section, we share some of the lessons we have learned from these projects.

3.1   Ask'nSeek

AsknSeek [2] is a two-player Web-based guessing game that asks users to guess the location of a hidden region within an image with the help of semantic and topological clues. One player (*master*) hides a rectangular region somewhere within a randomly chosen image, whereas the second player (*seeker*) tries to guess the location of the hidden region through a series of successive guesses, expressed by clicking at some point in the image. Rather than blindly guessing the location of the hidden region, the seeker asks the master for clues (i.e., indications) relative to the objects present in the image (e.g., "above the cat" or "partially on the dog" in Fig. 1). The game is

Fig. 1 Examples of object
detection and labeling results
obtained with AsknSeek:
two objects (*cat* and *dog*)
were detected and labeled

cooperative—i.e., both players score more points when the hidden region is found quickly—which leads to the master providing accurate clues, which are stored into the game logs for further processing.

The information collected from game logs is combined with the results from content analysis algorithms and used to feed a machine-learning algorithm that outputs the outline of the most relevant regions within the image and their names (Fig. 1). The approach solves two computer vision problems—object detection and labeling—in a single game and, as a bonus, allows the learning of spatial relations (e.g., "the dog is above the cat") within the image [8].

3.2 Click'n'Cut

Ask'nSeek was extended to address the object segmentation problem [9], where the game traces are combined with the ranked set of segments generated by the constrained parametric min-cuts (CPMC) algorithm [4].

In a parallel effort, an intelligent interactive tool for foreground object segmentation (*Click'n'Cut*) was created, where the users are asked to produce foreground and background clicks to perform a segmentation of the object that is indicated in a provided description (Fig. 2). Every time a user produces a click, the segmentation result is updated and displayed over the image with an alpha value of 0.5. This segmentation is computed using an algorithm based on object candidates [1] and aims at guiding the user to provide information (i.e., meaningful clicks) that will help improve the quality of the final segmentation result.

Several experiments were performed using the Click'n'Cut tool on a set of 105 tasks (100 images to be segmented plus 5 gold standard tasks, to control for errors), with two distinct groups of users:

- Experts: 15 computer vision researchers from academia, both students and professors.
- Workers: 20 paid workers from the platform https://microworkers.com/; each worker was paid 4 USD for annotating 105 images.

Click'n'Cut (1/105)

Extract the fish.

Left click on the Foreground
Right click on the background
To reset your clicks, please click "Clear Points"
Click on any point to remove it
Use the slider to modify the mask transparency
Once you are satisfied with the mask, click 'Done' to go to the next task

Clear Points

Transparency

Show points ● Yes ○ No

Done

Fig. 2 Click'n'Cut user interface

The results obtained by each group were also compared against the segmentation output produced by using clicks collected from 162 Ask'nSeek players (mostly students) who played the Ask'nSeek game on any number of images they wanted to.

The goal of such experiments was to assess: (i) the "crowdsourcing loss" between experts and less-skilled workers; and (ii) the "gamification loss" incurred by replacing an interactive crowdsourcing tool (Click'n'Cut) with a game (Ask'nSeek) as a generator of foreground and background clicks. Detailed results and discussions can be found in [3].

3.3 Guess That Face!

Guess That Face! [7] is a single-payer Web-based face recognition game that reverse engineers uses the human biological threshold for accurately recognizing blurred faces of celebrities under time-varying conditions (Fig. 3). The game combines a successful casual game paradigm with meaningful applications in both human- and computer-vision science. Results from preliminary user studies conducted with 28 users and more than 7,000 game rounds supported and extended preexisting knowledge and hypotheses from controlled scientific experiments, which show that humans are remarkably good at recognizing famous faces, even with a significant degree of blurring [10]. A live prototype is available at http://tinyurl.com/guessthatface.

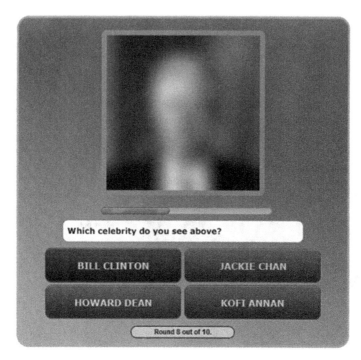

Fig. 3 Guess That Face! screenshot: The player is presented with a severely blurred image (de-blurring progress is indicated by the *green bar* below the image) and four options (*buttons*). Here, the correct answer has already been chosen (*green*)

3.4 Lessons Learned

The experience with the games and crowdsourcing projects described in this section has taught us several valuable lessons that we could not have anticipated if we had not developed and deployed the games and tools and conducted associated user studies and experiments, among them:

- Most of our users played the game for *extrinsic*—rather than *intrinsic*—reasons; for example, the highest response rates for *Ask'nSeek* came as a result of assigning bonus points in a course for students who played 30 or more rounds of the game.
- On a related note, for *Guess That Face!*, a different reward system was adopted, in which students placed in the "High scores" table were assigned extra bonus points. As a result, students played many more games than expected, driven by intrinsic reasons—placing themselves among the top scorers and beating their friends' scores. As a side effect of playing the game for many more rounds than originally expected, some players ended up "memorizing" the dataset, which led to additional work when processing and cleaning up the game logs.
- None of our games has reached a level of engagement remotely close to "viral" or "addictive."

4 Discussion

4.1 Problems

In this paper, I postulate that most of the efforts aimed at creating games for assisting in the solution of multimedia research problems suffer from two main problems:

1. The design process is often reversed, i.e., rather than designing a game with the player in mind, researchers (including the author) usually follow these steps: (i) start from a problem; (ii) think of a crowdsourcing solution; (iii) create a tool; and (at the end) (iv) "make it look like a game."
2. Our terminology is not exactly inspiring: expressions such as "serious games," "Games with a purpose (GWAP)," "Human-based computation games," or "Non-entertainment focused games" do not convey the idea that such games can (and should!) be fun to play.

4.2 Possible Solutions

I was recently asked to give a talk on this topic and provide advice for (young) researchers who are interested in the intersection of games and multimedia research. While reflecting upon the message I wanted to convey to the audience, I realized that the worst advice I could give was "Gamify everything!" Gamification has been overused and misused during recent years; it should not be seen as a "cure-all" solution, but rather a creative way to engage users in meaningful human computation tasks, while being driven by the intrinsic motivation evoked by a properly designed game.

These are some other pieces of advice that I hope will be helpful to readers of this paper:

- Do not try to gamify if you cannot see the world from a gamer's viewpoint. If you are not a gamer, learn more about it from resources such as the excellent book "Getting Gamers: The Psychology of Video Games and Their Impact on the People who Play Them" [6] and companion site (http://www.psychologyofgames. com) by Jamie Madigan.
- Select multimedia problems that are worth researching and can be modeled as tasks that fulfill at least two of the criteria stated earlier in this paper (and repeated here for convenience): (i) are easy for humans and hard for computers; (b) require intensive labor; (iii) enable noble scientific pursuits; and (iv) improve human life.
- Be mindful of (and try to incorporate, whenever possible) new devices and technologies, such as increasingly popular virtual reality (VR) kits, new sensors, and wearable devices and gadgets.

- Consider engaging in research on game effectiveness, e.g., creating experiments to find out if people are having fun while playing a certain game.
- Challenge the design workflow described in Sect. 4.1 (and turn it upside-down!)

5 Conclusion

In this paper, I have discussed the intersection between (serious) games and multimedia research and provided advice on how to use games to supplement traditional content analysis techniques and assist in the solution of hard multimedia problems.

As stated in an earlier paper [8], I believe that games should be designed such that the input collected from the users is as simple as possible, but carries as much meaningful information as possible. Moreover, we should not ignore the traditional approach of content analysis, which can be used to augment crowdsourcing, in order to reduce the number of participants needed to obtain meaningful results, using semi-supervised machine-learning approaches.

As a final reflection, for those readers who might want to consider engaging in this field of research and studies, a simplified SWOT analysis might be helpful:

- Strengths

  - There are many meaningful research problems waiting to be solved.
  - People love games!

- Weaknesses

  - Poorly designed games turn people away (quickly!).

- Opportunities

  - There are multiple game platforms to develop for—from Web to traditional consoles to mobile apps for iOS and Android devices.
  - There is a growing interest in games, and the trend is not likely to change any time soon.

- Threats

  - There may be better solutions for certain multimedia research problems (e.g., the increasingly popular use of *deep learning* techniques) that do not use games.

Acknowledgements The author would like to thank Prof. Sambit Bakshi and the ICACNI 2016 Program Committee for the invitation to prepare this paper.

References

1. P. Arbeláez, J. Pont-Tuset, J. T. Barron, F. Marques, and J. Malik. Multiscale combinatorial grouping. In *CVPR*, 2014.
2. A. Carlier, O. Marques, and V. Charvillat. Ask'nseek: A new game for object detection and labeling. In *Computer Vision–ECCV 2012. Workshops and Demonstrations*, pages 249–258. Springer, 2012.
3. A. Carlier, A. Salvador, F. Cabezas, X. Giro-i Nieto, V. Charvillat, and O. Marques. Assessment of crowdsourcing and gamification loss in user-assisted object segmentation. *Multimedia Tools and Applications*, pages 1–28, 2015.
4. J. Carreira and C. Sminchisescu. Constrained parametric min-cuts for automatic object segmentation. In *CVPR*, 2010.
5. E. L. Law, L. Von Ahn, R. B. Dannenberg, and M. Crawford. Tagatune: A game for music and sound annotation. *ISMIR*, 3:2, 2007.
6. J. Madigan. *Getting gamers: the psychology of video games and their impact on the people who play them*. Rowman & Littlefield, Lanham, Maryland, 2016. ISBN 978-1-4422-3999-9.
7. O. Marques, J. Snyder, and M. Lux. How well do you know tom hanks?: Using a game to learn about face recognition. In *CHI '13 Extended Abstracts on Human Factors in Computing Systems*, CHI EA '13, pages 337–342, New York, NY, USA, 2013. ACM. ISBN 978-1-4503-1952-2. doi: 10.1145/2468356.2468416. URL http://doi.acm.org/10.1145/2468356.2468416.
8. W. T. Ooi, O. Marques, V. Charvillat, and A. Carlier. Pushing the Envelope: Solving Hard Multimedia Problems with Crowdsourcing. *E-letter COMSOC Multimedia Communications Technical Committee*, 8(1):37–40, 2013. URL http://oatao.univ-toulouse.fr/13004/.
9. A. Salvador, A. Carlier, X. Giro-i Nieto, O. Marques, and V. Charvillat. Crowdsourced object segmentation with a game. In *Proceedings of the 2nd ACM international workshop on Crowdsourcing for multimedia*, pages 15–20. ACM, 2013.
10. P. Sinha, B. Balas, Y. Ostrovsky, and R. Russell. Face recognition by humans: Nineteen results all computer vision researchers should know about. *Proceedings of the IEEE*, 94(11):1948–1962, Nov 2006. ISSN 0018-9219. doi: 10.1109/JPROC.2006.884093.
11. L. von Ahn and L. Dabbish. Labeling images with a computer game. In *Proceedings of the SIGCHI Conference on Human Factors in Computing Systems*, CHI '04, pages 319–326, New York, NY, USA, 2004. ACM. ISBN 1-58113-702-8. doi: 10.1145/985692.985733. URL http://doi.acm.org/10.1145/985692.985733.
12. L. von Ahn, M. Kedia, and M. Blum. Verbosity: A game for collecting common-sense facts. In *Proceedings of the SIGCHI Conference on Human Factors in Computing Systems*, CHI '06, pages 75–78, New York, NY, USA, 2006a. ACM. ISBN 1-59593-372-7. doi: 10.1145/1124772.1124784. URL http://doi.acm.org/10.1145/1124772.1124784.
13. L. von Ahn, R. Liu, and M. Blum. Peekaboom: a game for locating objects in images. In *ACM CHI*, 2006b.

Part II
Advanced Image Processing Methodologies

Effective Image Encryption Technique Through 2D Cellular Automata

Rupali Bhardwaj and Vaishalli Sharma

Abstract Basic idea of encryption is to encrypt the pixels of an image in such a manner that the image becomes chaotic and indistinguishable. In this paper, we have done exploratory study on image encryption techniques through 2D cellular automata (single layer and double layer) which create chaos on pixels of considered image. A comparative analysis on effectiveness of scrambling technique is provided by scrambling degree measurement parameters, i.e., gray difference degree (GDD) and correlation coefficient. Experimental results showed that the 2D cellular automata (single layer)-based encryption gives better result than 2D cellular automata (double layer).

Keywords Encryption · Cellular automata · Game of Life · Gray difference degree · Correlation coefficient

1 Introduction

Trivial encryption is done over multimedia contents for their secure transmission, but with improvement in speed of computer and with the use of parallel processing, they can be easily deciphered by algorithms. Images are one of the kinds of multimedia technology; as far as understandability or expressive world is concerned, images play best role at that. Images are quite intuitive for believing or seeing and they are informative too. We need to provide secure way of transmitting images over Internet like any other media. Due to the lack of guaranteed secure transmission of images by traditional encryption scheme, transmission of image has become very big problem, so two encryption methods, viz. digital watermarking and digital scrambling, are proposed as a solution. Digital image scrambling means

R. Bhardwaj (✉) · V. Sharma
Department of Computer Science & Engineering, Thapar University, Patiala, Punjab, India
e-mail: rupali.bhardwaj@thapar.edu

V. Sharma
e-mail: vaishalli.sharma@gmail.com

© Springer Nature Singapore Pte Ltd. 2018   39
P.K. Sa et al. (eds.), *Progress in Intelligent Computing Techniques: Theory,*
Practice, and Applications, Advances in Intelligent Systems and Computing 518,
DOI 10.1007/978-981-10-3373-5_3

creating confusion in image elements to make it non-informative for the sake of protecting it from theft, manipulation of content and privacy measures. At present, there are many technologies for scrambling image such as classical method Arnold cat map, Hilbert curve transform, Josephus traversing and chaos sequence. A digital image scrambling method based on a 2D cellular automaton, specifically the well-known Game of Life, produces an effective image encryption technique.

The following paper contents are as follows: Sect. 2 features literature review. After a brief discussion of cellular automata and Game of Life in Sect. 3, Sect. 4 describes digital encryption algorithm using 2D cellular automata (double layer). Section 5 demonstrates experimental study, and Sect. 6 concludes the paper.

2 Literature Review

Digital image scrambling using 2D cellular automata is presented in this paper [1]. An evaluation method of image scrambling degree based on adjacent pixels is proposed on a 256×256 grayscale image [2]. Digital image scrambling using 2D cellular automata for periodic and null boundary is presented in this paper [3]. Arnold cat map is discussed and explored here [4]. Anti-Arnold transformation algorithm is presented here [5]. An algorithm based on Arnold cat map and logistic map for image encryption is described here [6]. A multiregion scrambling technique which splits non-square regions and scrambles each region is presented here [7]. 3D Arnold cat map-based scrambling technique and a histogram-based comparison between 2D and 3D Arnold cat map is presented here [8]. Image scrambling along with watermarking scheme based on orbits of Arnold transformation is presented here [9–11]. Watermarking algorithm based on cellular automata is described in this paper [12]. A new evaluation method of image scrambling is proposed here [13].

3 Scrambling Techniques

Images means collection of atomic units called pixel, which represents some color value. Images can be thought of as a matrix containing pixel values of each pixel for whole image. Primarily the basic idea behind scrambling of image is to change position of pixels of image through matrix transformation. In this way, high quality of confusion is created in image within few evolutions. Scrambling is used to create a non-intelligible image which prevents human or computer vision system from recognizing the true content. First image is scrambled or we can say encrypted, i.e., distortion is created; then, it is descrambled to get back actual image by using descrambling algorithm. Only authorized person is capable of descrambling the scrambled image as only he/she will be given all possible keys used while scrambling the image. Scrambling has attracted researchers in recent years, and many scrambling techniques such as Arnold cat map and 2D cellular automata have been proposed.

3.1 Cellular Automata

The term cellular automata is a discrete model studied in mathematics, physics, computability theory, complexity science and theoretical biology. Image enhancement, compression, encryption and watermarking are applications of CA in the field of digital image processing. Cellular automata mean a regular finite grid of cell, where each cell encapsulated an equal portion of the state, and arranged spatially in a regular fashion to form an n-dimensional lattice. At every time step t, upcoming state of a cell is being calculated based on its present state as well as local neighbor's present state. In 2D CA, Von Neumann neighborhood of range r is defined by Von Neumann as

$$N^v_{(x_0, y_0)} = \{(x, y): |x - x_0| + |y - y_0| \leq r\} \tag{1}$$

Here, (x, y) are possible neighbors of current cell (x_0, y_0), where r is the radius which represents neighborhood's dimension and the number of cells in each neighborhood is $2r(r + 1) + 1$. If $r = 1$, which is typically used, yields a total of five cells in the Von Neumann neighborhood, with neighbors to the top, bottom, left and right cell around current cell. Moore neighborhood of range r is defined by Moore as:

$$N^m_{(x_0, y_0)} = \{(x, y): |x - x_0| \leq r, |y - y_0| \leq r\} \tag{2}$$

Here, (x, y) are possible neighbors of current cell (x_0, y_0), where r is the radius which represents neighborhood's dimension and the number of cells in each neighborhood is $(2r + 1)^2$. Typically, $r = 1$ yields a total of nine cells in Moore neighborhood. In Moore neighborhood, diagonal cells to current cell are also considered; hence, it is a bit different than Von Neumann neighborhood.

3.1.1 Game of Life (2D Cellular Automata)

The universe of Game of Life is 2D cellular automata devised by J. H Conway in 1970 called as Conway's Game of Life. It is actually an infinite orthogonal grid which consists of cells, each of which is either dead (represented by zero) or alive (represented by one). According to local update rule, it is being asked from each cell their present state and state of their neighbor cells. For alive cell, it stays alive (survives) if and only if it has two or three alive neighbors; otherwise, it is dead because of loneliness or overcrowding. A dead cell becomes alive if and only if it has exactly three live neighbors. These rules are applied simultaneously to each cell. As the process is repeated over and over again, a dynamical system is obtained that persists amazingly complex behavior.

4 Digital Encryption

Digital encryption with 2D cellular life automata overcomes problem associated with Arnold cat map, like it does not have periodicity property and it can work on square as well as rectangular images too. We scrambled image through Game of Life rowwise as well as columnwise on image, with the help of scrambling algorithm given by Fasal Qadir [3] as follows:

1. First, read $X \times Y$ image that is to be scrambled and then convert it to grayscale as we have to perform experiment on grayscale image only.
2. Create a random array of rows and columns, respectively, X and Y, which stores random binary values in it call it BIN_INIT this is used to produce an initial state of each and every pixel at time $t = 0$.
3. For a particular number of iterations, repeat following steps

   - Calculate sum of neighbors of current pixel by Moore neighborhood or Von Neumann neighborhood.
   - Check this sum according to Game of Life's rule.
   - Store the next state of each cell in one-dimensional binary array, NEXT_STATE.
   - For all live cells, take out their gray value and store sequentially in a separate array and apply same procedure for dead cells and put it in another array.
   - Concatenate both arrays now this is our scrambled image.

4. Now taking transpose of scrambled image obtained after step 3 and apply again step 3 on this scrambled image.
5. For descramble the image, reverse steps of scrambling technique is applied.

5 Experimental Study

To calculate scrambling parameters, experimental study is performed through 2D CA (Moore neighborhood periodic boundary) scrambling/descrambling algorithms on a 172×172 arbitrary grayscale image.

5.1 Gray Difference Degree (GDD)

Scrambling degree (GDD) is used to evaluate effectiveness of scrambling algorithm. Generally, if the scrambling degree of scrambled/encrypted image is high, this simply means higher security. Therefore, scrambling degree measure is of great importance.

Consider a digital image I of size X × Y; first, compute the gray difference between the pixel (i, j) and its four neighboring pixels as follows:

$$GD(i,j) = \frac{1}{4} \sum_{i,j} \left[G(i,j) - G(i',j') \right]^2 \tag{3}$$

After computing the gray difference for whole image, except the pixels at the edges, we can compute the full image's mean gray difference as follows:

$$E(GD(i,j)) = \frac{\sum_{i=2}^{X-1} \sum_{j=2}^{Y-1} GD(i,j)}{(X-2)(Y-2)} \tag{4}$$

The scrambling degree is defined as follows:

$$GDD = \frac{[E'(GD(i,j)) - E(GD(i,j))]}{[E'(GD(i,j)) + E(GD(i,j))]} \tag{5}$$

where $E(GD(i,j))$ and $E'(GD(i,j))$ denote the average neighborhood gray differences of the original and the scrambled images, respectively. Range of GDD is in between $[-1, 1]$. If GDD value is close to 1, better scrambling effect is obtained [1]. Figure 1 shows scrambling degree evaluation, and Table 1 gives comparative study of maximum GDD for 2D cellular automata (single- and double-layer Moore periodic boundary) scrambling algorithms.

5.2 Correlation Coefficient

Correlation coefficient is based on gray value of adjacent pixels of current pixel. It tests gray relation value of adjacent pixels in cipher image. In any meaningful image, correlation of adjacent pixels will be high, but this strong correlation will be broken when image is scrambled. Zero value of correlation between original and scrambled image implies more effective scrambling is done. Correlation coefficient is defined in Eq. (6). Range of correlation coefficient is in between $[-1, 1]$. Correlation coefficient is given as:

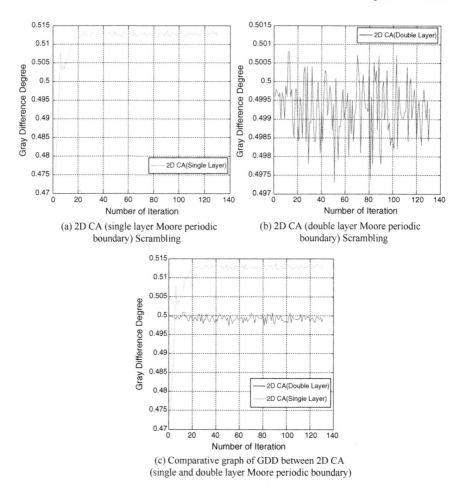

(a) 2D CA (single layer Moore periodic
boundary) Scrambling

(b) 2D CA (double layer Moore periodic
boundary) Scrambling

(c) Comparative graph of GDD between 2D CA
(single and double layer Moore periodic boundary)

Fig. 1 Scrambling degree

Table 1 GDD of different scrambling techniques, k = 131

	2D CA (single-layer Moore neighborhood)	2D CA (double-layer Moore neighborhood)
Maximum GDD	0.5154	0.5008

$$R = \frac{\sum_m \sum_n (A_{mn} - A')(B_{mn} - B')}{\sqrt{\sum_m \sum_n (A_{mn} - A')^2 (B_{mn} - B')^2}} \qquad (6)$$

Here, $R = 1$ implies that scrambled image is very similar to original one. $R = -1$ implies that scrambled image is a negative to original image and just by inverting it

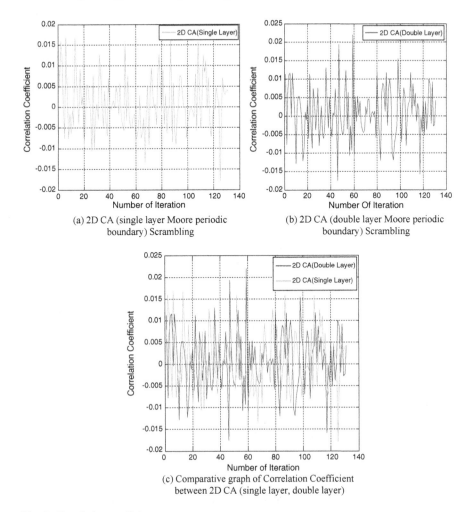

(a) 2D CA (single layer Moore periodic
boundary) Scrambling

(b) 2D CA (double layer Moore periodic
boundary) Scrambling

(c) Comparative graph of Correlation Coefficient
between 2D CA (single layer, double layer)

Fig. 2 Correlation coefficient

we can get original image. R = 0 implies there is no relation between scrambled and
original image and they are not at all correlated, it is an ideal condition for scrambling. Figure 2 shows correlation coefficient of 2D cellular automata (single- and
double-layer Moore periodic boundary) scrambling algorithms.

6   Conclusion

For secured transmission of images, we need approaches that guarantee robustness;
scrambling solves this issue and proved to be more robust than encrypting images.
Arnold cat map is widely used scrambling technique and very popular for its

simplicity, but it is less secure because of its periodic nature and condition to work upon square images only. 2D Cellular automata is an interesting and capable way of solving problems associated with Arnold cat map as it does not possess periodic nature and can work upon rectangular images too. We proved by our experimental study that 2D cellular automata (single layer) performs better scrambling effect as compare to 2D cellular automata (double layer).

References

1. Abdel Latif Abu Dalhoum, Ibrahim Aldamari "Digital Image Scrambling Using 2D Cellular Automata", 14th IEEE International Symposium on Multimedia, 2012.
2. Congli Wang, Zhibin Chen, Ting Li, "Blind Evaluation of Image Scrambling Degree based on the Correlation of Adjacent Pixels", Telkomnika, Indonesian Journal of Electrical Engineering, Vol. 11, No. 11, pp. 6556–6562, November 2013.
3. Fasel Qadir, M. A. Peer, K. A. Khan, "Digital Image Scrambling Based on Two Dimensional Cellular Automata", International Journal of Computer Network and Information Security, Vol. 2, pp 36–41, 2013.
4. Gabriel Peterson, "Arnold's Cat Map", Newyork, 1997.
5. Lingling Wu, Jianwei Zhang, Weitao Deng, Dongyan He, "Arnold Transformation Algorithm and Anti-Arnold Transformation Algorithm", 1st International Conference on Information Science and Engineering (ICISE2009), IEEE.
6. Mao-Yu Huang, Yueh-Min Huang, Ming-Shi Wang, "Image Encryption Algorithm Based on Chaotic Maps", 3rd IEEE International Conference on Computer Science and Information Technology (ICCSIT), 2010, IEEE.
7. Min Li, Ting Liang, Yu-jie He, "Arnold Transform Based Image Scrambling Method", 3rd International Conference on Multimedia Technology (ICMT 2013).
8. Pawan N. K., Narnaware, M., "Practical Approaches for Image Encryption/Scrambling Using 3D Arnolds Cat Map", CNC 2012, LNICST 108, pp 398–404, 2012.
9. Ruisong Ye, "A Novel Image Scrambling and Watermarking Scheme Based on Orbits of Arnold Transform", Pacific-Asia Conference on Circuits, Communications and System, 2009, IEEE.
10. http://en.wikipedia.org/wiki/Arnold%27s_cat_map.
11. V. I. Arnold, A. Avez, "Ergodic Problems in Classical Mechanics", New York: Benjamin, 1968.
12. Ruisong Ye, Huiliang Li, "A Novel Image Scrambling and Watermarking Scheme Based on Cellular Automata", International Symposium on Electronic Commerce and Security, 2008, IEEE.
13. Tan Yongjie, Zhou Wengang, "Image Scrambling Degree Evaluation Algorithm based on Grey Relation Analysis", International Conference on Computational and Information Sciences, 2010, IEEE.

Comparison of Different Renal Imaging Modalities: An Overview

Ravinder Kaur and Mamta Juneja

Abstract The kidneys play an important role in our health, and renal dysfunction is identified by a successive loss in renal functionality with passage of time. The term 'uroradiology' is used to describe imaging and interventional techniques involved in the examination of urinary tract. Imaging plays an important role for assessment of different kidney abnormalities. The choice of particular imaging technique is based on the radiation burden, cost involved, possible complications and the diagnostic yield. Various attempts have been made for improving the correctness of renal diagnosis using distinct medical imaging techniques. In this article, we explore the potential of different renal imaging techniques currently employed in clinical set-ups along with their advantages and disadvantages.

Keywords Renal biopsy · CT · Magnetic resonance imaging · OCT · KUB

1 Introduction

Kidneys are bean-shaped organ of human body which remove waste products from blood, help to maintain acid base balance, form urine and support other functions of the body. According to the World Health Organization (WHO), renal dysfunction is a worldwide health crisis and it is feasible to control the growth of renal diseases with early diagnosis and treatment [1–3]. Non-communicable diseases such as renal disorders or heart diseases have taken place of communicable diseases such as AIDS in increasing the mortality and morbidity rates globally. The common indications for renal imaging are abnormal renal function tests that suggest the need of imaging. Different types of imaging techniques are being used by clinicians to

R. Kaur (✉) · M. Juneja
Department of Computer Science and Engineering, University Institute of Engineering
and Technology, Panjab University, Punjab, India
e-mail: ravinder.kaur7@yahoo.com

M. Juneja
e-mail: mamtajuneja@pu.ac.in

© Springer Nature Singapore Pte Ltd. 2018
P.K. Sa et al. (eds.), *Progress in Intelligent Computing Techniques: Theory,
Practice, and Applications*, Advances in Intelligent Systems and Computing 518,
DOI 10.1007/978-981-10-3373-5_4

investigate different diseases. The choice of particular imaging technology depends on the type of disease to be diagnosed. In this paper, we discuss about different imaging techniques that are used to investigate different renal abnormalities.

2   Renal Imaging Modalities

In the last few decades, imaging techniques are consistently enforced for determination and analysis of renal disorders. There are different renal imaging modalities that are widely used such as CT, US, MRI, OCT, angiography and nuclear medicine. Each of these modalities has merits and demerits which will be explained in related section. The choice of particular imaging modality depends on the diagnostic output, pricing requirement, effect of radiation and complexity of disease. CT, US, MRI, US, DCE-MRI and angiography are susceptible to computerized analysis, and much of work is done on the automatic handling of images from these imaging techniques [4–6]. In contrast, less number of attempts has been made by researchers for computerized analysis of nuclear medicine and OCT. Here, we discuss distinct kidney imaging modalities which generate different images for analysis and detection of diseases.

2.1   Plain Abdominal Radiographs

Imaging work initially started with abdominal radiograph which is the first imaging modality that was used in 1972 [7]. Plain abdominal fluoroscopy or kidneys, ureters and bladder (KUB) X-ray, as shown in Fig. 1, is one of the important imaging modalities for detection of radio-opaque stones. However, the growing use of US, CT and MRI has confined the usage of abdominal radiographs; still, they are used in the administration of severe abdominal pain. The problem with the plain radiographs is that they are unable to distinguish phleboliths present in pelvis from the actual kidney stones. The limitation of technique is that 10% of stones (known as radiolucent stones) are not detected and smaller renal calculi are often overlooked.

2.2   Intravenous Urogram (IVU)

Intravenous urogram, also known as intravenous pyelography, is a procedure in which iodinated contrast medium is injected into the patient body and serial radiographs are taken that allows visualization of the renal tract as shown in Fig. 2. If there is any interruption in the one-sided flow of contrast agent through the renal pelvis and ureter, then it would be evidence of blockage. If flow is two-sided, then

Fig. 1 KUB X-ray of kidney (*arrow* pointing towards kidney)

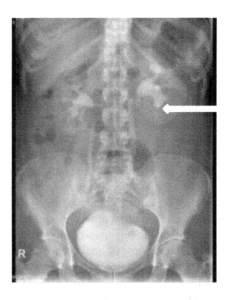

Fig. 2 IVU of kidney showing kidney drainage with contrast (*arrow* pointing towards kidney)

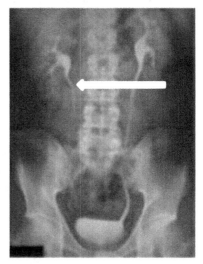

intrinsic problems, such as low renal perfusion or function, can be examined [8]. So the degree of delay in the passage of contrast agent can be used to find the degree of obstruction. Intravenous urogram (IVU) further assists radiologists in the ferreting out inborn anomalies of the human kidney, such as renal agenesis, renal hypoplasia and renal dysplasia. The limiting factor of IVU is that it cannot be used for pregnant women, paediatric patients or if patients have allergy due to contrast agent. The advantage of an IVU is it helps us to confirm obstructions present in collecting duct system of kidneys. The limitation of technique is that it requires more radiation dose as compare to plain radiographs and administration of contrast agent.

2.3 Ultrasonography (US)

Ultrasonography has taken place of IVU as the first-line analysis of renal diseases. It is non-invasive, quick and inexpensive imaging modality that provides 2D images of body internal organs and tissues using high-frequency sound waves. It evades the use of ionizing radiation and can precisely measure renal size without the enlargement effects associated with IVU. Figure 3 shows a normal human kidney using US imaging modality. The real-time images of US helps for assessment of renal biopsy, renal stones, renal cysts, renal mass lesions and interventional procedures. In chronic kidney failure, US show abnormal echo pattern and the kidneys become smaller. An irregularity in renal size is an indication of renal artery stenosis which can be confirmed using Doppler interrogation. US images usually possess a texture pattern termed as speckle. These texture patterns are dependent on the anatomy of the organ and distinctive imaging specifications. Speckle pattern is formed as an outcome of interference of sound waves from many dispersers that are small fragments of size proportionate to its wavelength, i.e. ultrasound wave wavelength. These disperser particles not only refract or reflect the wave although they produce a complicated diffused scattering that forms the primitive approach for formation of speckles. Speckle is an immanent attribute of US imaging modality that has multiplicative noise which makes the US images deceitful and therefore reduces its diagnostic value. Speckle tracking methods use same speckle patterns to evaluate internal tissue movement. Doppler ultrasound imaging is also employed for computation of renal blood flow (RBF), and when this technique is adapted to obtain images of tissue, it is known as tissue Doppler imaging (TDI) [9].

Fig. 3 Ultrasound of kidney (*arrow* pointing towards kidney)

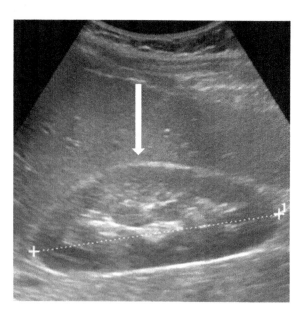

The benefit of using ultrasonography is that there is no exposure of subject to radiation and it is cost-effective. The limitation of technique is that it requires optimal image quality in corpulent patients and it is operator dependent.

2.4 Computed Tomography (CT)

Computed tomography imaging becomes invaluable means for diagnosis of any pathology affecting kidneys as it provides images with high contrast and good spatial resolution. In many departments, IVU is being replaced by CT imaging modality to obtain detailed images of body organs as they provide more anatomical and functional information than standard X-rays as shown in Fig. 4. CT scan is a non-invasive approach of imaging which provides different views of same body organ or tissue (often called slices) within seconds. Modern CT devices perform scan of abdomen within few seconds, thereby reducing breath-hold time, and data after acquisition are forwarded to allocated work stations for further diagnosis [9]. Renal multi-detector computed tomography (MDCT) has ability to capture pictures of the entire abdomen with thin slice collimation in one gantry rotation. Initial series of CT scanners allow taking only one or two slices in a single gantry revolution and undergo through the problems such as motion artefacts due to respiration and loss of IV contrast agents. Presently, multi-detector CT scanners moderately raised number of detectors such as 64, 128 or 256 detector CT scanners, thereby reducing acquisition time, and dual-source CT imaging is feasible with numerous gantry revolutions per second acquiring different CT slices (such as 128 slices for a 128 detector row scanner). CT angiography (CTA) is extremely responsive for detection of deep vein coagulation, renal artery stenosis and aneurysm. Kidney CT scanners are now extensively used in kidney diagnostic centres irrespective of its expense. New advances in CT include spectral (dual-energy) CT that has possibility to use

Fig. 4 CT scan of kidney (*arrows* pointing towards kidneys)

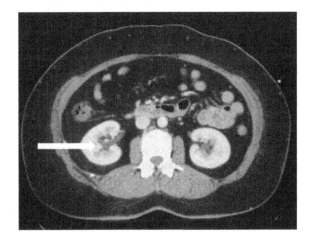

Fig. 5 MRI of kidneys
(*arrow* pointing towards
kidney)

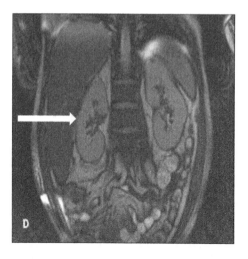

the spectrum of X-ray energies in order to carry out characterization of tissue and lesion detection [10]. The benefit of using CT modality is that it provides high-resolution images with good anatomical details and provides access to all abdominal insights. Limitation of using CT modality is that patient is exposed to radiations that may have side effects.

2.5 Magnetic Resonance Imaging (MRI)

Magnetic resonance imaging (MRI), as shown in Fig. 5, is a non-obtrusive imaging method which utilizes radio frequency waves and strong magnetic field to generate elaborative description of different organs and tissues. MRI is considered as non-invasive technique as it does not use ionizing radiation. However, contrast agents may be injected for augmentation of different parts of organ under consideration. MRI is contraindicated for patients having artificial valves, pacemaker, stunts or any other implanted medical devices. However, there are no known biological hazards, but the technique is expensive. MRI is an excellent procedure of imaging of tumour coagulum in renal veins as it can easily differentiate between angiomyolipoma, simple cyst, complex cysts and renal cell carcinoma. MRI is also helpful for the evaluation of possible renal donors prior to renal transplantation. In comparison with CT, MRI is suitable for pregnant and younger patients as it does not involve the use of ionizing radiations. It is also useful in those who are intolerant of iodinated agents because of allergy. In contrast to other imaging modalities, MRI provides assessment of human kidney perfusion and it allows you to acquire images with different orientations that do not require further image processing which makes it a good choice for imaging. An MRI imaging protocol named DCE-MRI uses gadolinium-based contrast agent which is injected into kidney and images are taken rapidly and repeatedly. One drawback of this technique is that due to fast scanning, spatial resolution of dynamic MR images is low and it suffers from motion introduced by patient breathing. MRI techniques that make use of contrast agent may be harmful to patient with kidney diseases. In order to avoid these problems, researchers investigated MRI technique known as diffusion MRI, which is non-invasive functional modality that relies on the movement of water molecules inside the organ as it provides indirect information about the structure of organ surrounding these water molecules. Diffusion-weighted MRI (DW-MRI) is being considered by researchers for assessment of kidneys because of its major role in filtering the water and blood of body [11]. DW-MRI is an emerging area and computer vision researchers have made efforts to check out its usefulness in the classification of kidney lesions [12–16], renal infections [17] and renal parenchymal disease.

As in case of CT, MRI does not involve any radiations while scanning and useful for detection of abnormalities in soft tissues. However, MRI scans are susceptible to patient movement and it is difficult to complete the procedure as patient feel claustrophobic based on the kind of study being accomplished.

2.6 Nuclear Medicine (Isotope Imaging)

Radio-isotope scanning approaches such as positron emission tomography (PET) and spectral positron emission tomography (SPECT) use gamma rays or radiotracer in order to investigate functional and qualitative details of kidney rather than anatomical details. PET images possess lower spatial resolution as compared to CT and MRI and it provide quantitative information about tissue abnormalities and structures by accumulating emitted photons from a radiotracer. Isotope imaging of renal falls in two broad categories, i.e. static and dynamic renal scans [8, 9]. Dynamic renography is useful for measurement of glomerular filtration rate (GFR) and differential renal blood flow (RBF) and provides the quantitative evaluation of renal function. It gives information about degree of obstruction and is used in assessment of the filtration, perfusion and drainage of renal transplants. Static renography is useful in children for detecting cortical scars in kidney with urinary tract infections (UTIs) (Fig. 6).

Among different isotopes, fluorodeoxyglucose (FDG) is the most frequently used PET radiotracer for the assessment of various neoplasms as well as in the planning of radiotherapy in kidney cancer. It usually recognizes abnormalities in initial stages of disease maturation—long before with any other diagnostic examination. Nuclear medicine scans are usually conducted by radiographers. Albeit nuclear studies involve some threats to the patient health, yet they turn out to be intensely capable with good specificity and sensitivity in the evaluation of subjects with renal artery disorder and in investigating the flaws related to perfusion. Nuclear imaging methods (specifically PET) are extravagant, and moreover, they provide images with low resolutions that may be the reason less work has been done on this modality by computer vision researchers [18]. Isotope imaging helps

Fig. 6 Nuclear medicine
scan of kidney (*arrow*
pointing towards kidney)

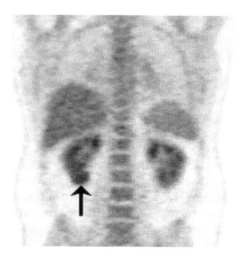

you to get both anatomical and functional details of organ under consideration, but these medicine scans are more sensitive as compare to other imaging modalities.

2.7 Optical Coherence Tomography (OCT)

Optical coherence tomography (OCT) is evolving scanning procedure as it provides real-time high-resolution cross-sectional images of tissues structures. An OCT scan of kidney is shown in Fig. 7. Optical coherence tomography is closely related to ultrasonography, with the exception that it works with echo deferment of light waves in place of sound waves to create pictures [19]. This modality has been employed in various biomedical applications incorporating, cardiology [20, 21],

Fig. 7 OCT image of the
human kidney in which
glomeruli (*G*), tubules (*T*) and
the kidney capsule (*C*) are
discernible

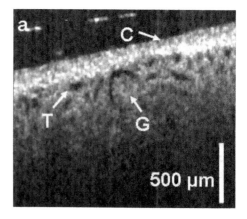

Table 1 Comparison of renal imaging modalities

	X-ray	IVU	US	CT	MRI	Nuclear medicine	OCT
Cost involved	Low priced	Low priced	Low priced	Moderate cost	High cost	High cost	High cost
Radiation risk	Less radiation is involved	Less amount of radiation is involved	No radiation is involved	Moderate radiation is involved	No radiation is involved	Less amount of radiation is involved	No radiation is involved
Time involved	Scanning takes less time duration	Scanning takes less time duration	Scanning takes less time duration	Scanning takes less time duration	Scanning takes more time duration	Scanning takes more time duration	Scanning takes less time duration
Spatial resolution	Low	Low	High	High	High	Low	High

gastroenterology [22, 23], dermatology [24], dentistry [25], urology [26] and others.

OCT has high-powered modality which integrates physical and functional information that can be utilized to investigate renal condition in vivo studies and at the time of medical procedures. OCT imaging of renal donor prior to renal transplant can help the surgeons to predict transplant outcomes. The adaptability of OCT imaging procedures makes it a perfect procedure to acquire images of the kidney in situ as it provides high-resolution facilities and increased depth examination, in comparison with other modalities [27]. Table 1 shows comparison of different renal imaging modalities based on cost, radiation risk, time involved and spatial resolution. Only few studies have been observed on assessment of kidney function using OCT, and it is new area for researchers to explore.

3 Conclusion

In the last two decades, we have seen major improvements in the diagnosis of kidney disorders with the improved imaging procedures. Different imaging modalities can be used to diagnose different kinds of renal disorders. Advancements in imaging techniques continue to evolve the role of imaging in decreasing the mortality and morbidity rates due to kidney disorders. Presently, researchers are investigating imaging techniques at the molecular level which will assist in understanding the nature of disease growth and development. It has been observed in the literature that one technique is not entirely suitable for investigation of all renal diseases. Thus, investigation is persistently being performed in order to upgrade the current imaging techniques and to invent new techniques.

References

1. Barsoum, RS.: Chronic kidney disease in the developing world. N. Engl. J. Med. (2006) 997–999.
2. Meguid, El., Nahas, A., Bello, AK.: Chronic kidney disease: the global challenge. Lancet (2005).
3. World Health Organization: Preventing Chronic Disease: A Vital Investment. Geneva, WHO (2005).
4. Webb, A.: Introduction to Biomedical Imaging. John Wiley and Sons Inc., NY, Hoboken (2003).
5. Sutton, D., Grainger, RG.: A Textbook of Radiology. E.S. Livingstone, Edinburgh (2002).
6. Myers, GL., Miller, WG., Coresh, J., Fleming, J., Greenberg, N., Greene, T. et al.: Recommendations for improving serum creatinine measurement: a report from the Laboratory Working Group of the National Kidney Disease Education Program. *Clinical Chemistry* (2006) 5–18.
7. Gans, SL., Stoker, J., Boermeester, MA.: Plain abdominal radiography in acute abdominal pain; past, present, and future. *International Journal of General Medicine* (2012) 525–533.
8. Sebastian, A., Tait, P.: Renal imaging modalities. Medicine (2007) 377–382.
9. Harvey, C., Hare, C., Blomley, M.: Renal Imaging Medicine. Imaging and Biopsy (2003).
10. Silva, AC., Morse, BG., Hara, AK., Paden, RG., Hongo, N., and Pavlicek, W.: Dual-energy (spectral) CT: applications in abdominal imaging. *Radiographics* (2011) 1031–1046.
11. Goyal, A., Sharma, R., Bhalla, AS., Gamanagatti, S., Seth, A.: Diffusion-weighted MRI in assessment of renal dysfunction. Indian J. Radiology Imaging (2012).
12. Squillaci, E., Manenti, G., Di Stefano, F., Miano, R., Strigari, L., Simonetti, G.: Diffusion weighted MR imaging in the evaluation of renal tumours. J. Exp. Clin. Cancer Res. (2004) 39–45.
13. Cova, M., Squillaci, E., Stacul, F., Manenti, G., Gava, S., Simonetti, G., et al.: Diffusion weighted MRI in the evaluation of renal lesions: Preliminary results. Br. J. Radiol. (2004).
14. Yoshikawa, T., Kawamitsu, H., Mitchell, DG., Ohno, Y., Ku, Y., Seo, Y., et al.: ADC measurement of abdominal organs and lesions using parallel imaging technique. Am. J. Roentgenol (2006)1521–30.
15. Taouli, B., Thakur, R., Mannelli, L., Babb, JS., Kim, S., Hecht, EM., et al.: Renal lesions: Characterization with diffusion–weighted imaging versus contrast–enhanced MR imaging. Radiology (2009) 398–407.
16. Sandrasegaran, K., Sundaram, CP., Ramaswamy, R., Akisik, FM., Rydberg, MR., Lin, C., et al.: Usefulness of diffusion–weighted imaging in the evaluation of renal masses. Am. J. Roentgenol. (2010) 438–45.
17. Verswijvel, G., Vandecaveye, V., Gelin, G., Vandevenne, J., Grieten, M., Horvath, M., et al.: Diffusion–weighted MR imaging in the evaluation of renal infection: Preliminary results. JBR–BTR (2002)100–103.
18. Fred, D., Mettler, A., Milton, J., Guiberteau, MD.: Essentials of Nuclear Medicine Imaging. Fifth ed., WB Saunders. Philadelphia (2005).
19. Andrews, PM. and Chen, Y.: Using Optical Coherence Tomography (OCT) to Evaluate Human Donor Kidneys Prior to and Following Transplantation. Nephrology & Therapeutics (2014).
20. Brezinski, M.: Characterizing arterial plaque with optical coherence tomography. Curr. Opin. Cardiol. (2002) 648–655.
21. Jang, IK., Bouma, B., MacNeill, B., Takano, M., Shishkov, M., et al.: In-vivo coronary plaque characteristics in patients with various clinical presentations using Optical Coherence Tomography. Circulation (2003) 373–373.
22. Bouma, BE., Tearney, GJ., Compton, CC., Nishioka, N.: High-resolution imaging of the human esophagus and stomach *in vivo* using optical coherence tomography. Gastrointest Endosc. (2000) 467–474.

23. Chen, Y., Aguirre, AD., Hsiung, PL., Desai, S., Herz, PZ., et al.: Ultra high resolution optical coherence tomography of Barrett's esophagus: preliminary descriptive clinical study correlating images with histology. Endoscopy (2007) 599–605.
24. Welzel, J., Lankenau, E., Birngruber, R., Engelhardt, R.: Optical coherence tomography of the human skin. J. Am. Acad. Dermatol. (1997) 958–963.
25. Otis, LL., Everett, MJ., Sathyam, US., Colston, BW.: Optical coherence tomography: a new imaging technology for dentistry. J. Am. Dent. Assoc. (2000) 511–514.
26. D'Amico, AV., Weinstein, M., Li, X., Richie, JP., Fujimoto, J.: Optical coherence tomography as a method for identifying benign and malignant microscopic structures in the prostate gland. Urology (2000) 783–787.
27. Qian, Li., Maristela, L., Onozato: Automated quantification of microstructural dimensions of the human kidney using optical coherence tomography (OCT), Optical Society of America (2009).

A Dynamic Model to Recognize Changes in Mangrove Species in Sunderban Delta Using Hyperspectral Image Analysis

Somdatta Chakravortty, Dipanwita Ghosh and Devadatta Sinha

Abstract Remote sensing and hyperspectral image data play an important role to map and monitor changes in mangrove forest mostly caused by natural disasters, anthropogenic forces and uncontrolled population growth. These have resulted in changes in the availability of natural resources and the physicochemical environment leading to a competition between the existing mangrove species for their survival. This paper applies hyperspectral data to obtain the fractional abundance of pure and mixed mangrove species at sub-pixel level and extracts the data to analyze inter-species competition in the study area over a period of time. The influence of different species traits such as rate of growth, rate of reproduction, mortality rate and processes that control coexistence of plant species that form the basis of competition analysis has been extracted from Hyperion time series image data of the study area. The image-extracted results have been verified through field visits. This study determines the dominance of certain mangrove species in a pixel area and examines its state of equilibrium or disequilibrium over a fixed time frame. This study has attempted to predict mangrove species dynamics within a mixed mangrove patch where they are interacting for coexistence or mutual exclusion.

Keywords Competition factor · Time series hyperspectral data · Eco-dynamic system · Mangrove species

S. Chakravortty (✉) · D. Ghosh
Department of Information Technology, Government College
of Engineering & Ceramic Technology, Kolkata, West Bengal, India
e-mail: csomdatta@rediffmail.com

D. Sinha
Department of Computer Science and Engineering, Calcutta University,
Kolkata, West Bengal, India

© Springer Nature Singapore Pte Ltd. 2018
P.K. Sa et al. (eds.), *Progress in Intelligent Computing Techniques: Theory,
Practice, and Applications*, Advances in Intelligent Systems and Computing 518,
DOI 10.1007/978-981-10-3373-5_5

1 Introduction

The Sunderban Biosphere Reserve that fringes the southern coastal belt of West Bengal is an ideal location where remote sensing can be successfully applied for survey and mapping of a wide range of mangrove plantation that are unique to this pristine ecological system. The Sunderban is the world's largest single-patch mangrove forest that offers an ecosystem of mangrove species, some of which are extremely rare and endangered in nature. Considering the enormous importance of this region, Sunderban is named as the "World Heritage Site" and "Ramsar site." Ironically, the Sunderban mangrove forest is now a threatened habitat that has turned into an "ecological hotspot" at recent times. Further, mangrove habitat experiences spatial and temporal changes because of the dynamicity that remains associated with the coastal environment. These disturbances cause ecological degradation which alters the state ecosystem permanently and changes the resource availability. Hence, an accurate up-to-date assessment of the nature, extent and distribution of mangrove species is very essential to frame up a management plan for survival of this dwindling ecosystem. The effect of physical disturbance makes species respond by initiating regeneration to retain their existence, thus promoting biodiversity in the ecosystem. In a similar way, the mangrove community interacts with each other. These interactions are competitive as well as mutual. This paper intends to analyze the interaction among the various species of mangrove community with the help of detailed information obtained from the hyperspectral image data [1–4]. There are studies in which the dynamic mangrove ecosystem has been modeled through physical monitoring of dense forests and ground truthing over several years. Dislich et al. [5] have used density-dependent mortality, and light-dependent colonization to analyze coexistence of species has graphically explained competition model on the basis of competitive ability and mortality [6]. There have been studies where changes in vegetation are monitored by observing satellite image data with respect to time. NDVI is calculated with varying time to measure the greenness of a particular area [7]. Tilman has measured competition between two species by colonization and dispersal rate of a species [8]. Changes over years have been mapped using techniques like image differencing by subtracting intensity values of same coordinates of two images acquired at different time periods [9]. Image ratioing of pixels have been used in studies to identify changes [10]. Change vector analysis has been done to calculate the magnitude of spectral changes among vectors of two images to monitor changes [11]. Studies have applied PCA, ICA and k-means algorithms for unsupervised change detection in hyperspectral data [12–16]. Studies have used linear predictors to predict a space-invariant affine transformation between image pairs and assume changes in hyperspectral image [17, 18]. The novelty of this study is the application of image-derived fractional abundance values obtained from spectral unmixing of hyperspectral data to analyze the eco-dynamic system of mangrove species. The abundance of mangrove species in time series hyperspectral data is used to analyze how they compete with each other to establish their dominance over other species with time. The fractional abundance values are also used to predict the outcome of competitive interaction with other species and whether it displays a state of equilibrium or instability.

2 Study Area

The mangrove habitats of Henry Island of the Sunderban Delta have been selected for the present study. The basis of selecting the study area is that the island comprises a variety of mangrove species which are rare and some which are gradually getting extinct. The island is dominated by pure and mixed patches of Avicennia Alba, Avicennia Marina, Ceriops Decandra, Bruguiera Cylindrica, Aegialitis and Excoecaria Agallocha.

3 Methodology

3.1 Acquisition of Data and Ground Survey

An Earth Observatory-1 satellite mounted Hyperion sensor-extracted hyperspectral imagery of the study area has been requested from the US Geological Survey, and a time series imagery was captured on May 27, 2011, and November 23, 2014. Field survey has been done to select sampling points and demarcate areas of pure and mixed mangrove species in the study area based on patterns visible in the hyperspectral image data. Radius of trees of mangrove species is measured to calculate area of each species which is used to parameterize the model. Ground data collection is used to validate the results obtained from hyperspectral data.

3.2 Endmember Detection and Derivation of Abundance of Sub-pixels

To identify distinct endmembers (unique signatures of mangrove species), an unsupervised target detection algorithm, N-FINDR, has been applied on the hyperspectral imagery [19]. The main principle of N-FINDR algorithm is to find pixels that can construct a maximum volume simplex, and these pixels are considered as endmembers. The pure signature of endmembers extracted from N-FINDR is fed into the fully constrained linear spectral unmixing (FCLSU) classifier to derive the fractional abundances of sub-pixels in the hyperspectral image [20].

3.3   Model Description

The model assumes that mangrove species spatially adjacent to each other compete with each other to survive within the limited resources. The fractional abundance values of each species derived from FCLSU for each pixel area (900 m^2) of Hyperion data represents its extent of occurrence within the area. *It is essential to calculate the survival capacity of the species over the pixel area to predict its dominance over others.* However, the values vary with time as they depend upon the type of species, competition and the physical environment around them. The survival capacity of mangrove species is dependent upon its rate of reproduction, rate of mortality and rate of growth over a period of time. The survival capacity K_i of species i is calculated as:

$$K_i = \frac{(a_i - \mu_i) * p_i}{a_i} \tag{1}$$

$$K_i = \frac{r_i * p_i}{a_i} \tag{2}$$

where a_i is the rate of reproduction, μ_i is the rate of mortality, r_i is the growth rate and p_i is the fractional abundance of species i in a pixel area over the years 2011 and 2014. r_i and μ_i have been calculated on the basis of increase and decrease in species population size (difference in fractional abundance occupied by species i) with time, respectively. Growth of species over a period of time has an assumption of minimum mortality rate (0.0005), and mortality has an assumption of minimum growth rate (0.0005). a_i is the summation of r_i and μ_i. Another important parameter to predict species response in competition with various species is the competition factor (CF). CF is the spatiotemporal probability of occurrence of species—imposed on species i by species j and vice versa. The competition factor is expressed as the importance value of each species (according to Curtis) and divided by 100.

$$\beta_{ij} = \frac{IV_{ij}}{100} \tag{3}$$

Importance value is an index which has been considered as the relative dominance of a species in the mangrove community among other species which may be expressed as:

$$RDO_i = \left(\frac{A_i}{\sum_{i=1}^{m} A} * 100 \right) \tag{4}$$

where RDO_i represents relative dominance of a mangrove species, A_i represents the fractional abundance area occupied by species i and m is the total number of species present within the pixel area. Depending upon the competition factor and survival capacity of species, competition coefficient is calculated based on the reproduction

Table 1 Inequalities used to interpret state of equilibrium of mangrove species

Inequalities	$k_i > (k_j/a_{ji})$	$k_i < (k_j/a_{ji})$
$k_j > (k_i/a_{ij})$	Unstable state	Species j wins
$k_j < (k_i/a_{ij})$	Species i wins	Stable state

k stands for survival capacity and a_{ij}, a_{ji} are the competition coefficients

rate of species to analyze interaction between various species. If competition coefficient between two species (i and j) is a_{ij} and reproduction rate of species is f_i, a_{ij} is expressed as:

$$a_{ij} = \frac{\left(2 * \beta_{ij} * k\right) + f_i}{f_i} \tag{5}$$

Combination of species interacting with each other may exhibit a state of equilibrium or disequilibrium due to competition among themselves or other external influences. Inequalities used to exhibit stability or instability due to pair-wise interactions between mangrove species have been used and are displayed in Table 1.

4 Result and Analysis

The proposed model applies time series hyperspectral image data of 2011 and 2014 to identify pure endmembers (mangrove species) with N-FINDR in the densely forested study area and derive its fractional abundance values using sub-pixel classification algorithm. FCLSU has been applied to obtain mangrove species abundance distribution maps for two years (Figs. 1 and 2). The increase or decrease in extent of abundance of mangrove species in certain regions of the study area in year 2011 and 2014 is shown in Table 2.

Fig. 1 Mangrove species distribution map algorithm of 2011

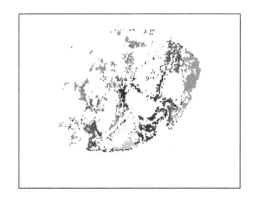

Fig. 2 Mangrove species
distribution map algorithm of
2014

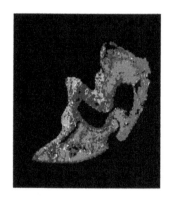

Table 2 Fractional abundance of mixed mangrove species obtained from hyperspectral image of 2011 and 2014

Name of species	Geographic location (2014 image)	Fractional abundance	Geographic location (2011 image)	Fractional abundance
Excoecaria agallocha	21.57702, 88.27503	0.6094	21.57702, 88.27503	0.5833
Avicennia officinalis		0.2183		0.2048
Avicennia alba		0.0164		0.1009
Brugeria cylindrica		0.0575		0.1110

A detailed analysis of mangrove species dominance has been made from spectral unmixing of hyperspectral data. Reproduction rate and survival capacities of individual mangrove species have been derived from the abundance estimates of endmembers in the sub-pixels. Growth rate has been calculated from the changes derived from the hyperspectral data of 2011 and 2014. It is assumed that the mortality rate of mangrove species does not change with time. The growth rate, mortality rate and reproduction rate of certain locations are shown in Table 3. Equilibrium analysis of the mangrove ecosystem has been expressed by varying survival capacities (K_i and K_j) and interaction coefficients (a_{ij} and a_{ji}). Importance value (IV) is another parameter which indicates dominance of a particular species over another. The higher the IV of a species, the more it dominates in the area. The importance value and survival capacity of mangrove species present in certain locations of the study area are given in Table 4.

Analysis of Table 3 indicates that *Excoecaria agallocha* is the most dominant and important species on 900 m^2 pixel area of geographic location 21.57702, 88.27503. *Avicennia officinalis* is the next dominant species in that region. The lesser important species are *Avicennia alba* and *Brugeria cylindrica*. The matrix

Table 3 Growth rate, mortality rate and reproduction rate of important mangrove species

Geographic location	Name of species	Growth rate	Mortality rate	Reproduction rate
21.57702, 88.27503	*Excoecaria agallocha*	0.0087	0.0005	0.0092
	Avicennia officinalis	0.0045	0.0005	0.005
	Avicennia alba	0.0005	0.0282	0.0287
	Brugeria cylindrical	0.0005	0.0178	0.0183

*Minimum mortality rate: 0.0005; minimum growth rate: 0.0005

Table 4 Survival capacity and importance values of mixed mangrove species

Geographic location	Name of species	Survival capacity	Importance value
21.57702, 88.27503	*Excoecaria agallocha*	0.5763	67.5909
	Avicennia officinalis	0.1965	24.2125
	Avicennia alba	0.0003	1.8190
	Brugeria cylindrica	0.0016	6.3776

Table 5 Competition coefficients of mangrove species at location 21.576933, 88.2752

	Excoecaria agallocha	*Avicennia officinalis*	*Bruguiera cylindrica*	*Avicennia alba*
Excoecaria agallocha		1.0029	1.0555	1.3606
Avicennia officinalis	1.0000		1.0010	1.0063
Bruguiera cylindrica	1.0005	1.0275		1.0337
Avicennia alba	1.0344	2.9575	1.3682	

representing competition coefficients between different species existing in location 21.576933, 88.2752 is displayed in Table 5. After further analysis of survival capacity values and thereafter evaluating the coexistence inequalities (Table 6), it is apparent that certain degrees of disturbances produce changes in equilibrium status.

At geographic location 21.57702, 88.27503, it is observed that *Excoecaria agallocha* and *Avicennia officinalis* have the same strength to compete with each other. An unstable state at this location indicates that there is a possibility of dominance of either species in this region. However, competition between *Excoecaria agallocha-Avicennia alba* and *Excoecaria agallocha-Bruguiera cylin-drica* indicates that *Excoecaria agallocha* is the stronger species among them. *Avicennia officinalis* is the second most persistent species in the area followed by *Bruguiera cylindrica* and *Avicennia alba*. It is thus observed that on analysis of competition between mangrove species over the study area using time series data of

Table 6 Pair-wise competition among species and their outcomes

Geographic location	Species	Competing species	Species 1 wins	Species 2 wins	Unstable	Stable
21.57702, 88.27503	*Excoecaria agallocha*	*Excoecaria agallocha + Avicennia officinalis*			Y	
	Avicennia officinalis	*Excoecaria agallocha + Avicennia alba*	Y			
	Avicennia alba	*Excoecaria agallocha + Bruguiera cylindrica*	Y			
	Brugeria cylindrica	*Avicennia officinalis + Avicennia alba*	Y			
		Avicennia officinalis + Bruguiera cylindrica	Y			
		Avicennia alba + Bruguiera cylindrica		Y		

2011 and 2014, *Excoecaria agallocha*, *Avicennia officinalis* and *Avicennia alba* have a better success rate of survival over other mangrove species in the study area.

5 Conclusion

This competition model has been uniquely implemented on hyperspectral data and provides connections between competing mangrove species in the study area. This model has helped predict mangrove species which dominate over others and will survive with time. This paper has extracted features of hyperspectral imagery to estimate fractional abundances of target objects and used the values as input to derive parameters such as relative dominance, importance value of individual mangrove species and their survival capacity in the study area. The model has successfully identified the dominant and dominated species of the pixel areas. On analysis of results, it has been observed that *Excoecaria agallocha*, *Avicennia alba* and *Ceriops decandra* are the most dominant followed by *Phoenix* and *Bruguiera cylindrica*. It is noted that existence of dominant mangrove species such as *Excoecaria agallocha* and *Avicennia officinalis* in the same area leads to an unstable condition as they have the same strength to compete with each other to survive. These results will help in formulating plan for conservation and management of mangroves and preparing a framework for their regeneration and restoration in the study area.

References

1. Demuro, M., Chisholm, L: Assessment of Hyperion for Characterizing Mangrove Communities. Proc. of the Internat. Conf. AVIRIS Workshop 18–23 (2003).
2. Vaiphasa, C: Innovative Genetic Algorithm for Hyperspectral Image Classification. Proc. Intern. Conf. MAP ASIA [http://www.gisdevelopment.net/technology/ip/ma03071abs.htm] (2003).
3. Vaiphasa, C., Ongsomwang, S: Hyperspectral Data for Tropical Mangrove Species Discrimination. Proc. 25th ACRS Conf 22–28 (2004).
4. Vaiphasa, C., Ongsomwang, S: Tropical Mangrove Species Discrimination Using Hyperspectral Data: A Laboratory Study. Est., Coas. Shelf Sc. 65(1–2) 371–379 (2005).
5. Dislich C., Johst K., Huth A: What enables coexistence in plant communities? Weak versus strong Traits and the role of local processes. Helmoltz Center for Environmental Research UFZ Leipzig, Ecological Modelling, Leipzig, Germany 221(19) 2227–2236 (2010).
6. Adler, F., Mosquera, J: Is space necessary? Interference competition and limits to biodiversity, Ecology 3226–3232(2000).
7. John, R., Dattaraja, H., Suresh, H., Sukumar, R: Density-dependence in Common Tree Species in a Tropical Dry Forest in Mudumalai. J. Veg. Sci. 13 45–56 (2002).
8. Tilman, D: Competition and Biodiversity in Spatially Structured Habitats. Ecology 75 2–16 (1994).
9. Hussain M., Chen D., Cheng A., Wei H., Stanley D: Change Detection from Remotely Sensed Images: From Pixel-based to Object-based Approaches. ISPRS Journal of Photogrammetry and Remote Sensing. 8091–106 (2013).
10. Shaoqing Z., Lu X: The Comparative Study of Three Methods Of Remote Sensing Image Change Detection. The International Archives of the Photogrammetry, Remote Sensing and Spatial Information Sciences. 37(B7) (2008). Beijing.
11. Singh S., Talwar R: A Comparative Study on Change Vector Analysis based Change Detection Techniques. Indian Academy of Sciences. 39(6) 1311–1331 (2014).
12. Kumar V., Garg K.D: A Valuable approach for Image Processing and Change Detection on Synthetic Aperture Radar Data. International Journal of Current Engineering and Technology 3(2) (2013).
13. Celik T: Unsupervised Change Detection in Satellite Images Using Principal Component Analysis and K-Means Clustering. IEEE Geoscience and Remote Sensing Letters 6(4) (2009).
14. Vikrant Gulati, Pushparaj Pal: Enhancement of ICA Algorithm Using MatLab for Change Detection in Hyperspectral Images, IJESRR, Volume 1, Issue 5 (2014).
15. Gulati V., Pal P.A., Gulati: Survey on Various Change Detection Techniques for Hyper Spectral Images. International Journal of Advanced Research in Computer Science and Software Engineering 4(8) 852–855 (2014).
16. Benlin X., Fangfang L., Xingliang M., Huazhong J: Study On Independent Component Analysis: Application In Classification And Change Detection Of Multispectral Images. The International Archives of the Photogrammetry, Remote Sensing and Spatial Information Sciences. 37(B7). Beijing (2008).
17. Eismann M.T., Meola J. and Hardie R.C: Hyperspectral Change Detection in the Presence of Diurnal and Seasonal Variations, IEEE Transactions on Geoscience and Remote Sensing. 46(1) (2008).
18. Wu C., Du B., and Zhang L: A Subspace-based Change Detection method for Hyperspectral Images. Selected Topics in Applied Earth Observations and Remote Sensing, 6(2) 815–830 (2013).
19. Plaza A., Martínez P., Pérez R. and Plaza J: A Quantitative and Comparative Analysis of Endmember Extraction Algorithms from Hyperspectral Data. IEEE Trans. on Geo. and Rem. Sens., 42(3) (2004).
20. Keshava N. and Mustard: J. Spectral Unmixing. IEEE Signal Processing Magazine. 44–57 (2002).

A Content-Based Visual Information Retrieval Approach for Automated Image Annotation

Karthik Senthil, Abhi Arun and Kamath S. Sowmya

Abstract Today's digital world is filled with a vast multitude of content such as text and multimedia. Various attempts are being made to develop modern and powerful search engines in order to support diverse queries on this large collection of textual and multimedia data. For supporting intelligent search, particularly for multimedia data such as images, additional metadata plays a crucial role in helping a search engine handpick the most relevant information for a query. A common technique that is used to generate pertinent metadata for visual multimedia content is by the process of annotation. Automating the annotation process given the large volume of visual content available on the Web is highly advantageous. In this paper, we propose an automated image annotation system that employs a content-based visual information retrieval technique using certain features of the image. Experimental evaluation and analysis of the proposed work have shown promising results.

Keywords Content-based information retrieval · Automatic image annotation · Image search · Search engines

1 Introduction

Automatic image annotation refers to the process by which a computer processes an image and automatically assigns captions or keywords to accurately describe it. These annotations are stored as metadata for images and can be utilized by other applications (like search engines and indexers). The unprecedented growth of the

K. Senthil (✉) · A. Arun · K.S. Sowmya
Department of Information Technology, National Institute of Technology Karnataka,
Surathkal, Mangalore, Karnataka, India
e-mail: karthik.senthil94@gmail.com

A. Arun
e-mail: abhiarun94@gmail.com

K.S. Sowmya
e-mail: sowmyakamath@nitk.ac.in

© Springer Nature Singapore Pte Ltd. 2018   69
P.K. Sa et al. (eds.), *Progress in Intelligent Computing Techniques: Theory,*
Practice, and Applications, Advances in Intelligent Systems and Computing 518,
DOI 10.1007/978-981-10-3373-5_6

Internet and World Wide Web has led to an increase in importance of automatic annotation of images. Identifying relevant images from the Web or in massive image databases is a non-trivial task as most of these images exist without any metadata. Many approaches have been proposed to solve this problem, like using non-textual queries based on color, texture, shape, etc., as can be seen during image search using the popular Google Image Search. But, it is also a fact that users find it difficult to represent information using abstract image features and mostly prefer textual queries. Thus, automatic image annotation is a popular way of solving this problem.

The challenges that are commonly faced in the field of automatic image annotation include incorporation of multidimensional information into learning models, adequate use of shape information, and the need for extensive training in various semantic concepts which may be computationally inefficient. Several techniques aim to deal with these non-trivial problems through the use of a combination of high-level and low-level features of the image and based on the actual visual content of the image. Content-based visual information retrieval (CBVIR), also called content-based image retrieval (CBIR), refers to image search approach that uses the actual content of the image rather than its metadata such as keywords, tags, or descriptions. "Content" here refers to actual properties of the image such as color, shape, and texture.

Any CBVIR-based technique is associated with some important terms—*image descriptor* (used to describe an image in the search space, e.g., color histograms, Hu moments); *feature* (the output of the *image descriptor*); and the *distance metric* (measure of how similar/dissimilar two image feature vectors are, e.g., chi-squared distance and Euclidean distance). A CBVIR system that can address the associated challenges of large-scale image retrieval, with high accuracy and in good time is considered an effective system. A very famous CBIR system for reverse image search is TinEye [7]. TinEye claims to be the first image search engine that enables users to search images by submitting a query image itself. When a query image is submitted, TinEye processes it by applying proprietary algorithms and matches it with indexed images. TinEye is capable of also matching and returning edited versions of the query image, but other similar images are usually not included in the results.

In this paper, we present a CBVIR system for the purpose of automatic annotation of unknown images, to support text-based natural language queries from users. The system is fully automated categorizing unannotated images based on the classification performed on a training (or proof of concept) dataset of tagged images. The users of the system can also evaluate the accuracy of the algorithm by means of a Web portal. The rest of the paper is organized as follows: In Sect. 2, some related works in the area of automatic image annotation is discussed. Section 3 describes the proposed work and methodology used for developing the automatic image annotation system. In Sect. 4, we discuss in detail the CBVIR-based algorithm for the system. Section 5 presents the results and analysis of the technique used for automatic annotation. Section 6 provides the conclusions and discusses possible future work.

2 Related Work

Several approaches for automated content-based image retrieval are available in literature, of which some significant works in the area of automatic image annotation are discussed here. Yang et al. [1] proposed an active-tagging scheme which combines humans and computers to automatically assign representative tags to images. Their iterative algorithm uses tag prediction and sample selection as two major components. A set of images are selected in each iteration as per predefined criteria (such as ambiguity, citation, and diversity) and manually tagged by the human annotators. These are added to the tagged dataset, after which, tag prediction is performed by the computer for the remaining images. This process is performed iteratively for the entire image dataset.

He et al. [2] propose an approach in which the annotation of images is regarded as multiclass image classification. The training set consists of a set of pre-annotated images which are segmented into a collection of regions associated with a keyword. For each region, a number of low-level visual descriptors are extracted and clustered, and the cluster center is calculated. For a new image region to be annotated, the distance between the instance and every cluster center is measured and the nearest categorys' keyword is chosen for annotation. Similarly, Wan [3] presents a technique that divides the image into regions with similar visual features, extracting the image visual information of each region. It then uses an improved decision tree algorithm for automatic annotation of images.

Li et al. [4] proposed a search-based framework which combines the two methods of social user tagging and automatic image tagging. Annotation of images is done by retrieving the visual neighbors of the images. In this process, k-nearest neighbors of the image from a user-tagged image database are found. Then, using tags of neighbor images, the most relevant images are selected. The unreliability and sparsity of user tagging is tackled using a joint-modality tag relevance estimation method which efficiently addresses both textual and visual clues. Jiang et al. [5] propose a machine learning-based approach for automatic image annotation. It combines Bayesian learning algorithm with the decision tree algorithm for this purpose.

It can be seen that different methods have been used for image annotation purpose and image features such as color histograms, image regions, similarity with other images have been considered. In this paper, we consider the shape of the image for the annotation process. We use chi-squared distance metric [6] to compare the feature vectors. We also define the terms *true positive* and *false positive* which allow the user to decide whether the annotation tags match the particular image or not. Finally, a crowdsourced application, where people can provide relevance feedback on the accuracy of the image annotation is used.

3 Proposed Work

Figure 1 gives an overview of the proposed work and methodology adopted for the CBIR-based algorithm for automatic tagging of images.

3.1 Retrieving a Dataset of Tagged Images

In this phase, we use *flickr.com* API [8], the online photo-sharing application, to scrape images and use them for training our CBIR application. All images provided in this Web site are tagged by humans as a crowdsourced initiative. The *flickr.com* API is used to obtain 60 images (6 categories X 10 images per category). The information for the categories and image URLs are stored using the JSON format in *tags.json*. For each search category, the corresponding JSON from this file is loaded and the image is fetched from each specified URL. Then, the image name and tags are stored in a database *tagged_images.db*.

3.1.1 Proposed CBIR Algorithm

The proposed algorithm is intended for automatically analyzing the content features of input images to generate metadata in the form of natural language tags. Here, the algorithm is first trained on a tagged dataset, and then, it is tested on an untagged dataset. The detailed design and working of the algorithm is discussed in Sect. 4.

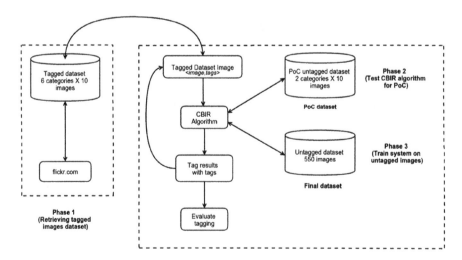

Fig. 1 Methodology for the proposed framework

3.1.2 Proof of Concept Testing for Proposed Algorithm

In this phase, the correctness of our CBIR-based auto-annotation algorithm is verified by applying it to a proof of concept (PoC) dataset. The procedure followed is outlined in Algorithm 1, where *image* refers to the image object stored in the database while *tags* refers to an array of tags associated with an *image* object.

Algorithm 1 Procedure for PoC testing of CBIR algorithm

1: **procedure** POC_TESTING
2:       **for** each record in tagged images dataset **do**
3:             Determine the parameters <*image,tags*>
4:             Run CBIR algorithm with *image* on "PoC dataset"
5:             Tag the results with *tags*
6:       **end for**
7:       Evaluate the tagging
8: **end procedure**

3.1.3 Training the CBIR Algorithm on Final Untagged Dataset

After the verification from the previous phase, the proposed algorithm is applied to the final untagged dataset of 550 images. The procedure followed is exactly as described in Algorithm 1 with the exception of using the final untagged dataset in Step 4.

4 Proposed CBVIR Algorithm

The main image processing function *CBIR* discussed in this paper is illustrated using the workflow diagram depicted in Fig. 2. The proposed CBVIR algorithm is designed to use certain parameters as below:

1. *Image descriptor or feature extracted by the algorithm*: In the proposed algorithm, we use the **shape** feature characterized by statistical moments (Hu moments [9]) to describe the image. This feature was decided based on search domain of the framework. The dataset involved inanimate objects of fixed shape and animate objects with an approximately similar outline. Other alternatives like color histogram would not produce accurate results given the diversity in color between images of the same category.

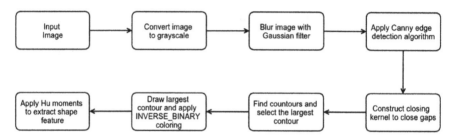

Fig. 2 Workflow of the *CBIR* function

2. *Distance metric/similarity function used to compare features of images*: The **chi-squared distance** metric was adopted to compare feature vectors. The Pearson's chi-squared statistic is commonly used for features with probabilistic distributions and since Hu moments is calculated based on a probability distribution function, this metric is an ideal choice. The formula represents the following probability distribution function

$$M_{ij} = \frac{\sum_x \sum_y x^i y^j I(x, y)}{\sum_x \sum_y I(x, y)} \tag{1}$$

where $I(x, y)$ represents pixel intensities of the image.

3. *Threshold value for tolerance in the distance metric*: Based on the output from the distance metric function, we sort the images in the dataset in ascending order of the distance from the query image's shape feature. Using heuristic observations and PoC testing, we assigned **threshold = 5.1e-10** to allow maximum number of true positives[1] while maintaining a low false-positive count.

The proposed CBIR-based technique for auto-tagging is described using two procedures defined in Algorithm 2. As shown, the procedure *CBIR_TRAINING* takes a tagged image dataset as input and uses it to annotate the untagged image dataset. In this procedure, *index* refers to an index file containing features of all images in the dataset. The *SEARCH* procedure extracts the shape feature from an input image (*queryImg*) and identifies all images in untagged dataset that have a similar feature. The similarity is quantified using *chi2_distance* metric function. These similar images are collected in *results* array.

---

[1]Terms "true positive" and "false positive" are defined in Sect. 5.4.

Algorithm 2 CBIR-based algorithm for auto-tagging of images

```
 1: procedure CBIR_TRAINING(tagged_images.db)
 2:      dataset = untagged_images
 3:      for each image in tagged_images.db do
 4:          queryImg = image
 5:          tags = image.tags
 6:          results = SEARCH(queryImg, index, dataset)
 7:          for each resultImg in results do
 8:              resultImg.tags ≪ tags
 9:          end for
10:          Add results to cbir_results.db
11:      end for
12: end procedure
```

```
 1: procedure SEARCH(queryImg, index, dataset)
 2:      queryImg_feature = CBIR(queryImg)
 3:      for each i in index do
 4:          distance = chi2_distance(i.feature, queryImg_feature)
 5:          if distance ≤ threshold then
 6:              Add i.image to results
 7:          end if
 8:      end for
 9:      return results
10: end procedure
```

5 Experimental Results and Analysis

In this section, the experimental results obtained for the CBIR-based technique for auto-tagging of images is discussed in detail. The CBIR technique was implemented and tested on a single core of an Intel(R) Core(TM) i5-3337U CPU @ 1.80 GHz processor.

- Tolerance threshold value for the distance metric = 5.1e-10
- Total images tagged or identified = 499 out of 550 (90.73%)
- Total tags used from tagged dataset of images = 96 out of 98 (97.96%)
- Total runtime of algorithm = 45 min.

The above results show the range and coverage of the CBIR algorithm. By using a relatively small training dataset of 60 images, the technique achieves nearly 90% coverage indicating the capability of the technique. We also observed that the execution is not particularly efficient with regard to the time taken for analyzing the complete untagged dataset of images.

5.1   Image Category-Wise Results

Using the above-mentioned experimental conditions, the CBIR algorithm was applied to untagged images spanning all considered categories. The results are summarized below.

5.1.1   Number of Images Tagged or Identified per Category

Table 1 indicates the coverage of the technique for each category of the search domain. We can observe that inanimate objects such as airplanes, cars, and motorbikes give a high percentage of coverage while animate objects such as cats, puppies, and flowers result in a relatively lower coverage.

5.1.2   Number of Tags per Category

Table 2 indicates the tags assigned for images in each category. Again, we observe that inanimate objects such as motorbikes, airplanes, and cars get relatively more number of tags than objects such as puppies, cats, and flowers.

These results will be further validated by the evaluation of CBIR algorithm where we used a Web application to assess the tagging. This is an ongoing stage of the work since the approach is based on the crowdsourcing and the results from this will be tabulated when evaluations for all 499 images are complete.

Table 1   Results for number of images tagged per category

Category	Number of images identified	Total number of images	Percentage
Airplanes	95	100	95
Cars	89	100	89
Cats	83	100	83
Flowers	83	100	83
Motorbikes	100	100	100
Puppies	49	50	98

Table 2   Results for number of tags per category

Category	Number of tags
Airplanes	18
Cars	18
Cats	17
Flowers	17
Motorbikes	20
Puppies	13

5.2 Proof of Concept (PoC) Testing of CBIR Algorithm

The above experiment and results provide a numerical evaluation of the algorithm. We performed a proof of concept testing to verify the correctness of our algorithm, on a small scale. The experimental setup is laid out below:

- Number of untagged images = 20
- Pre-evaluation of untagged dataset = 2 categories (airplanes and cars)
- Training dataset = 5 tagged images of airplanes.

The results for this test was evaluated manually. We discerned the results as "true positives" and "false positives" as explained in Sect. 5.4. Outcomes of the PoC testing are stated below:

- Total number of airplane images in POC untagged dataset = 10
- Total number of car images in POC untagged dataset = 10
- Query = 1 image of an airplane
- Results returned = 13 images
- Number of airplanes in the results = 10
- Therefore, number of true positives = 10 out of 10 (100%)
- Number of non-airplanes (cars) in the results = 3
- Therefore, number of false positives = 3 out of 10 (33%).

We can observe that the algorithm successfully detected all 10 airplanes from the untagged dataset, but misidentified 3 images of cars to be airplanes (false positive).

5.3 Threshold Heuristic Analysis

We also performed a heuristic analysis of the influence of the threshold parameter on the coverage of the CBIR algorithm. The program was executed on an untagged image dataset of 550 images and results obtained are shown in Fig. 3. The graph indicates that the number of tagged images increases with increase in threshold value but the efficiency of the algorithm also proportionally decreases. The execution time was noticed to increase with increase in threshold. The tendency to get more false positives also increases.

5.4 Crowdsourcing-Based Evaluation of CBIR Algorithm

The output of the CBIR-based technique for tagging images is evaluated using a crowdsourced approach. We developed a Web application where a set of human users were shown the image and the corresponding tags assigned to it by the CBIR algorithm. The user can then discern the result to be either a "true positive" or "false positive":

Fig. 3 Results of threshold
heuristic analysis

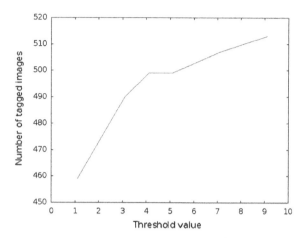

- **true positive**: Result is true positive only if all the tags assigned to image belong to the same search domain category of image
- **false positive**: Result is false positive if some tags of the image do not belong to the same category as image.

Crowdsourced evaluation of CBIR based auto-tagging of images

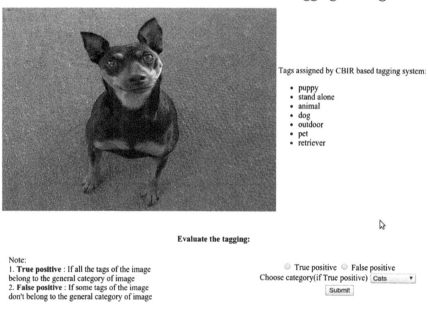

Fig. 4 Evaluation of CBIR algorithm

The correctness of the algorithm is determined by maximizing the number of true positives while at the same time minimize the false-positive count. Figure 4 illustrates the working of the Web application developed for the human assessment of the proposed CBIR algorithm.

6   Conclusion and Future Work

In this paper, a content-based image retrieval system for automatic tagging of images is proposed. The CBIR system uses object shape as a feature in an input image and the chi-squared distance metric for comparing the features of images. Experimental results showed that the algorithm produces promising and conducive outcomes. The PoC testing shows that the technique produces 100% accuracy in detecting true positives. We can also conclude that the threshold heuristic analysis helps in identifying the feasible value for the "threshold" parameter in the algorithm. Thus, the CBIR-based technique can be certainly used to create a well-tagged image database.

As part of future work, the quality of tagging can be further improved. It is currently not completely accurate, for example, many images of flowers get tagged as "pink" even though they are of different colors. This is an inherent disadvantage of using object shape and contours as features for CBIR. One possible solution for this could be to use **color histograms** as a second feature and reprocess the results from the current algorithm. We can also improve the tagging for large-scale datasets by incorporating machine learning techniques. The execution time of the algorithm is also critically high especially for a large dataset. There is more scope for improving this performance through techniques from high-performance computing.

References

1. Kuiyuan Yang, Meng Wang, Hong-Jiang Zhang.: Active tagging for image indexing. In: IEEE International Conference on Multimedia and Expo (ICME), pp. 1620–1623, New York (2009)
2. Dongjian He, Yu Zheng, Shirui Pan, Jinglei Tang.: Ensemble of multiple descriptors for automatic image annotation. In: 3rd International Congress on Image and Signal Processing (CISP), vol. 4, pp. 1642–1646, Yantai (2010)
3. Shaohua Wan.: Image Annotation Using the SimpleDecisionTree. In: Fifth International Conference on Management of e-Commerce and e-Government (ICMeCG), pp. 141–146, Hubei (2011)
4. Xirong Li, Snoek C.G.M., Worring Marcel.: Annotating images by harnessing worldwide user-tagged photos. In: IEEE International Conference on Acoustics, Speech and Signal Processing (ICASSP 2009), pp. 3717–3720, Taipei (2009)
5. Lixing Jiang, Jin Hou, Zeng Chen, Dengsheng Zhang.: Automatic image annotation based on decision tree machine learning. In: International Conference on Cyber-Enabled Distributed Computing and Knowledge Discovery (CyberC '09), pp. 170–175, Zhangijajie (2009)
6. Greenwood, Priscilla E., Michael S. Nikulin.: A guide to chi-squared testing. John Wiley & Sons (1996)
7. TinEye Reverse Image Search, https://www.tineye.com/ *Last accessed on 7 February 2016*

8. Flickr. Flickr Service, https://www.flickr.com/services/api/ *Last accessed on 7 February 2016*
9. Ming-Kuei Hu.: Visual pattern recognition by moment invariants. In: IRE Transactions on Information Theory, vol. 8, no. 2, pp. 179–187 (1962)

Hand Gesture-Based Stable PowerPoint Presentation Using Kinect

Praveen Kumar, Anurag Jaiswal, B. Deepak and G. Ram Mohana Reddy

Abstract Recent trends in the development of interactive devices provide a better human–computer interaction (HCI) experience in various domains, e.g., academics, corporate world, teaching-assistant tools, and gaming. Presently sensors and camera-based applications are the area of interest of many researchers. In this paper, we mainly focus on developing natural hand gesture-based presentation tool for controlling PowerPoint presentation with more efficiency and stability. In order to provide stability during presentation using Microsoft Kinect, this paper proposes a novel idea of locking and unlocking for recognition of gestures. A comparative study has been done by considering three parameters namely viewer's interest, frequency of movements, and overall stability during presentation. Results show that gesture-based presentation system provides better experience to both presenter and viewer as compared to traditional system. Further, there is no need of any extra device such as mouse or keyboard while presenting. This tool can also be used for development of other related applications.

Keywords Human–computer interaction (HCI) · Hand gesture recognition · Kinect depth sensor

P. Kumar (✉) · A. Jaiswal · B. Deepak · G.R.M. Reddy
National Institute of Technology Karnataka,
Surathkal, Mangalore 575 025, Karnataka, India
e-mail: agrawalpraveen241@gmail.com

A. Jaiswal
e-mail: anuragraj.90@gmail.com

B. Deepak
e-mail: daredreamer1990@gmail.com

G.R.M. Reddy
e-mail: profgrmreddy@gmail.com

© Springer Nature Singapore Pte Ltd. 2018 81
P.K. Sa et al. (eds.), *Progress in Intelligent Computing Techniques: Theory,
Practice, and Applications*, Advances in Intelligent Systems and Computing 518,
DOI 10.1007/978-981-10-3373-5_7

1 Introduction

Presently, the development of game console based on physical interfaces has become the trend for many companies such as Sony and Microsoft. In the History of Video game Consoles, the "Sony Play Station Move" was the first motion-sensing gaming device, "Wii" from Nintendo focused on integration of controllers with movement sensors along with joystick, then later on Microsoft joined the scene by launching "Kinect." Unlike the other two, Kinect does not use any controller rather makes the user as a controller for the game.

Based on depth-sensing camera and built in mic, Kinect provides flexibility to users to interact and control through their computer or console through natural gesture recognition and spoken commands without any external game controller. For real-time user interaction, Microsoft Kinect provides an easy and interactive way. The Kinect Software Development Kit (SDK) Driver Software with Application Programming Interfaces (API) released by Microsoft gives access to skeleton tracking and raw sensor data streams. Table 1 summarizes the comparison of different technologies used in motion sensing.

Although it is acceptable that all applications cannot be suitable based on gesture recognizing interfaces for daily life task, because user may feel tired quickly if the application requires lots of actions and it would be difficult to achieve it through gesture only. Still there are some tasks which can be performed even better through gestures.

Presentations are frequently being used in several areas such as academics, business, meetings, workshops, and training to convey thoughts and ideas in a simple and direct manner. Presentation using traditional style needs access of mouse, keyboard, holding a remote or controller and also may need an extra person for help. This may be a poor interaction with viewer and is less effective. Hence, there is a need for the development of an effective way of presentation, in which a presenter

Table 1 Comparison of existing motion-sensing devices [1]

Technology	Merits	Demerits
Kinect	Specific SDK, drivers, open-source libraries. Capable for processing both color and depth images	Little bit costly in comparison to regular Webcams. Dependent on light. Image processing is needed
EyeToy	Lower in cost and easy to develop using Webcams, available open-source image processing libraries	Based on the resolution and the frame rate quality varies and often more dependent on light
Move	Camera tracking along with accelerometer	Controller dependent. No availability of open-source libraries
Wii	Accelerometer-based tracking along with available open-source libraries, makes image processing unnecessary	Dependent on controller

can use his hands gesture while talking and walking through the presentation slides for next, previous, play, pause, full screen, start, stop etc.

Use of gesture for presentation is a challenging task due to continuous recognition of gesture by sensors. Sometimes it also amplifies the recognized gesture even if presenter does not want while existing works [6] have the advantage of using gestures for PowerPoint presentation, but these works suffer from the weakness of stability. This motivates us to devise a method for achieving stability during the presentation using gestures.

To overcome the above-mentioned problem, in this paper, we propose the idea of locking and unlocking for the recognition of gestures. By using the gestures, the presenter can lock and unlock the recognition capability of the sensor. This eliminates the unnecessary recognition and thus provides stability along with effectiveness during presentation for both presenter and viewer.

The core idea behind the proposed work is based on the computation of geometrical absolute difference between the detected ovals for both palms of the presenter for both locking and unlocking capabilities of the sensor.

The Key Contributions of this paper are as follows:

- Design and development of a stable and robust presentation system through locking and unlocking of hand gestures.
- A comparative study of effectiveness between traditional and gesture-based presentation style on three parameters.

The rest of this paper is organized as follows: Sect. 2 deals with the related work; Sect. 3 explains the proposed methodology; Sect. 4 deals with the results and analysis; and finally, concluding remarks and future work are included in the Sect. 5.

2 Related Work

There are several existing work in the area of hand gesture recognition using Kinect.

Yi Li [2] proposed multiscenario gesture recognition for hand detection using contour tracing algorithm and angle-based finger identification algorithms but suffers from the accuracy during recognition of gestures.

Meenakshi Panwar [3] discussed the significance of shape parameters in recognition of hand gestures based on image segmentation and orientation detection, but it does not follow any systematic approach and made assumption for most of the taken parameters.

Omer Rashid et al. [4] proposed the integration of gesture and posture recognition using HMM (hidden Markov model) and SVM (support vector machine). Marek Vanco et al. [5] analyzed gesture identification for system navigation in 3D scene, but it is not suitable for recognition of dynamic gestures and is limited for static gestures only. This work is an extension of the earlier work [6] for PowerPoint presentation using Kinect.

Table 2 Literature survey of existing work

Technology	Merits	Demerits
Harvard graphics [7]	Used to incorporate slides, texts, information graphics, and charts	Printed transparencies needed to be placed on an overhead projectors, switching between slides is difficult, orientation of slides had to be remembered
PowerPoint by Microsoft [8]	Most widely known and used, overcame the limitation of using transparencies	Traditionally controlled by keyboard and mouse, remote controllers
Sideshow presenter for macintosh computer [9]	Controlled by hand gestures and data gloves	Problem remains the use of data gloves
Laser pointer [10]	Use of laser pointer as a pointing tool for interacting through slides	Required a laser pointer and clickable buttons on slides

Outcome of Literature Survey

Moreover, existing works do not provide the consistent, stable, and robust capabilities to presenter which sometimes become irritable and less effective for interaction during presentations, which leads toward the idea and motivation for the development of an improved and stable version of PowerPoint presentation system using hand gestures which can play a very significant role in modern lifestyle, look and feel to users as compared to traditional approach. These applications make user free from being holding mouse, keyboard, or any extra controllers and provide flexibility to users. Table 2 summarizes the merits and demerits of existing works in the development of presentation systems using different techniques.

3   Proposed Methodology

The research focuses upon the work to propose a stable and efficient presentation system using gesture recognition through Kinect.

3.1   Gesture Detection

In the context of development of any HCI system based on the gestures [11], most important aspect is to track the presenter's body before any hand detection could occur. This recognition of actions can be done by the skeleton tracking features of Kinect SDK along with the NUI (Natural User Interface) library.

Fig. 1 Skeleton tracking

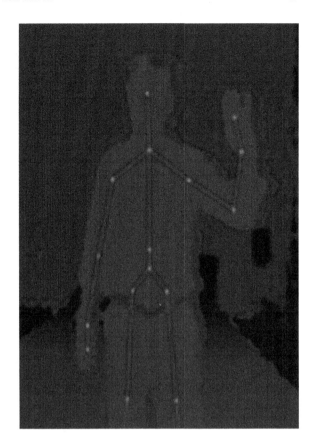

According to documentation of Kinect by Microsoft, up to six users can be recognized by the infrared (IR) camera; furthermore, in detail only two users can be tracked. The considerable point here is that Kinect sensor itself does not recognize a person and that it sends the information of depth image obtained to the host device either a computer or an Xbox. Application software running on host machine then processes this depth image based on the logic and recognizes the presence of human according to elements in the image with characteristic human shapes. In the background, it is being able to done because the system has been trained on various data of human shapes, positions, movements, etc. Alignment of the body, its movement plays a vital role in order to track them.

For a person standing in front of Kinect sensor, 20 joints including head, shoulder, arm, elbow, and wrist can be tracked using skeleton tracking application as shown in Fig. 1. After tracking of skeleton, the position of every joint is returned by the NUI library in the form of X, Y, and Z coordinates in 3D space expressed in meters based on skeleton coordinate system.

Fig. 2 Proposed
methodology

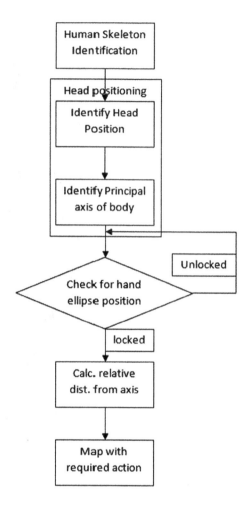

3.2 *Work Flow of the Proposed Approach*

We proposed a part-based algorithm which localizes the position of hands relative to
the distance from principal axis of human body. This principal axis has been decided
by the position of head. Figure 2 shows the work flow of our proposed approach for
the development of a stable application for the PowerPoint presentation.

3.2.1 Kinect Setup

The development of this application involves reference of Microsoft.Kinect.dll
assembly in the C# program and inclusion of namespace directive Microsoft.Kinect
after the installation of required driver software and hardware (Kinect). Also a

connection is needed to be established by using the object of the KinectSensor class available in Kinect SDK. This object is responsible for processing the commands generated through developer's program and also for generating error messages [12].

3.2.2 Data Collection from Kinect Sensor

As per the reference of Kinect, the application should likewise have the capacity to get the stream information (e.g., color, depth, and skeleton) from the sensor which is vital for the effective utilization of this application. This is done by enabling the ColorStream part of the Kinect which gives the RGB video stream, 3D representation of image in front of sensor is provided by the DepthStream, and for obtaining the skeleton information SkeletonStream has to be enabled.

3.2.3 Skeleton Tracking for Joints and Data Processing

Once enabling of skeleton Stream is done, for each time of a new skeleton data, it generates an event. An event handler is connected to the FrameReady() methods for obtaining skeleton [13] frame data.

The methods sensorSkeletonFrameReady() and sensorColorFrameReady() are connected with the frame ready events and called each time the skeleton tracking code complete the analysis of human skeleton in front of the Kinect sensor. When these methods are called with roviding arguments (sender object and SkeletonFrameReadyEventArgs), it gives the access to the skeleton frame data.

3.3 Stability of Application

Joints such as head, right hand, left hand, left foot, and right foot have been considered for tracking and thereby developing this application. Kinect cannot track any joints by itself until these are initiated, a developer is needed to initiate those joints in a program which are needed to be tracked.

When a presenter stands in front of the Kinect sensor, position of his head and both hands get tracked and this is represented as a red oval shown in Fig. 3. The program considers the head position as the central point and creates a logical central axis. Also the movement of hands get tracked and relative distance from the central axis has been calculated. Based on a threshold value, if the calculated distance between hand and central axis is greater than the threshold value, required right or left button of keyboard can be triggered. Similarly, other keys, e.g., left arrow, esc (close), and f5 (full screen), have been mapped for performing actions accordingly during the presentation.

The major issue and challenge in this system was to eliminate unwanted detection of movements of hands during presentation. Because of the continuous tracking of

Fig. 3 Gesture recognition

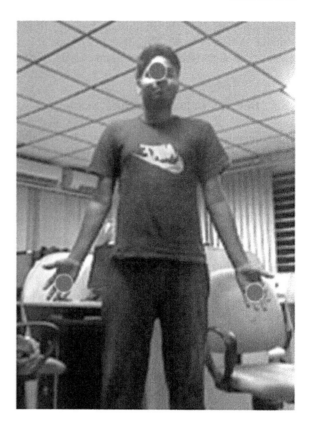

gestures through the Kinect sensor, sometimes it was creating irritation to the presenter and audience both.

This instability fault and limitation has been overcome through "locking and unlocking" concept as shown in Figs. 4 and 5. Left or right movement of hands initiates left or right movement, respectively, of slides as shown in Fig. 6. The snippet of locking/unlocking condition of code is shown in Fig. 7.

For this, we calculated the absolute difference between both the ovals of palm and when its approximately superimposes each other then tracking of gesture become locked. For unlocking, again presenter needs to put both palms together to superimpose both ovals to each other. This makes the application very stable and more interactive.

Red-colored small oval shows inactive and non detecting scenario, whereas green and large oval show the active and detecting scenario during presentation as shown in Figs. 4 and 5 respectively.

Fig. 4 Locking and
unlocking

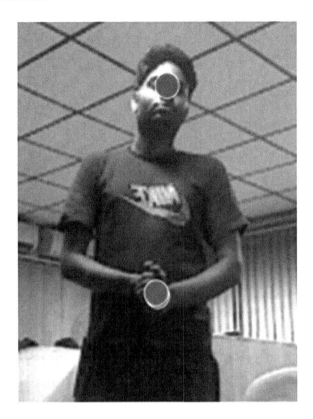

4 Results and Analysis

Experimental Setup The experimental setup to develop this stable and interactive
presentation system includes the following:

- Kinect Sensor device
- Kinect SDK installed OS-Window 7 64 bit
- Visual Studio 2010
- CPU-Core i7-2600@3.40 Ghz with 8 GB RAM

The experiment with Kinect device for the development of gesture-based Power-
Point presentation tool (as discussed in Sect. 3) was conducted on a total of 20 people,
and feedback for the 3 parameters, namely: (1) viewer's interest, (2) frequency of presenter's movement, and (3) stability in slide movements, was taken. Based on these
parameters, a comparative study between traditional and gesture-based presentation
style has been performed.

Fig. 5 Unlocked position of hands

Fig. 6 Forwarding or backwarding of slides

```
if (Math.Abs(rightHand.Position.X - leftHand.Position.X) + Math.Abs(rightHand.Position.Y - leftHand.Position.Y) <= 0.02 )
//if ((rightHand.Position.X == leftHand.Position.X) && (rightHand.Position.Y ==leftHand.Position.Y))
{
    if (flag == true)
    {
        Console.WriteLine("Locked");
        flag = false;
    }
    else
    {
        Console.WriteLine("UnLocked");
        flag = true;
    }
}
```

Fig. 7 Lock/unlock condition

Fig. 8 Viewer's interest
using traditional style

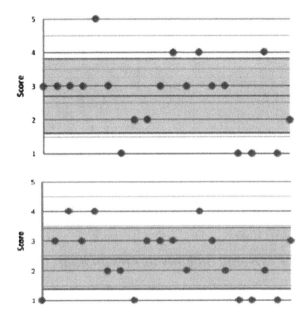

Fig. 9 Viewer's interest
using gesture-based
presentation

4.1 Viewer's Interest

The first test performed was about the comparison of viewer's understandings and their interest regarding remembrance of pictures and contents. Figures 8 and 9 show the obtained viewer's interest score on scale 5.0 with the mean values 2.3 and 2.8 for traditional and gesture-based presentation respectively, which clearly describe that using gestures for presentation imposes a great impact on viewers.

4.2 Frequency of Presenter's Movement

Similarly second test was performed about frequency of presenter's body movement during presentation. The obtained results (mean movement score 1.5 and 3.06,

Fig. 10 Movement frequency during traditional style

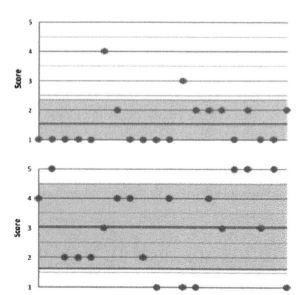

Fig. 11 Movement frequency during gesture-based presentation

respectively, for traditional and gesture based on the scale of 5.0) are shown in Figs. 10 and 11 and it is observed that using the traditional style of presentation, lesser body movements occurs as compared to gesture-based presentation.

4.3 Overall Stability in Presentation

After using the concept of locking and unlocking of recognition of movement of presenter this application became very stable and error free. Figures 12 and 13 show that there is no significant difference between traditional and gesture-based presentation styles in terms of stability. Both are nearly same, this is a tremendous success of our method that maintains the stability nearly same as compared to traditional style. Still gesture-based presentation is able to put more interactive and effective remarks on users.

Fig. 12 Stability of presentation using traditional style

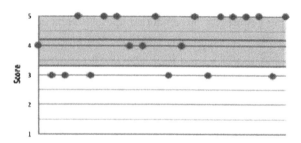

Fig. 13 Stability using locking/unlocking of gesture

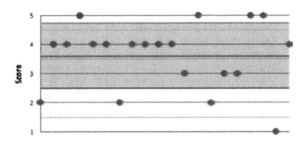

Table 3 Obtained ratings (mean on scale of 5.0) for all three experimental parameters

Parameters	Traditional	Gesture-based
Viewer's interest	2.3	2.8
Frequency of presenter's movement	1.5	3.06
Stability in presentation	4.2	3.7

Table 3 summarizes the obtained results of these above-mentioned three parameters for both traditional and gesture-based presentations.

5 Conclusion and Future Work

In this paper, we developed a stable and robust gesture-based PowerPoint presentation system using Kinect and performed a comparative study of traditional and gesture-based presentation style considering three parameters. Obtained results clearly demonstrate that although both gesture-based and traditional style are nearly same in terms of stability (as desired); furthermore, gesture-based presentation is more effective regarding viewer's interest. Still there are several scope of future extensions for making it more flexible or stable as per the need of user. Any actions related to keyboard or mouse can be added to this work as per requirements. Additional actions or operations can also be carried out using voice recognition feature of Kinect. This needs some training and device-independent voice system for more refined voice-based operations to be carried out in the near future.

References

1. Kolbjrnsen, Morten. "A Comparison of Motion-Sensing Game Technologies for use in Physical Rehabilitation." (2012).
2. Li, Yi. "Multi-scenario gesture recognition using Kinect." Computer Games (CGAMES), 2012 17th International Conference on. IEEE, 2012.

3. Panwar, Meenakshi. "Hand gesture recognition based on shape parameters." Computing, Communication and Applications (ICCCA), 2012 International Conference on. IEEE, 2012.
4. Rashid, Omer, Ayoub Al-Hamadi, and Bernd Michaelis. "A framework for the integration of gesture and posture recognition using HMM and SVM." Intelligent Computing and Intelligent Systems, 2009. ICIS 2009. IEEE International Conference on. Vol. 4. IEEE, 2009.
5. Van o, Marek, Ivan Minrik, and Gregor Rozinaj. "Gesture identification for system navigation in 3D scene." ELMAR, 2012 Proceedings. IEEE, 2012.
6. Kinect PowerPoint Control link: https://kinectpowerpoint.codeplex.com/.
7. Rufener, Sharon L. (May 26, 1986), "Harvard Graphics Is Easy to Learn and Use", InfoWorld, pp. 4748.
8. Rogers, Lee F. "PowerPointing." American Journal of Roentgenology 177.5 (2001): 973–973.
9. Baudel, Thomas, and Michel Beaudouin-Lafon. "Charade: remote control of objects using free-hand gestures." Communications of the ACM 36.7 (1993): 28–35.
10. Sukthankar, Rahul, Robert G. Stockton, and Matthew D. Mullin. "Self-calibrating camera-assisted presentation interface." Proceedings of International Conference on Automation, Control, Robotics and Computer Vision. Vol. 2000.
11. Cassell, Justine. "A framework for gesture generation and interpretation." Computer vision in human-machine interaction (1998): 191–215.
12. Miles, Rob. Start Here! Learn the Kinect API. Pearson Education, 2012.
13. [Untitled Photograph of Kinect Sensors on Human Body] Retrieved March 10, 2013 from: http://gmv.cast.uark.edu/uncategorized/working-with-data-from-the-kinect/attachment/kinect-sensors-on-human-body/.

Wavelet Statistical Feature Selection Using Genetic Algorithm with Fuzzy Classifier for Breast Cancer Diagnosis

Meenakshi M. Pawar and Sanjay N. Talbar

Abstract Breast cancer diagnosis at its early stage is achieved through mammogram analysis. This paper presents a genetic fuzzy system (GFS) for feature selection and mammogram classification. Mammogram image is decomposed into sub-bands using wavelet transform. Wavelet statistical features are obtained from 100 biggest wavelet coefficients from each sub-band. From each level of decomposition, 20 WSFs are extracted. Therefore, total 80 WSFs are extracted from four levels of decomposition. At first level, 20 WSFs are given to GFS, which selects five features with classification accuracy of 60.94%. For second level, 18 features are selected from 40 features and classification accuracy of 80.66% is obtained. Further, at third level, 18 features are selected from 60 features with classification accuracy of 85.25%. At last, for fourth level, 21 features are selected from 80 features and classification accuracy improved to 93.77%.

Keywords Breast cancer diagnosis · Feature selection · Wavelet statistical feature

1 Introduction

Breast cancer is second largest cancer and leads to death among the women of 35–40 age group. ACS has estimated new cases of cancer as 231,840 among women and 2,350 cases in men and also 40,730 deaths due to breast cancer during

M.M. Pawar (✉)
Electronics and Telecommunication Engineering, SVERI's College of Engineering, Pandharpur, Solapur 413304, Maharashtra, India
e-mail: minakshee2000@gmail.com

S.N. Talbar
Electronics and Telecommunication Engineering, S.G.G.S.I.E. & T, Nanded, Maharashtra, India
e-mail: sntalbar@sggs.ac.in

© Springer Nature Singapore Pte Ltd. 2018
P.K. Sa et al. (eds.), *Progress in Intelligent Computing Techniques: Theory, Practice, and Applications*, Advances in Intelligent Systems and Computing 518, DOI 10.1007/978-981-10-3373-5_8

the 2015 [1]. Therefore, it is necessary to detect breast cancer in its early stages, which will reduce the number of deaths. Digital mammography is proved to be effective for early detection of breast cancer worldwide [2]. Radiologist finds some clues on mammogram, viz. mass/tumor, microcalcification spots, size and shape to identify type of abnormality. Screening of large volume of mammogram may result in extreme tiredness due to which important clues can be overlooked by radiologist. Studies showed that 10–30% cases of the noticeable cancers undergo undetected and 70–80% biopsy cases are unnecessarily performed [2]. Hence, computer-aided diagnosis (CAD) system would assist radiologist as second reader for automatic detection of mammograms. CAD system works efficiently in diagnosis process for which researchers have demonstrated that cancer detection rate has improved from 4.7% to 19.5% compared to radiologist [3].

The CAD system is comprised of: (1) preprocessing, (2) computation of features, (3) selection of appropriate features and (4) classification stages. Region of interest (ROI) image is obtained in preprocessing stage. Researchers have studied various feature extraction techniques based on wavelet and curvelet transforms, demonstrated their usefulness in mammogram image analysis [2–9]. However, methods introduced in [2] take feature size of 1238, 150, 5663 and 333, for [3] have selected 10–90% wavelet coefficients and in [7] take 400 coefficients from each level of decomposition which is quite large feature size and computationally demanding. Therefore, it is challenging problem to reduce feature size. However, in this regard various feature selected approaches have been suggested, Eltoukhy et al. [8] proposed curvelet moments for breast cancer detection and achieved 81% accuracy with eight features for malignancy detection. Dhahbi et al. [9] used genetic algorithm for feature selection and AUC = 0.90 is obtained. In above study, only severity of breast abnormalities has been considered, i.e., classification between benign and malignant tumor.

Fuzzy logic (FL) tool is better for mammogram analysis [10] and advantageous over other tools as it requires less computation and has a better interpretability [11–13]. Therefore, this paper proposes wavelet-based moment extraction from mammograms of Mammographic Image Analysis Society (MIAS) database [14] and GFS algorithm for choosing appropriate features as well as classification. The remaining paper is described as follows: Sect. 2 specifies feature extraction techniques; Sect. 3 explains GFS algorithm for feature selection and classification. Section 4 gives result and discussion, while conclusion is given in Sect. 5.

2 Discrete Wavelet-Based Feature Extraction

2.1 Feature Extraction

This section gives detail explanation about extraction of wavelet transform-based statistical features from mammogram image.

2.1.1 Data Set

In this study, MIAS database is used for the experimentation work. MIAS data set has been chosen as it includes mammograms that were labeled by expert radiologist and also its various cases as shown in Table 1. MIAS database comprised of 322 mammogram images of left and right breast, obtained from 161 patients in which 64 cases reported as benign, 51 as malignant and 207 normal cases.

2.1.2 ROI Selection

The size of mammogram from MIAS database is 1024 × 1024 pixels, containing 50% background with lots of noise. The first step of classification is ROI (region of interest) selections as shown in Fig. 1; the mammogram is cropped using the physical annotation provided with database itself. Thus, almost all background noise is being cut off with tumors left with ROI. One hundred and forty-two ROIs of size 128 × 128 pixels are extracted, and some of these are presented in Fig. 2.

Table 1 Detailed description of MIAS data set

Breast abnormality	Benign	Malignant	Total
Normal	–	–	27
Microcalcification (CALC)	12	13	25
Ill-defined mass (MISC)	7	7	14
Circumscribed mass (CIRC)	19	4	23
Spiculated mass (SPIC)	11	8	19
Asymmetry (ASYM)	6	9	15
Architectural distortion (ARCH)	9	10	19
Total	64	51	142

Fig. 1 **a** Original mammogram image. **b** Region of interest (ROI) image

(a)   **(b)**

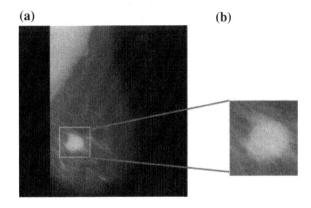

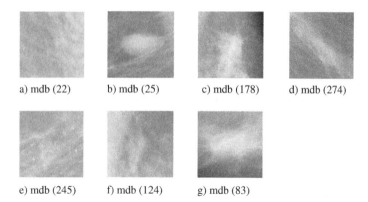

a) mdb (22)   b) mdb (25)   c) mdb (178)   d) mdb (274)

e) mdb (245)   f) mdb (124)   g) mdb (83)

Fig. 2 Samples of ROIs of mammogram images: **a** Normal. **b** CIRC. **c** SPIC. **d** MISC. **e** CALC. **f** ARCH. **g** ASYM

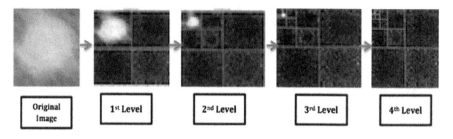

| Original Image | 1st Level | 2nd Level | 3rd Level | 4th Level |

Fig. 3 Four level decomposition of mammogram image using wavelet transform

2.2 Discrete Wavelet Transform

Wavelet filter bank is used to decompose mammogram image into wavelet sub-bands with the help of low-pass and high-pass filters. In each level of mammogram decomposition, 2D-DWT-filter bank structure uses approximate coefficients from upper level as shown in Fig. 3. The wavelet function db8 is used for mammogram decomposition based on previous studies [8–10, 13]. Here, wavelet statistical features (WSFs) were extracted from wavelet coefficients obtained from each level of DWT decomposition.

Wavelet Statistical Features (WSFs): Wavelet coefficients are obtained by arranging each sub-band coefficients in a row with descending order such as $C_k(i)$ where $1 \leq i \leq N$ and $N = M \times M$. Statistical moments have been calculated from biggest 100 wavelet coefficients. Following statistical moments are obtained by using equation as:

$$\text{Mean:} \quad f_1 = \mu_k = \frac{1}{N} \sum_{i=1}^{N} C_k(i)_{A, D_{1,2,3,4}} \tag{1}$$

$$\text{Standard Deviation:} \quad f_2 = \sigma_k = \left[\frac{1}{N} \sum_{i=1}^{N} \left(C_k(i)_{A, D_{1,2,3,4}} - \mu_k \right)^2 \right]^{1/2} \tag{2}$$

$$\text{Skewness:} \quad f_3 = skew_k = \frac{\frac{1}{N} \sum_{i=1}^{N} \left(C_k(i)_{A, D_{1,2,3,4}} - \mu_k \right)^3}{\left(\sqrt{\frac{1}{N} \sum_{i=1}^{N} \left(C_k(i)_{A, D_{1,2,3,4}} - \mu_k \right)^2} \right)^3} \tag{3}$$

$$\text{Kurtosis:} \quad f4 = kurt_k = \frac{\frac{1}{N} \sum_{i=1}^{N} \left(C_k(i)_{A, D_{1,2,3,4}} - \mu_k \right)^4}{\left(\sqrt{\frac{1}{N} \sum_{i=1}^{N} \left(C_k(i)_{A, D_{1,2,3,4}} - \mu_k \right)^2} \right)^2} \tag{4}$$

$$\text{Energy:} \quad f_5 = E_k = \frac{1}{N} \sum_{i=1}^{N} C_k(i)^2_{A, D_{1,2,3,4}} \tag{5}$$

where $C_k(i)$ is the ith element of the wavelet coefficient of kth DWT sub-band of dimension N. Here, statistical moments are denoted as wavelet statistical features (WSFs).

3 GFS Algorithm for Feature Selection and Classification

Feature selection is an important step that improves classifier performance, selects optimal number of highly discriminative features and reduces the number of computations. Therefore, combining fuzzy logic and genetic algorithm provides an automatic feature selection and called as a genetic fuzzy system (GFS).

This section explains briefly fuzzy classifier and development of GFS algorithm for breast abnormality detection.

3.1 Fuzzy Classifier (FC)

Fuzzy logic has been applied in many areas like control system, pattern recognition, artificial intelligence, image processing. The schematic of fuzzy classifier is shown in Fig. 4. Typical multi-input single-output (MISO) FC system maps input feature vector $f = \{f_1, f_2, \ldots, f_n\}$ to output class W using four units, viz. fuzzification, inference engine, rule base and defuzzification.

The FC system can be represented as

$$F : f \in R^n \rightarrow W \in R \tag{6}$$

where $f = f_1 \times f_2 \times \cdots \times f_n \in R^n$ act as input space and $W \in R$ as output space.

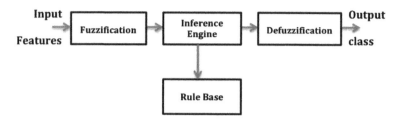

Fig. 4 Schematic of fuzzy classifier (FC)

The selection of the best rule set is difficult in fuzzy system. Selection of rules and membership functions are important in the view of capturing accurate relation of input and output variable.

Fuzzification: Input features (WSF) are applied to fuzzy logic classifier and act as crisp variables. Fuzzification maps input crisp variables into fuzzy sets with the use of membership function. Therefore, fuzzy set (A) is defined by degree of membership function $U: \mu(x) \rightarrow [0\ 1]$ and can be written as

$$\mu(f) = e^{-0.5\left(\frac{f-mean}{sd}\right)^2} \tag{7}$$

where mean and standard deviation are obtained from features of each group of samples of above-mentioned abnormality.

Inference Engine and Rule Base: Rules for fuzzy system are derived from corresponding fuzzy sets (F). Thus, ith rule can be written as

$$R_i: IF\ f_1\ is\ F_{i1}\ AND\ f_2\ is\ F_{i2}\ AND\ \dots\ f_n\ is\ F_{in}\ THEN\ y = C_i$$
$$i = 1, 2, \dots, L \tag{8}$$

where n and L are amount of input features and rules or amount of output class. For each input feature corresponding to mammogram abnormality, membership value for corresponding fuzzy set is calculated. Here, each rule is obtained from each mammogram abnormality with n number of input features. Fuzzy product implication for each rule can be written as

$$\mu_{R_i}(f, y) = \mu_{F_1}(f_1) * \mu_{F_2}(f_2) * \dots * \mu_{F_i}(f_n) * \mu_{C_i}(y) \tag{9}$$

where * is a min or product of f and y. In the above rules, membership values in fuzzy set F will change from class to class.

Defuzzification: In this study, maximum matching method is used as a defuzzification method. Matching of pth pattern in the antecedent part (IF part) of ith rule can be written as

$$D_p^i = \prod_{l=1}^{m} \mu_{li} \qquad (10)$$

where μ_{li} is the membership value of image feature k for ith rule in fuzzy region. The maximum identical rules (rules $i = 1, 2, \ldots, N$) for generating class C_i

$$D_p^{\max}(C_{i*}) = \max_i D_p^i(C_i) \qquad (11)$$

The classification accuracy (CA) of the FC system is calculated as

$$CA = \frac{N_C}{N_T} * 100 \qquad (12)$$

where N_C denotes correctly classified samples and N_T is total sample set for each class.

3.2 Proposed GFS Algorithm

The aim of this study is to formulate GFS algorithm for feature selection and classification. The GFS is shown in Fig. 5.

Input and output environment: Original mammogram image is cropped into 128×128 ROI image. WSFs are calculated from ROI image. The features are represented as of $f = \{f_1, f_2, \ldots, f_n\}^T$, and seven classes of abnormalities are

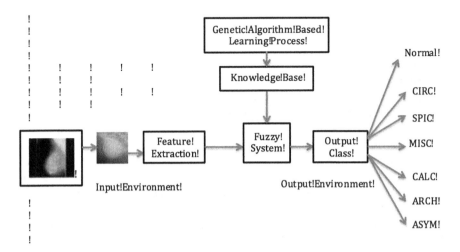

Fig. 5 Schematic for genetic fuzzy classifier (GFC) system

denoted as $C = \{C_1, C_2, \ldots, C_L\}^T \in W$ where n and L are number of features and number of output class labels, respectively.

Optimization problem for GFS: Genetic algorithms (GA) are evolutionary algorithms that uses notion of Darwin's theory of evolution. Genetic algorithm performs search in a "natural" way for solutions to problems. At start, all possible solutions for a problem are generated called as population where chromosomes represented by binary strings. GA performs intelligent search using operators such as crossover and mutation to obtain better solution. Goodness of solution is decided with the help of fitness function. The optimization problem for GFS using genetic optimization tool can be written as

$$
\begin{aligned}
&Minimize \ \sum_{i=1}^{L} \beta(i) \\
&subject\ to: \alpha_{ij} \in \{0\ 1\} \\
&\mu_{ij} = 1, \quad \alpha_{ij} = 0; \\
&\mu_{ij} = \mu_{ij}, \quad \alpha_{ij} = 1; \\
&j = 1, 2 \ldots, n
\end{aligned}
\tag{13}
$$

where L denotes the number of rules of fuzzy system, β is the classification error $(\beta = 1 - CA)$, and μ_{ij} is the membership value of ith rule for jth feature. GFS uses classification error as a fitness function. The membership value is tuned by using solution obtained from genetic algorithm. If feature is selected, i.e., $\alpha_{ij} = 1$, membership value will remain as it is. However, if feature is not selected, i.e., $\alpha_{ij} = 0$, then $\mu_{ij} = 1$. According to this, the knowledge base of fuzzy system is generated and rules are modified subsequently.

4   Results and Discussion

The experimentation work is carried on mammogram images from MIAS database. The mammogram image is converted into wavelet sub-band coefficients using db8 wavelet function. From each sub-band coefficients, five statistical features were computed. Therefore, total 20 features were extracted from each level and 80 features from four levels of decomposition. The performance of GFS algorithm is tested using WSFs at level of decomposition. At first level, 20 WSFs were given to GFS, which selects five features with an improved classification success rate from 26.67 to 60.94%. Further, 40 features from second level of decomposition are given to GFS, out of which it select 12 features and improves success rate from 18.73 to 80.66%. Subsequently, 60 features up to third level of decomposition are applied to GFS where 18 optimal features are selected and classification is improved from 16.31 to 85.25%. At last, total 80 features from fourth level of decomposition are given to GFS; it selects 21 most discriminative features with an improved classification rate from 21.61 to 93.77%. Figure 6 shows that fitness value is decreased from 0.39 to 0.0623 as we move from Level1 to Level4. The reduction in fitness value increases the classification accuracy. Therefore, the classification accuracy is improved from 60.94 to 93.77% in breast abnormality classification.

Table 2 depicts the number of feature selected in each level of decomposition. It is observed that for Level1 to Level3 of mammogram decomposition the contribution of mean feature is more as compared to other features. At fourth level, contribution of skewness feature is more. From Table 2, it is clearly stated that GFS algorithm successfully selects optimal features.

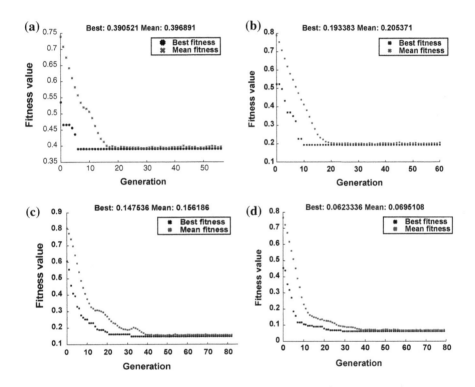

Fig. 6 Fitness curve and individual feature selected in each generation

Table 2 Detailed description of MIAS data set

DWT level of decomposition name of feature	Level1	Level2	Level3	Level4
Mean	02	04	07	06
Std. dev.	01	02	02	04
Skewness	01	01	04	07
Kurtosis	–	03	03	04
Energy	01	02	02	–
Total features selected	05	12	18	21

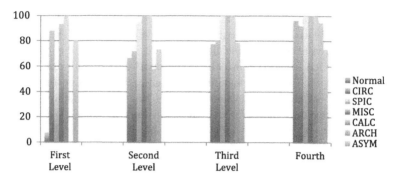

Fig. 7 Success rate at each level for each individual class of abnormality

Figure 7 represents overall classification accuracy at each level of decomposition for breast abnormalities. It is observed that as we move from Level1 to Level4, average success rate is improved from 60.94 to 93.77%. At fourth level, SPIC, MISC and CALC types of abnormality are classified with 100% accuracy, whereas ARCH and ASYM are classified with success rate of 94.74% and 73.33%, respectively. Normal class is classified with 96.3% success rate.

5 Conclusion

In the present study, the proposed CAD system is based on wavelet statistical features with GFS algorithm for feature selection as well as classification of breast abnormalities. The performance of CAD system is demonstrated using MIAS database. The mammogram is transformed into wavelet coefficients using db8 wavelet function. WSFs are computed from 100 biggest wavelet coefficients from each level. GFS algorithm performance is analyzed from WSFs from each level of decomposition. GFS algorithm selects 25 discriminative features from four level of decomposition, and breast abnormalities are classified with success rate of 93.77%. The results show that breast abnormalities are classified successfully from optimal features selected by GFS algorithm.

References

1. American Cancer Society, Internet site address: ⟨http://www.cancer.org/Research/Cancer-FactsFigures/index, 2015.
2. Mohamed Meselhy Eltoukhy, Ibrahima Faye, Brahim Belhaouari Samir. "A statistical based feature extraction method for breast cancer diagnosis in digital mammogram using multiresolution representation." Computers in Biology and Medicine 42 (2012) 123–128.

3. Essam A. Rashed, Ismail A. Ismail, Sherif I. Zaki. "Multiresolution mammogram analysis in multilevel decomposition." Pattern Recognition Letters 28 (2007) 286–292.
4. J. Dheeba, N. Albert Singh, S. Tamil Selvi. "Computer-aided detection of breast cancer on mammograms: A swarm intelligence optimized wavelet neural network approach." Journal of Biomedical Informatics 49 (2014) 45–52.
5. Cristiane Bastos Rocha Ferreira, Díbio Leandro Borges. "Analysis of mammogram classification using a wavelet transform decomposition." Pattern Recognition Letters 24 (2003) 973–982.
6. Shradhananda Beura, Banshidhar Majhi, Ratnakar Dash , "Mammogram classification using two dimensional discrete wavelet transform and gray-level co-occurrence matrix for detection of breast cancer" Neurocomputing 154 (2015) 1–14.
7. Rodrigo Pereira Ramos, Marcelo Zanchetta do Nascimento, Danilo Cesar Pereira. "Texture extraction: An evaluation of ridgelet, wavelet and co-occurrence based methods applied to mammograms." Expert Systems with Applications 39 (2012) 11036–11047.
8. Mohamed Meselhy Eltoukhy, Ibrahima Faye, Brahim Belhaouari Samir. "A comparison of wavelet and curvelet for breast cancer diagnosis in digital mammogram." Computers in Biology and Medicine 40 (2010) 384–391.
9. Sami Dhahbi, Walid Barhoumi, Ezzeddine Zagrouba, Breast cancer diagnosis in digitized mammograms using curvelet moments Computers in Biology and Medicine 64 (2015) 7990.
10. A. Vadive, B. Surendiran. "A fuzzy rule-based approach for characterization of mammogram masses into BI-RADS shape categories." Computers in Biology and Medicine 43 (2013) 259–267.
11. Pawar P M. Structural health monitoring of composite helicopter rotor blades, Ph.D. Thesis, Indian Institute of Science, Bangalore, India, 2006.
12. Prashant M. Pawar, Ranjan Ganguli, "Genetic fuzzy system for damage detection in beams and helicopter rotor blades." Comput. Methods Appl. Mech. Engrg. 192 (2003) 2031–2057.
13. Shailendrakumar M. Mukane, Sachin R. Gengaje, Dattatraya S. Bormane. "A novel scale and rotation invariant texture image retrieval method using fuzzy logic classifier." Computers and Electrical Engineering 40 (2014) 154–162.
14. http://peipa.essex.ac.uk/ipa/pix/mias.

Least Square Based Fast Denoising Approach to Hyperspectral Imagery

S. Srivatsa, V. Sowmya and K.P. Soman

Abstract The presence of noise in hyperspectral images degrades the quality of applications to be carried out using these images. But, since a hyperspectral data consists of numerous bands, the total time requirement for denoising all the bands will be much higher compared to normal RGB or multispectral images. In this paper, a denoising technique based on Least Square (LS) weighted regularization is proposed. It is fast, yet efficient in denoising images. The proposed denoising technique is compared with Legendre-Fenchel (LF) denoising, Wavelet-based denoising, and Total Variation (TV) denoising methods based on computational time requirement and Signal-to-Noise Ratio (SNR) calculations. The experimental results show that the proposed LS-based denoising method gives as good denoising output as LF and Wavelet, but with far lesser time consumption. Also, edge details are preserved unlike in the case of total variation technique.

Keywords Hyperspectral denoising · Least square · Legendre-Fenchel · Wavelet · Total variation · Signal-to-noise ratio

1 Introduction

With the improvement in spectral resolution of imaging sensors, the technology that evolved is the hyperspectral imagery, in which, a location or a scene is captured in hundreds of bands. It is one of the most important technological developments in the field of remote sensing. With hyperspectral sensors, contiguous or noncontiguous bands of width around 10 nm over the region of 400–2500 nm [1] in the

S. Srivatsa (✉) · V. Sowmya · K.P. Soman
Center for Excellence in Computational Engineering and Networking (CEN),
Amrita School of Engineering, Amrita Vishwa Vidyapeetham, Amrita University,
Coimbatore, Tamil Nadu, India
e-mail: srivatsa-s@live.com

S. Srivatsa · V. Sowmya · K.P. Soman
Amrita Vishwa Vidyapeetham, Amrita University, Coimbatore, Tamil Nadu, India
e-mail: v_sowmya@cb.amrita.edu

© Springer Nature Singapore Pte Ltd. 2018
P.K. Sa et al. (eds.), *Progress in Intelligent Computing Techniques: Theory,
Practice, and Applications*, Advances in Intelligent Systems and Computing 518,
DOI 10.1007/978-981-10-3373-5_9

electromagnetic spectrum can be obtained. Since lot of spectral and spatial data are present in hyperspectral image, many specialized algorithms are being developed for extracting the great deal of information present in them. Different land cover types can be precisely discriminated using the abundant spectral information. This also has the potential to detect minerals, aid in precision farming, urban planning, etc.

However good the sensors be, presence of noise in captured images is unavoidable. Noise tends to degrade the performance of image classification, target detection, spectral unmixing, and any such application. So, there is a need for denoising hyperspectral images. Thus, hyperspectral image (HSI) denoising has become an important and essential pre-processing technique.

In the previous years, many denoising techniques have been employed. Ting Li [2] used total variation algorithm, Hao Yang [3] used wavelet-based denoising method. Legendre-Fenchel transformation-based denoising was employed by Nikhila Haridas [4]. But there are some inherent limitations in any of these techniques, like total variation causes the denoised image to lose the edge information. Wavelet-based denoising technique's major success is due to its satisfactory performance for piecewise smooth function for signals of one dimension. But, this ability to smooth is lost when it is applied to two or higher dimension signals [5].

Each hyperspectral data consists of around 150–200 or even more bands. When there are huge number of captured images, denoising all bands in all the images takes very high computation time. So, there is a need to implement a fast and robust technique of denoising. The proposed least square-based method is a solution for this.

2 Methodology

2.1 Hyperspectral Denoise Model

Denoising can be modelled as—construct a clean image X out of a noisy image Y. Let each pixel be represented by x_{ijb} for X and y_{ijb} for Y. The indices i, j and b represent the row number, column number, and the band number, respectively. Considering the image to be additive zero-mean Gaussian and represented as w_b and with a standard deviation σ, we can express Y as

$$y_{ijb} = x_{ijb} + w_b \tag{1}$$

Here, the noise w_b is band-dependent. The solution to the Eq. 1, which is a problem of estimating $\hat{X}$ from Y, with a L2 fidelity constant, i.e., $\hat{X} = argmin_X ||Y - X||_2^2$, in general is not unique. Having a prior knowledge about the ideal image aids in finding better solution, which can be used to regularize the denoising problem.

2.2 Least Square Model

Application of least square for one-dimensional denoising was proposed by Ivan W. Selesnick [8]. Here, the usage of least square for two-dimensional signals, i.e., images are explained. For any single dimensional signal, let y be the input noisy signal and x represents the unknown denoised signal. Then, the problem formulation for least square is given by

$$min_x \|y - x\|_2^2 + \lambda \|\mathbf{D}x\|_2^2 \tag{2}$$

where, min_x represents a minimization function that describes that the given equation is to be minimized with respect to x; λ is a control parameter; $\|.\|_2$ represents L2 norm; D is a second-order difference matrix.

The main idea behind any norm-based denoising is to generate a signal or image that does not contain any noise but still resemble the information present in the original noisy data. This similarity is taken care by the first term of the Eq. 2, whose minimization forces output to be similar to the input. The second term captures the degree to which smoothness is achieved, thus reducing the noise.

Solving the Eq. 2 results in a formulation for least square signal denoising as

$$x = (\mathbf{I} + \lambda \mathbf{D}^T \mathbf{D})^{-1} y \tag{3}$$

The control parameter, $\lambda > 0$, controls the trade-off between smoothing of y and preserving similarity of x to y. The Eq. 3 can be directly applied for denoising one-dimensional (1D) signals. For two-dimensional (2D) signals (i.e., images), this needs to be first applied on the columns of the 2D matrix, then to its rows. Thus, whole procedure consists of simple matrix operations.

2.3 Experimental Procedure

An image (in gray scale) is a two-dimensional matrix, where each element of the matrix represents corresponding pixel value. An image which has been captured in multiple wavelengths is similar to a three-dimensional matrix, where the third dimension represents the spectral bands of the image. For example, a color image has three bands, namely red, green, and blue. Similarly, a hyperspectral image consists of hundreds of bands. Most of the pre-processing techniques, also denoising, involve manipulation over each band separately.

Assessment of denoising may be done either by simulating noise or by identifying already noisy bands. The latter seems to be more appropriate, because it may not be always possible to simulate noise to match the real-world situations. The choice of λ is crucial in least square-based denoising. For computational time calculation, all the bands in each dataset are denoised and total time requirement is considered.

Least square-based denoising of an image is a matter of simple matrix operations and hence must be faster than most other denoising method involving complex differential equations. Legendre-Fenchel transform-based denoising uses the concept of duality [7]; wavelet technique uses Discrete Wavelet Transform (DWT) and inverse DWT; total variation method is solvable through primal-dual method, and the solution is non-trivial. When comparing all the above-mentioned denoising methods, we can perceptually tell that proposed LS technique is simple and must run relatively faster in any system or software.

2.4   Data Sources

For experimentation, following standard hyperspectral image data [4, 6, 7] were used: Salinas scene, Pavia University, and Indian Pines. Salinas scene and Indian Pines were captured by NASA AVIRIS Sensor. Salinas scene was collected in 224 bands over Salinas Valley, California, and is characterized by high spatial resolution (3.7-m). The area covered comprises 512 lines by 217 samples. Indian Pines scene was gathered over the Indian Pines test site in North-western Indiana and consists of 145×145 pixels and 220 spectral reflectance bands in the wavelength range 0.4–$2.5\,\mu m$. Pavia University image was captured by ROSIS sensor during a flight campaign over Pavia, Northern Italy. The image has 103 bands with dimensions 610×340. But some of the bands contain no information and needs to be discarded during experimentation.

3   Results and Analysis

3.1   Visual Interpretation

The proposed method is tested on original data without simulating noise. The evaluation of denoise quality is primarily based on visual analysis and a posterior measure—Signal-to-Noise Ratio (SNR) calculation. For better analysis of least square technique, noise simulation is done on ground truth of Indian Pines image and denoising is done.

Denoising effectiveness is analyzed by varying control parameter λ and the obtained outputs are shown in the Fig. 1. It can be observed that, as the value of λ increases, noise removal is better but the image becomes smoother and edge information gradually reduces. It shows that there is a trade-off between degree of noise removal and the amount of smoothness of the image. So, based on the amount of noise and precision of information required from the noisy image, the control parameter λ is to be set.

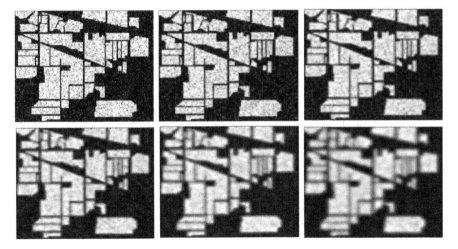

Fig. 1 Least square denoised images with different λ values

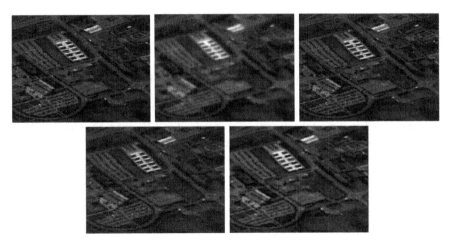

Fig. 2 *Clockwise from Top-Left*: Noisy band (band 15) of Pavia University; TV denoised; LF denoised; proposed LS denoised; wavelet denoised

The Figs. 2 and 3 show the comparison of different denoise methods applied over Pavia University (band 15) and Salinas scene (band 2) datasets. Visual perception shows that LS can perform as good as LF and better than Wavelet and TV denoising techniques.

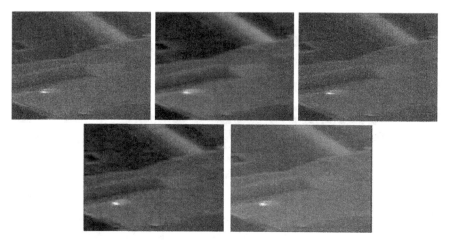

Fig. 3 *Clockwise from Top-Left*: Noisy band (band 2) of Salinas scene; TV denoised; LF denoised; proposed LS denoised; wavelet denoised

3.2 SNR Calculation

The numerical approach for quality measurement of denoised images is the signal-to-noise ratio or the SNR, which is a measure of signal power to that of noise power. It is expressed in decibels (dB). Each band has been captured in different wavelength, and every element has different reflectance properties for different wavelengths. Thus, any single band cannot be used as a reference band for calculating SNR. A new approach for SNR calculation of hyperspectral images is used by Linlin Xu [7]. For a given band, SNR may be calculated as

$$SNR = 10 \log_{10} \frac{\Sigma_{ij} \hat{x}_{ijb}^2}{\Sigma_{ij} (\hat{x}_{ijb} - m_b)^2} \tag{4}$$

where, $\hat{x}_{ijb}$ is the denoised pixel and m_b is the mean value of $\{\hat{x}_{ijb}\}$ in an area where the pixels are homogeneous. So, the estimation of the SNR relies on the selection of the homogeneous area. Class labels can be used to identify homogeneous area, since the pixels belonging to the same class are more similar to each other. Calculating SNR for all the pixels has an added advantage that the approach reduces the chance of bias that could occur by the selection of a particular homogeneous area.

The Table 1 shows the calculated SNR values after denoising particular bands in each dataset. The Table 2 shows the SNR values obtained after all the bands are denoised using various denoise models. SNR values shown in the table represent the average of SNR of all bands.

The inference that can be made by observing these two tables is that the proposed least square denoising has SNR values that are comparable with other denoise

Table 1 Signal-to-noise ratio for particular bands of various datasets

Denoise technique	Signal-to-noise ratio (dB)		
	Indian Pines (Band 102)	Pavia University (Band 15)	Salinas scene (Band 2)
Original noisy image	31.83	6.41	19.87
Total variation	35.85	7.71	24.40
Wavelet	33.03	6.47	24.06
Legendre-Fenchel	33.23	7.24	23.35
Least square	**34.45**	**7.56**	**23.12**

Table 2 Signal-to-noise ratio after denoising all bands for various datasets

Denoise technique	Signal-to-noise ratio (dB)		
	Indian Pines	Pavia University	Salinas scene
Total variation	32.47	10.65	16.31
Wavelet	29.96	9.14	15.48
Legendre-Fenchel	31.07	9.06	15.77
Least square	**30.52**	**9.87**	**15.41**

techniques. As known, increasing λ value removes noise to much higher extent, giving higher SNR values in both TV and the proposed LS denoising models. But, higher λ values have the effect of depleting edge information. For this reason, it is necessary that image quality after denoising must be interpreted both numerically and visually.

3.3 Computational Time Requirements

The main advantage of least square is that its model is very simple and thus needs lesser time for computation. This has been proved by showing average time requirement for denoising all the bands in each of the three datasets (Table 3). It is clearly visible that the time taken by total variation and Legendre-Fenchel techniques is very high compared to wavelet and least square techniques. But LS is still far faster than wavelet. Low time requirement becomes prevalent in real-time applications where huge amounts of data are being collected by the sensors every minute and needs to be pre-processed. In such situations, techniques like LS gain more importance than slow-processing techniques.

Table 3 Time requirement of different denoise techniques

Denoise technique	Time requirement (s)		
	Indian Pines	Pavia University	Salinas scene
Total variation	10.3	15.6	27.6
Wavelet	3.1	2.9	6.1
Legendre-Fenchel	14.9	26.5	40.5
Least square	**1.6**	**1.4**	**3.2**

4 Conclusion

This paper explains a novel method of denoising hyperspectral images that can be used as an alternative to many denoising methods proposed till date. The proposed algorithm has the ability to denoise images to a satisfactory level with very less time consumption. The denoising technique has been tested on different datasets for analyzing its denoising capability and time consumption to have more precise conclusion. The approach used for noise level approximation is simple enough but gives much better idea about SNR of each band and also for whole dataset. This gives a clear picture of the denoise quality. But, there is a need for some other technique that can measure the loss of edge information.

As a future work, improving denoise quality by altering difference matrix, D will be considered. Further, denoising hyperspectral images with LS technique prior to classification can be done to assess the improvement in classification accuracy.

References

1. Qiangqiang Yuan, Liangpei Zhang: Hyperspectral Image Denoising Employing a Spectral Spatial Adaptive Total Variation Model. IEEE Trans. Geosci. Remote Sens., vol 50, issue 10, pp 3660–3677, 2012.
2. Ting Li, Xiao-Mei Chen, Bo Xue, Qian-Qian Li, Guo-Qiang Ni: A total variation denoising algorithm for hyperspectral data. Proceeding of SPIE, The International Society for Optical Engineering, Nov 2010.
3. Hao Yang, Dongyan Zhang, Wenjiang Huang, Zhongling Gao, Xiaodong Yang, Cunjun Li, Jihua Wang: Application and Evaluation of Wavelet-based Denoising Method in Hyperspectral Imagery Data. Springer Berlin Heidelberg, CCTA 2011, Part IIIFIP AICT 369, pp 461–469, 2012.
4. Nikhila Haridas, Aswathy. C, V. Sowmya, Soman. K.P.: Hyperspectral Image Denoising Using Legendre-Fenchel Transform for Improved Sparsity Based Classification. Springer International Publishing. Advances in Intelligent Systems and Computing, vol 384, pp 521–528. Aug 2015.
5. Xiang-Yang Wang, Zhong-Kai Fu: A wavelet-based image denoising using least squares support vector machine. Engineering Applications of Artificial Intelligence. vol 23, issue 6, Sep 2010.
6. Aswathy. C, V. Sowmya, Soman. K.P.: ADMM based Hyperspectral Image Classification improved by Denoising Using Legendre Fenchel Transformation. Indian Journal of Science and Technology. Indian Journal of Science and Technology, vol 8(24), Sep 2015.

7. Linlin Xu, Fan Li, Alexander Wong, David A. Clausi: Hyperspectral Image Denoising Using a Spatial-Spectral Monte Carlo Sampling Approach. IEEE Journal of Selected Topics in Applied Earth Observations and Remote Sensing, vol 8, issue 6, Mar 2015.
8. Selesnick, Ivan.: Least squares with examples in signal processing. [Online], Mar 2013.

Image Forgery Detection Using Co-occurrence-Based Texture Operator in Frequency Domain

Saurabh Agarwal and Satish Chand

Abstract Image tampering/forgery can be accomplished easily using precise image editing software. However, different type of traces is introduced during image forgery process that may helpful in image forgery detection. This paper is an attempt to propose a robust blind image tampering detection technique in accordance with wavelet transform and texture descriptor. Image crucial details are highlighted using shift-invariant stationary wavelet transform. This transform provides more information about image internal statistics in comparison with DWT. Further to convert this information in terms of the feature vector, co-occurrence-based texture operator is used. The linear kernel SVM classifier is utilized to distinguish between pristine and forged images. The effectiveness of the proposed technique is proved on three image forgery evaluation databases, i.e., CASIA v1.0, Columbia and DSO-1.

Keywords Image forgery detection · Texture operator · Splicing

1 Introduction

Nowadays, images are popular over text due to high-speed internet and mobile devices. Many highly precise image editing softwares such as Photoshop and Corel Photo paint are available that can be misused for image forgery. Image forgery can be performed by changing its content. Conventionally, changes are carried out by splicing or copy–move forgery. To appeal forged image as genuine, different operations are performed like the boundary of the inserted object may be modified that disturbs the image underlying statistics. Digital watermark and digital signature

S. Agarwal (✉) · S. Chand
Department of Computer Engineering, Netaji Subhas Institute of Technology,
Sector-3, Dwarka, New Delhi, India
e-mail: saurabhnsit2510@gmail.com

S. Chand
e-mail: schand20@gmail.com

© Springer Nature Singapore Pte Ltd. 2018
P.K. Sa et al. (eds.), *Progress in Intelligent Computing Techniques: Theory,
Practice, and Applications*, Advances in Intelligent Systems and Computing 518,
DOI 10.1007/978-981-10-3373-5_10

can detect this disturbance easily, but it needs to be present in the image. Many times source of the image is unknown, and no prior information about the image is available. For this unknown scenario, we are discussing an approach that is called blind/passive image forgery detection. In passive image forgery detection, the only image itself is available for analysis. The changes in boundary/edges and important content of an image can be seen clearly in wavelet domain high-pass filtered images. Motivated by this fact, different types of features are extracted in wavelet domain [1–3]. For example, features are extracted using various wavelet characteristic functions moments and 2-D phase congruency information [1], by generalized Gaussian model in wavelet transform domain [2], using Markov model in DWT domain [4, 5], using LBP operator [6] in wavelet domain [7], and they provide satisfactory results. Encouraged by these papers, we extract features using a robust texture operator, i.e., rotation-invariant co-occurrence, among adjacent LBPs (RICLBP) [8] into more informative stationary wavelet transform (SWT) domain. Due to SWT shift-invariant nature, abrupt changes are highlighted in a better way. Further, the proposed technique is described in Sect. 2, afterward experimental setup and results are given in Sect. 3. At last, the outcomes of this paper are concluded in Sect. 4.

2 Proposed Method

Passive image forgery detection is difficult to achieve. In this paper, a technique is proposed that tries to offer better detection accuracy. As YCbCr color model performs better than RGB and other color models [7], we convert the images in YCbCr color model. Image internal details are better visualized in the frequency domain in comparison with spatial domain [1–5, 7]. Wavelet and Fourier transform both provide important details in the frequency domain. The Wavelet transform is preferred over Fourier transform due to its computational simplicity. The wavelet transform domain-based features also provide good results in denoising, edge detection, texture classification, etc. Generally, DWT is used as a wavelet transform. Its translation-invariant version SWT can perform better because it is the necessary property required for change detection. Image forgery detection goal is to detect changes in forged images, so SWT better suits for it. SWT also reduces deviations in energy distribution between coefficients at different levels and provide better information. The SWT modifies the high and low pass filters at each level of decomposition and eliminates down sampling by using á trous algorithm as shown in Fig. 1. In DWT, coefficients are down-sampled by 2 and reuse the same pair of filters in different decomposition levels. SWT costs a little bit more in comparison with DWT but provides more information (undecimated) that increases image forgery detection accuracy.

Further, we apply texture operator to extract the feature vector from SWT high-pass filtered images, i.e., LH, HL and HH. The LBP operator variant RICLBP [8] is computationally simple yet powerful is used as a texture operator. It is used

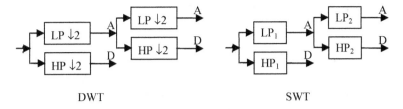

DWT   SWT

Fig. 1 Filter bank of DWT and SWT with 2-level decomposition

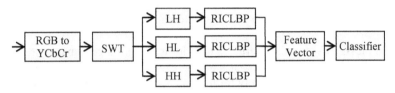

Fig. 2 Block diagram for suggested method

successfully in many applications such as image retrieval, face recognition and cell classification due to its strong extraction capability of important internal statistical information of the images. RICLBP provides global structural information between different LBP pairs. Here, brief introduction of RICLBP operator is given. First, LBP's are generated for an arbitrary pixel g_c with neighbors $g_i (i = 0, 1, .., I - 1)$. In an image by using thresholding,

$$\text{Threshold}(g_c) = S(g_i - g_c), \quad S(x) = \begin{cases} 1, & x \geq 0 \\ 0, & x < 0 \end{cases} \tag{1}$$

Further, LBP pairs are divided into two rotation equivalence classes for a particular distance R, (i) for $\theta = 0$, $\pi/4$, $\pi/2$, $3\pi/4$ and (ii) for $\theta = k$, $k + \pi$ are considered rotation equivalence classes. Using this criteria mapping, table T is created, and rotation-invariant LBP pair at distance R is labeled by utilizing this table. The feature vector of length 136 is generated by taking a histogram of $P_\theta^{RI}(R)$.

$$P_\theta^{RI}(R) = T(P_\theta(R + \Delta R_\theta)) \tag{2}$$

where $\Delta R_\theta = (R \cos \theta, R \sin \theta)$

Finally, SVM classifier of linear kernel is utilized for training and testing using feature vector generated by the above-defined process. Important steps of the proposed technique can be visualized in Fig. 2. We apply SWT on each color component individually, i.e., luminance component (Y) and Chrominance components (Cb and Cr).

Experimental environment, image databases, classifier and results are discussed in subsequent section.

3 Experimental Setup and Results

Detection of image forgery is a challenging task. Three different types of databases, i.e., CASIA v1.0 (CASIA) [9], Columbia [10] and DSO-1 [11], have been used to assess the discussed technique performance in Sect. 2, in which CASIA database have 800 pristine, 469 spliced and 452 copy–move forged images. Some forged images of this database are undergone with common post-processing operations to give them a realistic appearance. Columbia database contains 180 pristine and 183 spliced images that are not undergone to any post-processing operations. DSO-1 database has 100 original and 100 spliced high-resolution images. Forged images are created very precisely such that their credibility cannot be decided by naked eyes.

SVM classifier with linear kernel is used for training. Forged images are labeled as 1 and pristine images as 0. For each database, 30 sets are created, and in each set, 50% images are selected randomly for training and 50% left over images for testing of both labels. Results are shown using Accuracy (ACC), True Positive Rate (TPR) and True Negative Rate (TNR) parameters by taking the average of 30 sets results.

We denote the proposed technique as SRICLBP. For the sake of completeness, we also compare SRICLBP with DRICLBP and ORICLBP. In DRICLBP technique, features are extracted in DWT domain unlike SWT domain as in SRICLBP, and rest process is similar to SRICLBP. ORICLBP simply denotes the technique in which directly uses RICLBP texture operator, i.e., images are not undergone to any wavelet transform, in YCbCr color model. Single level wavelet decomposition is considered in SRICLBP and DRICLBP. For RICLBP operator, (1,2), (2, 4) and (4, 8) configurations are considered in which the first coordinate denotes LBP size and the second coordinate indicates the interval between LBP pairs. This configuration of RICLBP is applied on each high-pass filtered decompositions, i.e., HH, HL and LL of DRICLBP and SRICLBP. The results, i.e., TPR, TNR, ACC, are illustrated graphically in Fig. 3. As evident from Fig. 3a, SRICLBP outperforms [7] and gives 96.81% detection accuracy, and its sensitivity and specificity are also highest. The ORICLBP performs better than DRICLBP on CASIA. In Columbia database (Fig. 3b) and DSO-1 database (Fig. 3c), SRICLBP is the clear winner and the performance of DRICLBP is better than ORICLBP.

The standard deviation of accuracies for 30 sets is shown with average accuracy in Table 1. The Columbia and DSO-1 databases have less number of images in comparison with CASIA v1.0 that is the reason for higher SD of DSO-1. We achieve 92.31% detection accuracy in Columbia database using SRICLBP, and it is better than techniques proposed in [1–3]. On DSO-1 database, we achieve 85.51% accuracy that is comparable with [11].

Figure. 3 and Table 1 show that features extracted in SWT domain provide better results. Now, we conclude the paper in next section.

Fig. 3 Performance evaluation on different databases

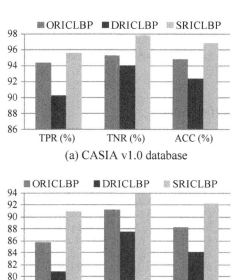

(a) CASIA v1.0 database

(b) Columbia database

(c) DSO-1 database

Table 1 Detection accuracy in percentage (%) and standard deviation (SD)

Techniques/ databases	CASIA v1.0		Columbia		DSO-1	
	ACC	SD	ACC	SD	ACC	SD
ORICLBP	94.84	0.64	88.32	3.53	80.83	4.46
DRICLBP	92.32	1.25	84.06	3.17	83.83	3.05
SRICLBP	96.81	0.57	92.31	2.34	85.51	3.34

4 Conclusion

In this paper, a robust passive image tampering detection technique has been discussed, and it provides satisfactory results in comparison with some available techniques. We have utilized robust texture operator based on co-occurrences of

LBP pairs for feature extraction in SWT domain. The SVM binary classifier is utilized to label tampered and pristine images. The validity of proposed technique has been evaluated on three different image forgery evaluation databases.

References

1. Chen, W., Shi, YQ, Su, W.: Image splicing detection using 2-d phase congruency and statistical moments of characteristic function, in Proc. of SPIE, San Jose, CA, (2007) 6505
2. Shao-Jie, S. U. N., W. U. Qiong, L. I. Guo-Hui: Detection of image compositing based on a statistical model for natural images. Acta Automatica Sinica, 35(12), (2009) 1564–1568
3. Zhang, P., Kong, X.: Detecting image tampering using feature fusion. in Proc. International Conference on Availability, Reliability and Security, ARES, (2009) 335–340
4. He, Z., Lu, W., Sun, W., and Huang, J.: Digital image splicing detection based on Markov features in DCT and DWT domain. Pattern Recognition, 45(12), (2012) 4292–4299
5. Zhao, X., Wang, S., Li, S., Li, J. and Yuan, Q.: Image splicing detection based on noncausal Markov model. In Proc. IEEE International Conference on Image Processing (ICIP), (2013) 4462–4466
6. Ahonen, T., Hadid, A., Pietikainen, M.: Face description with local binary patterns: Application to face recognition. IEEE Transactions on Pattern Analysis and Machine Intelligence, 28(12), (2006) 2037–2041
7. Muhammad, G., Al-Hammadi, M., Hussain, M., Bebis, G.: Image forgery detection using steerable pyramid transform and local binary pattern. Machine Vision and Applications, (2013) 1–11
8. Nosaka, R., Ohkawa, Y., Fukui, K.: Feature extraction based on co-occurrence of adjacent local binary patterns. in Proc. of fifth Pacific Rim Conference on Advance Image Video Technology, (2012) 82 –91
9. CASIA Tampered Image Detection Evaluation Database, http://forensics.idealtest.org
10. Ng, T-T, Jessie H., and Shih-Fu C.: Columbia image splicing detection evaluation dataset. (2009)
11. De Carvalho, T.J., Riess, C., Angelopoulou, E., Pedrini, H., de Rezende Rocha, A.: Exposing digital image forgeries by illumination color classification. IEEE Transactions on Information Forensics and Security, 8(7), (2013) 1182–1194

Multimodal Optical Image Registration Using Modified SIFT

Sourabh Paul, Ujwal Kumar Durgam and Umesh C. Pati

Abstract Although a variety of feature-based remote sensing optical image registration algorithms has been proposed in past decades, most of the algorithms suffer from the availability of a sufficient number of matched features between the input image pairs. In this paper, a modified version of scale-invariant feature transform (SIFT) is proposed to increase the number of matched features between images. Initial matching between the input images is performed by modified SIFT algorithm with cross-matching technique. Then, matched features are refined by using random sample consensus (RANSAC) algorithm.

Keywords Image registration · Scale-invariant feature transform · Random sample consensus

1 Introduction

The multimodal optical image registration is the process of aligning remote sensing images taken by different optical sensors. Images captured in two modalities contain different information. The main application of the multimodal image registration is to integrate the image information containing in different modalities. LANDSAT, SPOT, IKONOS, QUICKBIRD, ORBVIEW, etc., are the well-known remote sensing satellites, which capture optical images in the daytime. Image registration is the fundamental step to integrate the information contained in these optical images. Image registration process is broadly classified as intensity-based

S. Paul (✉) · U.K. Durgam · U.C. Pati
Department of Electronics and Communication, National Institute of Technology,
Rourkela 769008, Odisha, India
e-mail: sourabhpaul26@gmail.com

U.K. Durgam
e-mail: ujwalkumar06@gmail.com

U.C. Pati
e-mail: drumeshchandrapati@gmail.com

© Springer Nature Singapore Pte Ltd. 2018   123
P.K. Sa et al. (eds.), *Progress in Intelligent Computing Techniques: Theory,*
Practice, and Applications, Advances in Intelligent Systems and Computing 518,
DOI 10.1007/978-981-10-3373-5_11

methods and feature-based methods [1]. The accuracy of a feature-based registration algorithm depends on the number of correctly matched features. If the number of correctly matched features is more, the registration accuracy also becomes high.

Scale-invariant feature transform [2] is a very effective approach for registration of remote sensing optical images. But, standard SIFT algorithm suffers from the availability of a sufficient number of extracted features in some of the remote sensing optical images. Therefore, less number of matched features are obtained for these images when standard SIFT algorithm is used. Z. Yi et al. [3] presented gradient orientation modification SIFT (GOM-SIFT) and scale restriction (SR-SIFT) algorithms to increase the number of correct matches. Li et al. [4] proposed a robust scale-invariant feature matching algorithm to improve the correct rate of matched features in remote sensing images. Goncalves et al. [5] developed a feature-based registration algorithm by using image segmentation and SIFT. In [5], initial matched features between the input images are obtained by using standard SIFT and then refined by the bivariate histogram. Sedaghat et al. [6] proposed a uniform robust scale-invariant feature transform (UR-SIFT) to increase the number of extracted features in remote sensing images. In [6], uniform SIFT features are extracted from optical images and then matched through cross-matching technique.

Although a variety of SIFT-based registration algorithms has been proposed to register the remote sensing multimodal optical images, the availability of a sufficient number of correctly matched feature is still a challenging task. In this paper, modified SIFT algorithm is proposed to increase the number of matched feature between multimodal optical images. Initial matching candidates obtained from modified SIFT algorithm are further refined by RANSAC. Rest of the feature points are used to estimate the transformation parameters calculation.

Rest of the paper is organized as follows: Sect. 2 provides the details of the proposed algorithm. Section 3 discusses the comparison of the simulation results between the proposed algorithm and other methods. Finally, Sect. 4 offers a conclusion.

2 Proposed Method

In this section, the details of the proposed algorithm are described. Our proposed method consists of two steps: modifications of SIFT (Sect. 2.1) and matching candidates refinement (Sect. 2.2).

2.1 Modified SIFT

SIFT algorithm is an effective approach for feature extraction and feature matching in remote sensing optical images. Standard SIFT algorithm contains five major steps: scale-space extrema detection, feature point localization, orientation assignment, feature point descriptor formation, and feature point matching. In multimodal image registration as the modality is different for two sensors, sufficient matching candidates are unevenly obtained in standard SIFT algorithm. To improve the number of matches, modifications of SIFT algorithm are presented here.

(1) *Feature Selection*: Generally, first two steps of standard SIFT algorithm are used to select the strong features. In scale-space extrema detection step, Gaussian pyramid is formed by passing the image through Gaussian filter with different scales. Then, adjacent Gaussian filtered images are subtracted to form the difference of Gaussian (DOG) images. Feature points are selected in DOG images by comparing each pixel to its eight neighbors of current scale, and nine neighbors in the scale below and above. In [2], Lowe suggested a fixed threshold value ($T_s = 0.03$) for eliminating the low contrast features. Sedaghat et al. [6] showed that if this fixed value is used for feature extraction in remote sensing optical images, then sometimes the availability of the extracted becomes very less and in some cases, an excessive number of features is extracted. In [6], effects of different contrast values (T_s) on the number of extracted features are presented for various images. According to [6], 10 percent of the contrast range was used as threshold value (T_s). However, we have examined the effects of different percentages of the contrast range on the number of extracted features. This is presented in Fig. 1. The images used for the experiment are shown in Figs. 2a, b and 3a, b, respectively.

From Fig. 1, it can be observed that if the value of the percentage of contrast range is taken very low, the number of extracted features is very high for Figs. 2a and 3a. But, a very high value of the percentage of contrast range gives a very small number of extracted features for Figs. 2b and 3b. So, in our method, we have also found that 10 percent of the contrast range can provide a reasonable number of features for all the input images.

Fig. 1 Effects of the different percentages of contrast range on the number of extracted features

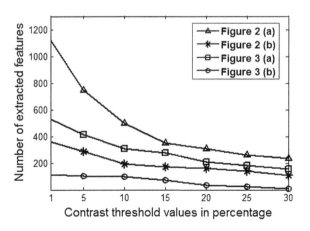

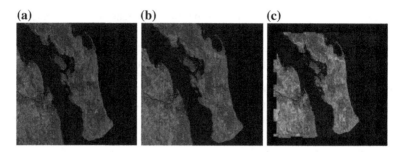

Fig. 2 Images of Campbell River in British Columbia. **a** Reference image, **b** sensed image, **c** checkerboard mosaiced image of (**a**) and (**b**)

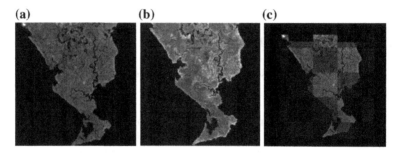

Fig. 3 Images of Chesapeake Bay in USA. **a** Reference image, **b** sensed image, **c** checkerboard mosaiced image of (**a**) and (**b**)

(2) *Orientation Assignment*: In standard SIFT, one or more dominant orientations are assigned to every feature points. Dominant orientations are calculated by forming a 36-bin orientation histogram. But, Fan et al. [7] have found out that orientations assigned to every feature point are not stable enough and reduce the number of matches in input image pair. Therefore, in our proposed method, we have skipped the orientation assignment to increase the number of matching candidates.

(3) *Feature Point Descriptor Formation*: In standard SIFT algorithm, the gradient magnitude and orientation of 16×16 neighborhoods around the feature points are calculated. Every 16×16 location grid is again divided into 16 subsections of size 4×4.

For each subsection, an orientation histogram is formed which contains 8 bins. Therefore, 128 elements descriptor is formed for every feature point by concatenating the orientation histograms entries of 16 subsections. We have used a comparatively larger supported region of size 24×24 in our proposed method to form the descriptor [9].

(4) *Feature Matching*: Feature matching operation in SIFT approach is performed by calculating the minimum Euclidean distance between the descriptors of input images. But, the minimum distance criterion gives many false matches.

So, Lowe [2] proposed a d_{ratio} parameter which is defined as the ratio of the distance of the first neighbor to the second neighbor. If the value of d_{ratio} is less than 0.8, then the corresponding pair is considered as a correct match. But, still many wrong correspondences are obtained in the case of multimodal image registration. The correct rate can be increased by taking a lower d_{ratio} value, but it yield to reduce the number of matches. So, we have used the cross-matching technique [6] to match the features between image pair. The cross-matching technique provides more matches compared to using a fixed d_{ratio} value.

2.2 Matching Candidates Refinement

The matching candidates obtained in the last step contain some outliers for remote sensing images. Outliers are removed by the RANSAC algorithm [8] to generate the final refined matches. Affine transformation model is used as a mapping function for RANSAC. So, the matched features obtained by RANSAC are used to estimate the value of transformation parameters.

3 Simulation and Analysis

In order to show the effectiveness of the proposed algorithm, two pairs of multi-modal remote sensing optical images (http://earthexplorer.usgs.gov) are presented here. The detail information of the images is provided in Table 1. The Image pairs are shown in Figs. 2 and 3, respectively. We have compared our proposed method with the three methods named as cross-correlation (CC) [1], Mutual information (MI) [1], and IS-SIFT [5] algorithms. In MI- and CC-based registration methods, the input images are divided into 4 square sections of equal sizes to obtain a set of control points between images.

Table 2 shows the matching quality comparison between CC, MI, IS-SIFT, and the proposed method for different image pairs. The performance of these registration algorithms is analyzed by the parameters RMS_{all} (root-mean-square considering all the matching points), RMS_{LOO} (root-mean-square by leaving one out), N_{red}

Table 1 Information of the test images

–	Images	Sensors	Resolution	Band	Size	Date of acquisition	Location
Pair 1	Ref. Image	Landsat TM	30	4	500 × 500	07/08/1989	Campbell
	Sen. Image	Landsat ETM	30	4	500 × 500	26/06/2000	Campbell
Pair 2	Ref. Image	Landsat TM	30	5	264 × 264	30/07/1988	Chesapeake
	Sen. Image	EO-1 ALI	30	9	264 × 264	13/05/2013	Chesapeake

Table 2 Matching quality comparison between different methods

Pair	Method	N_{red}	RMS_{all}	RMS_{LOO}	$BPP(1)$	MI
1	CC [1]	4	0.25	0.60	0	1.31
	MI [1]	4	0.25	0.57	0	1.34
	IS-SIFT [5] ($d_{ratio} = 0.5$)	9	0.79	0.91	0.22	1.12
	Proposed method	**20**	**0.22**	**0.24**	**0**	**1.46**
2	CC [1]	4	2.37	5.52	1	0.57
	MI [1]	4	1.83	4.26	1	0.60
	IS-SIFT [5] ($d_{ratio} = 0.7$)	12	0.94	0.99	0.16	0.71
	Proposed method	**21**	**0.71**	**0.75**	**0**	**0.96**

(number of matched points), and *BPP* (bad point proportion) proposed in [10]. From Table 2, we can observe that our proposed method provides more matched points compared to other algorithms. It also gives 0 bad point proportion (BPP) for two image pairs, and RMS_{all}, RMS_{LOO} values are less compared to other algorithms. In the case of first pair of images, CC and MI methods give higher accuracy compared to IS-SIFT method as the images are taken in the same frequency band. When different frequency bands are used (in the case of the second pair), CC and MI fail to give an accurate result. As the IS-SIFT method uses the d_{ratio} factor for matching the SIFT features, the number of matches is less compared to the proposed method. Moreover, the orientation assignment of the extracted features eliminates a number of correctly matched pairs.

4 Conclusion

In this paper, we have proposed a feature-based multimodal registration algorithm to increase the number of matches between the input images. Our proposed modified SIFT gives more matches compared to the other methods. To increase the number of matches, the cross-matching technique is performed between the image pairs. But, some outliers still present in the corresponding matches. RANSAC algorithm is implemented to refine the matches. The simulation results represent that the proposed method provides comparatively higher accuracy than the other existing algorithms.

References

1. Zitova, B., Flusser, J.: Image Registration Methods: A Survey. Image and Vision Computing 21, 977–1000 (2003)
2. Lowe, D.G.: Distinctive image features from scale-invariant keypoints. International journal of computer vision 60(2), 91–110 (2004)

3. Yi, Z., Zhiguo, C., Yang, X.: Multi-spectral remote image registration based on SIFT 44 (2),107–108 (2008)
4. Qiaoliang, L., Wang, G., Liu, J., Chen, S.: Robust Scale-Invariant Feature Matching for Remote Sensing Image Registration. IEEE Geoscience and Remote Sensing Letters 6(2), 287–291 (2009)
5. Goncalves, H., Corte-Real, L., Goncalves J. A.: Automatic image registration through image segmentation and SIFT. IEEE Geoscience and Remote Sensing Letters 49(7), 2589–260 (2011)
6. A. Sedaghat, M. Mokhtarzade, and H. Ebadi: Uniform robust scale-invariant feature matching for optical remote sensing images. IEEE Transactions Geoscience and Remote Sensing 49 (11), 4516–4527 (2011)
7. Fan, B., Wu, F., Hu, Z.: Aggregating gradient distributions into intensity orders: A novel local image descriptor. In: Conference on Computer Vision and Pattern Recognition (CVPR), pp. 2377–2384. IEEE (2011)
8. Fischler, M.A., Bolles, R.C.: Random sample consensus: a paradigm for model fitting with applications to image analysis and automated cartography. Communications of the ACM 24 (6), 381–395 (1981)
9. Hasan, M., Pickering, M. R., Jia, X.: Modified SIFT for multimodal remote sensing image registration. In: Conference on Geoscience and Remote Sensing Symposium, pp. 2348–2351 (2012)
10. Goncalves, H., Corte-Real, L., Goncalves J. A.: Measures for an objective evaluation of the geometric correction process quality. IEEE Geoscience and Remote Sensing Letters 6(2), 292–296 (2009)

Color Components Clustering and Power Law Transform for the Effective Implementation of Character Recognition in Color Images

Ravichandran Giritharan and A.G. Ramakrishnan

Abstract The problem related to the recognition of words in the color images is discussed here. The challenges with the color images are the distortions due to the low illumination, that cause problems while binarizing the color images. A novel method of Clustering of Color components and Power Law Transform for binarizing the camera captured color images containing texts. It is reported that the binarization by the proposed technique is improved and hence the rate of recognition in the camera captured color images gets increased. The similar color components are clustered together, and hence, Canny edge detection technique is applied to every image. The union operation is performed on images of different color components. In order to differentiate the text from background, a separate rectangular box is applied to every edge and each edge is applied with Power Law Transform. The optimum value for gamma is fixed as 1.5, and the operations are performed. The experiment is exhaustively performed on the datasets, and it is reported that the proposed algorithm performs well. The best performing algorithm in ICDAR Robust Reading Challenge is 62.5% by TH-OCR after pre-processing and 64.3% by ABBYY Fine Reader after pre-processing. The proposed algorithm reports the recognition rate of 79.2% using ABBYY Fine Reader. This proposed method can also be applied for the recognition of Born Digital Images, for which the recognition rate is 72.5%, earlier in the literature which is reported as 63.4% using ABBYY Fine Reader.

Keywords Color component clustering · Canny edge detection · Power law transform · Binarization · Word recognition

R. Giritharan (✉)
E.G.S. Pillay Engineering College, Nagapattinam, Tamil Nadu, India
e-mail: rvenkkatprabu@gmail.com

A.G. Ramakrishnan
Medical Intelligence and Language Engineering Laboratory,
Indian Institute of Science, Bengaluru, Karnataka, India

© Springer Nature Singapore Pte Ltd. 2018
P.K. Sa et al. (eds.), *Progress in Intelligent Computing Techniques: Theory, Practice, and Applications*, Advances in Intelligent Systems and Computing 518, DOI 10.1007/978-981-10-3373-5_12

1 Introduction

The recognition of texts in the image is an active research area. From the literature, it is reported that there are many different techniques proposed for the recognition of the text in the images. In those, only the grayscale images had been considered, and there involves two different processes. They are binarization and recognition. Binarization is an important process in the recognition, since optimum pre-processing is necessary for the proper recognition of the images. There are varieties of kind of OCRs available for Roman Scripts [1, 2]. But it is not necessary; the images submitted to those OCRs should produce good results. Basically, there are two different procedures involved in word recognition. The former one is detection or localization, and the latter one is recognition [3]. The best performing algorithm produced the result of recognition rate of 52% in ICDAR 2003 [4] and of 62.5% using TH-OCR and 64.3% using ABBYY Fine Reader in ICDAR 2011 [5]. As stated already, the recognition involves two stages: binarization followed by recognition. For recognition, there are training datasets or OCR Engines for Roman Scripts [7–10]. Hence, the latter part needs no more research expertise. It is only the earlier part; that is, the binarization process needs to be improved. Since the literature reported different algorithms for the binarization of the color images, none of them produced proper results. Hence, a novel algorithm of combing the method of color component clustering [6, 7] and Power Law Transform [8] is applied for the binarization of the images. Here the clustering of similar color components is a

Fig. 1 Some color image dataset considered

separate algorithm, and PLT is yet another different algorithm proposed separately for the binarization. But a novel algorithm is proposed by combining these two algorithms. Here we have considered certain scanned images containing different colors like pamphlets and posters are being considered (Fig. 1).

Here initially clustering of the color components is applied to the image, and the algorithm of PLT is applied. Since different color components are present in an image, the clustering algorithm gives the more efficient results. Since most of the characters seem to be linked with each other, Power Law Transform reduces the stroke width and hence separates the characters. By combining these two algorithms, the recognition rate would further be increased.

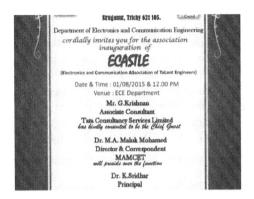

Fig. 2 Images in which PLT fails

Fig. 3 Binarized images by the PLT

2   Related Work

There are various algorithms proposed in the literature for the binarization of the color images. They include Global Threshold Algorithm [11], Adaptive threshold algorithm [8], Power Law Transform method [12], Clustering Algorithm [13–15] etc. All the algorithm have certain flaws, which reduce the recognition rate. Of these, power law transform is the most simple algorithm, with very good efficiency [8]. But in certain cases, the power law transform fails to binarize.

Figures 2 and 3 show the cases in which PLT algorithm fails to produce proper results.

3   Color Components Clustering

Text is one of the most important concentrations for the image which is taken into consideration. Since normally the colored images are multi-colored, the binarization becomes a challenging phenomenon. Hence, a novel approach of clustering different color components is considered. The major color planes of red, green, and blue color planes are considered. They are first extracted and the Canny edge detection [16–18] is performed on those extracted images independently and finally the union operation is performed between those three edge detected images. The edge map is obtained as shown below:

Fig. 4 Images considered and the subsequent images showing the edge detected *red*, *green*, and *blue* plane images and the final image representing the addition of all the three images (Color figure online)

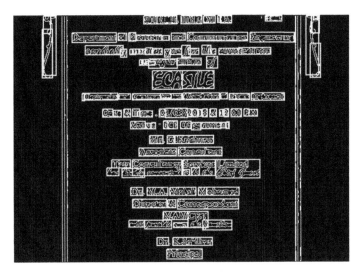

Fig. 5 Edge map image with edge box (EB)

$$E = E_R \lor E_G \lor E_B$$

Here E_R, E_G, and E_B are Canny edge detected image of red, green, and blue plane images, respectively, and $\lor$ is the union indicating the OR operation. Consider the same image taken in which the PLT had failed (Fig. 4).

Once this is done, based on 8 connected component labeling independent rectangular box is applied. Each of these rectangular boxes is called as edge box (EB). There are certain constraints for these EBs. The aspect ratio is fixed within the range of 0.1–10. The aspect ratio is assigned for eliminating the elongated areas. The edge boxes' size should be more than 10 pixels and less than 1/5th of the total image size. Else they should be subjected to further pre-processing (Fig. 5).

Here there is a possibility of the edge boxes to be overlapped. But the algorithm is designed in such a way that only the internal edge boxes are taken into consideration and others are ignored.

4 Power Law Transform

After the clustering of the similar color components and creating independent edge boxes for every text character, the Power Law Transform is applied within every edge box so as to perform independent binarization within those edge boxes, the common thresholding technique of Global Thresholding, i.e., Otsu's method can be applied after the Power Law Transformation [11]. Yet another problem is arriving the optimum threshold value. The histogram is split into two parts using the threshold value k. Hence, the optimum threshold value is arrived by maximizing the following objective function.

$$\sigma^2(k^*) = \max_k \frac{[\mu_T\,\omega\,(k) - \mu\,(k)]^2}{\omega\,(k)\,[1 - \omega\,(k)]}$$

where,

$$\omega(k) = \sum_{i=1}^{k} p_i; \quad \mu(k) = \sum_{i=1}^{k} ip_i; \quad \mu_T = \sum_{i=1}^{k} ip_i$$

Here, the no. of gray levels is indicated by 'L' and 'p$_i$' is the normalized probability distribution of the image. One of the most important advantages of the Power Law Transform is that it increases the contrast of the image and creates well-distinguishable pixels for improved segmentation.

$$s = cr^\nu$$

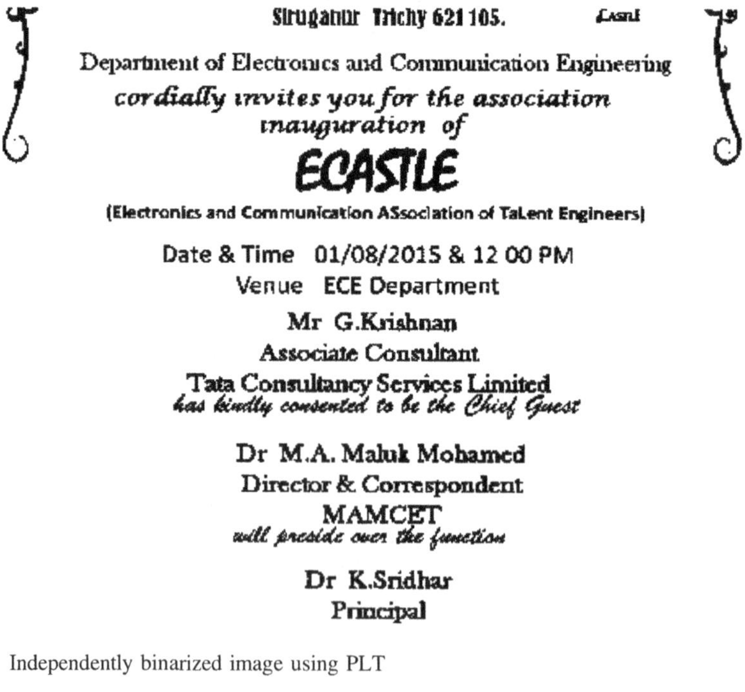

Sirug5anur Trichy 621 105.

Department of Electronics and Communication Engineering
cordially invites you for the association inauguration of

ECASTLE

(Electronics and Communication ASsociation of TaLent Engineers)

Date & Time   01/08/2015 & 12 00 PM
Venue   ECE Department

Mr G.Krishnan
Associate Consultant
Tata Consultancy Services Limited
has kindly consented to be the Chief Guest

Dr M.A. Maluk Mohamed
Director & Correspondent
MAMCET
will preside over the function

Dr K.Sridhar
Principal

Fig. 6 Independently binarized image using PLT

Here r is output intensity, s is input intensity, and c, ʋ are positive constants. Here the exponential component in Power Law is gamma. In experimentation, the gamma value is predefined as 1.5. To avoid another rescaling stage after PLT, the c is fixed as 1. With the increase in gamma value, the number of samples increases gradually, and for optimum segmentation, the gamma value for the PLT of images is fixed to 1.5 (Fig. 6).

5 Proposed Algorithm

As said above of the two important stages of recognition of text in images, binarization is an important task. And in case of color images, a novel method of combining the Clustering technique and PLT is utilized here, which shows more improved and universal results. Here the color components labeling is followed by the edge detection in separate clusters, edge map formation by union operation, and binarization using PLT method.

6 Recognition

Finally, the binarized image with text is given to OCR. Many OCRs are available such as OmniPage, Adobe Reader, ABBYY Fine Reader, and Tesseract. All the above-mentioned OCRs are relatively good, and hence, any of them can be used for the recognition of the binarized color images. Here, the train version of ABBYY Fine Reader is used. Performance evaluation is done with the ABBYY Fine Reader OCR's result.

7 Results

There are 2589 word images extracted from the color images as the training datasets. The testing dataset is of 781 word images. Table 1 compares the word recognition rate of proposed color clustering and PLT algorithms with those of other methods. Figure 7 shows certain images taken into consideration and their responses to the proposed algorithm.

Table 1 Performance evaluation of proposed color cluster and PLT algorithm on the color image dataset

S. No	Algorithm	Word recognition rate (%)
1	Color cluster an PLT	79.20
2	TH-OCR	62.50
3	Baseline method	64.30

Fig. 7 Other examples showing the binary results obtained using the color clustering and PLT algorithm

8 Conclusion

A novel algorithm with the combination of color components clustering and PLT technique is proposed for improving the recognition of color word image dataset. From experimentation, it is obtained that the algorithms proposed in the literature are not sufficient for the word recognition in the color images. Also in case power law transform, one of the most simple and effective algorithm performs appreciably well on many image dataset. But it fails in certain cases as shown above. Hence, a novel method is proposed by utilizing the effective algorithm PLT along with the connected component algorithm. The results obtained show that the proposed algorithm performs even more well universally for all kinds of image datasets, even for the images which PLT failed to produce results, and consequently enhance the recognition rate.

References

1. Abbyy Fine reader. http://www.abbyy.com/.
2. Adobe Reader. http://www.adobe.com/products/acrobatpro/scanning-ocrto-pdf.html Document Analysis and Recognition, pp. 11–16, September 2011, 2011.
3. S. M. Lucas et.al, "ICDAR 2003 Robust Reading Competitions: Entries, Results, and Future Directions", International Journal on Document Analysis and Recognition, vol. 7, no. 2, pp. 105–122, June 2005.
4. A. Mishra, K. Alahari and C. V. Jawahar, "An MRF Model for Binarization of Natural Scene Text", Proc. 11th International Conference of Document Analysis and Recognition, pp. 11–16, September 2011, 2011.
5. A. Shahab, F. Shafait and A. Dengel, "ICDAR 2011 Robust Reading Competition—Challenge 2: Reading Text in Scene Images", Proc. 11th International Conference of Document Analysis and Recognition, pp. 1491–1496, September 2011, 2011.
6. Thotreingam Kasar, Jayant Kumar and A.G. Ramakrishnan, "Font and Background Color Independent Text Binarization", Proc. II International Workshop on Camera-Based Document Analysis and Recognition (CBDAR 2007), Curitiba, Brazil, September 22, 2007, pp. 3–9.
7. T. Kasar and A. G. Ramakrishnan, "COCOCLUST: Contour-based Color Clustering for Robust Text Segmentation and Binarization," Proc. 3rd workshop on Camera-based Document Analysis and Recognition, pp. 11–17, 2009, Spain.
8. Deepak Kumar and A. G. Ramakrishnan, "Power-law transformation for enhanced recognition of born-digital word images," Proc. 9th International Conference on Signal Processing and Communications (SPCOM 2012), 22–25 July 2012, Bangalore, India.
9. J. N. Kapur, P. K. Sahoo, and A. Wong. A new method for gray-level picture thresholding using the entropy of the histogram. Computer Vision Graphics Image Process., 29:273–285, 1985.
10. C. Wolf, J. Jolion, and F. Chassaing. Text localization, enhancement and binarization in multimedia documents. ICPR, 4:1037–1040, 2002.
11. N. Otsu. A threshold selection method from gray-level histograms. IEEE Trans. Systems Man Cybernetics, 9(1):62–66, 1979.
12. Deepak Kumar, M. N. Anil Prasad and A. G. Ramakrishnan, "NESP: Nonlinear enhancement and selection of plane for optimal segmentation and recognition of scene word images," Proc. International Conference on Document Recognition and Retrieval(DRR) XX, 5–7 February 2013, San Francisco, CA USA.
13. Deepak Kumar, M. N. Anil Prasad and A. G. Ramakrishnan, "MAPS: Midline analysis and propagation of segmentation," Proc. 8th Indian Conference on Vision, Graphics and Image Processing (ICVGIP 2012), 16–19 December 2012, Mumbai, India.
14. D. Karatzas, S. Robles Mestre, J. Mas, F. Nourbakhsh and P. Pratim Roy, "ICDAR 2011 Robust Reading Competition—Challenge 1: Reading Text in Born-Digital Images (Web and Email)", Proc. 11th International Conference of Document Analysis and Recognition, pp. 1485–1490, September 2011, 2011. http://www.cv.uab.es/icdar2011competition.
15. D. Kumar and A. G. Ramkrishnan, "OTCYMIST: Otsu-Canny Minimal Spanning Tree for Born-Digital Images", Proc. 10th International workshop on Document Analysis and Systems, 2012.
16. J.Canny. Acomputationalapproachtoedgedetection. IEEE trans. PAMI, 8(6):679–698, 1986.
17. B. Epshtein, E. Ofek and Y. Wexler, "Detecting text in natural scenes with stroke width transform", Proc. 23rd IEEE conference on Computer Vision and Pattern Recognition, pp. 2963–2970, 2010.
18. P. Clark and M. Mirmhedi. Rectifying perspective views of text in 3-d scenes using vanishing points. Pattern Recognition, 36:2673–2686, 2003.

Grayscale Image Enhancement Using Improved Cuckoo Search Algorithm

Samiksha Arora and Prabhpreet Kaur

Abstract Meta-heuristic algorithms have been proved to play a significant role in the automatic image enhancement domain which can be regarded as an optimization question. Cuckoo search algorithm is one such algorithm which uses Levy flight distribution to find out the optimized parameters affecting the enhanced image. In this paper, improved cuckoo search algorithm is proposed which is used to achieve the better optimized results. The proposed method is implemented on some test images, and results are compared with original cuckoo search algorithm.

Keywords Image enhancement · Cuckoo search · Meta-heuristics · Gauss distribution · Improved cuckoo search algorithm

1 Introduction

Image enhancement means the technique of improving the visual appearance of an image so that resultant image is more accurate and comprehensive as compared to the original one, as well as, is appropriate for human and machine discernment. This process has application in various fields such as aerial/satellite imaging, bio-medical imaging, remote sensing. It is done to enhance the image's quality and appearance.

Image enhancement techniques mainly belong to two domains: spatial domain and frequency domain [1–3]. In this paper, image enhancement is considered in spatial domain. Spatial domain methods operate at the individual pixel intensities of the image. Main advantage of this method is that they are conceptually simple, comprehensive and lacks complexity which makes them useful in real-time applications. There are two generally used techniques: One is histogram equalization,

S. Arora (✉) · P. Kaur
Guru Nanak Dev University, Amritsar 143001, Punjab, India
e-mail: arorasamiksha1@gmail.com

P. Kaur
e-mail: prabhsince1985@gmail.com

© Springer Nature Singapore Pte Ltd. 2018   141
P.K. Sa et al. (eds.), *Progress in Intelligent Computing Techniques: Theory,*
Practice, and Applications, Advances in Intelligent Systems and Computing 518,
DOI 10.1007/978-981-10-3373-5_13

and other is linear contrast stretching (LCS). Both of them are used to improve images with low contrast. LCS can be accomplished using different techniques as max–min LCS, percentage LCS and piece-wise LCS [4, 5].

Frequency domain methods [2, 3] deal with analysis of mathematical functions with respect to frequency and manipulate transform coefficients such as Fourier Transform. These methods have big advantage like less computationally complex, robust and easy to observe and control frequency spectra of the image. The basic problem with these methods is that these methods cannot be easily automated and cannot enhance all parts of the image. In all these methods, there is subjectivity in judgment about how the image has improved. All these methods require the human interpreter is required to evaluate whether the image has improved for the required task.

This problem of subjectivity in judgment was finally removed by the introduction of genetic and various meta-heuristic algorithms [6–9]. In nature-inspired and meta-heuristic algorithms, image enhancement can be treated as an optimization problem which tends to generate the output image from given input image by using global or local transformation and also maximize the required fitness function. Transformation function used attempts to map the image in nonlinear fashion. Fitness function is an objective function which mathematically defines the quality of the image and hence makes enhancement procedure objective. Genetic algorithms started the optimization of image enhancement that overcomes the human judgment disadvantage of previous enhancement techniques [3, 10, 11].

Genetic algorithms (GA) were succeeded by particle swarm optimization (PSO) [9, 12]. Firefly [13] and differential evolution [14] algorithms are some other algorithms that have been implemented for image enhancement. Yang and Deb [11, 15] proposed other meta-heuristic algorithm as cuckoo search algorithm which is implemented for image enhancement. Results show that it outperforms many of the previous algorithms such as genetic algorithm and PSO [16]. Also cuckoo search can be implemented in domain of fingerprint contrast enhancement [17]. In this paper, an improvement over existing algorithm has been proposed for grayscale image enhancement that uses Gauss distribution [18]. Results obtained using proposed method come out to be better than original cuckoo search algorithm using criteria as: fitness and visual quality. The algorithm has been implemented using MATLAB software (R2010b) [19].

2   Cuckoo Search Algorithm

Cuckoo search algorithm (CS) is a meta-heuristic algorithm which was developed by Yin She Yang in 2009 [15]. This algorithm was based on reproductive behavior of cuckoo birds. Cuckoo never makes its own nest and depends on host nest for laying its eggs. Host bird takes care of the eggs taking them as their own eggs. If eggs are identified by host nest that they are not his eggs, then it either throws the alien eggs or discards the current nest and makes a fresh nest at some other place.

Eggs in the nest represent candidate solutions, and best egg is the new solution. There are three assumptions of this algorithm as follows:

1. Eggs are laid by cuckoo, and these eggs are put in a randomly selected nest.
2. Only eggs satisfying the criterion function will be taken forward in the subsequent iteration.
3. Count of existing host nests is kept constant. Probability of detection of egg by host bird p_d is taken as variable in the range [0, 1] [11]. Quality of eggs is determined by using a fitness function.

For generation of new solutions S (t + 1) for any cuckoo i, Levy flight distribution is used which is usual flight pattern of many flying animals and insects [15].

$$S_i^{t+1} = S_i^t + s \odot \text{Levy}\,(x). \tag{1}$$

Here, s(s > 0) represents size of the step made by above-mentioned distribution. The symbol $\odot$ depicts entry-wise multiplication. Levy flights impart a random walk which is derived from Levy distribution as:

$$\text{Levy} \sim z = t^{-1}, \; (1 < l \le 3). \tag{2}$$

that exhibits unbounded mean and variance. This random walk process is tracked by step length distribution governed by power law method. In case of nest discard, new nest is build using random walk and mixing at new location. Nests are discarded with probability p_d.

Image enhancement takes a transformation function that maps each input pixel to new intensity output pixel that generates the enhanced image. According to the literature [6–8], automatic image transformation can be done using Eq. (3) that takes into consideration global as well as local properties of an image.

$$o\,(x, y) = \frac{k * M}{s\,(x, y) + q} [i\,(x, y) + r * m\,(x, y)] + m\,(x, y)^p. \tag{3}$$

where i(x, y) is the gray value of (x, y)th pixel of the original image. o(x, y) is the intensity value of (x, y)th pixel of the output image. m(x, y) and s(x, y) are the mean and standard deviation of (x, y)th pixel over an n x n neighborhood, respectively. Global mean of the image is taken to be as M. p, q, r, k are the constraints to control the variation in the output image.

To eliminate the need of human interpreter to judge the output image, fitness function is defined. A fine image should have edges with sharp intensities, good number of pixels on the edges (edgels) and entropy measure [6].

$$F\,(I_e) = \log\,(\log\,(E\,(I_e) * \frac{\text{edgels}\,(I_e) * H\,(I_e)}{P * Q}. \tag{4}$$

In above equation, I_e is the enhanced image. E (I_e) is the sum of P × Q pixel intensities of the edge image, and edgels (I_e) is the number of edge pixels. Here, Sobel edge detector is used to find out edge image of output image. H (I_e) is the entropy value of the enhanced image.

3 Proposed Method

Levy flight distribution provides random walk for large steps using Eq. (2) which has infinite mean and variance. Hence, cuckoo search algorithm will always give the optimum solution. But random walk governs the entire search; hence, fast convergence cannot be guaranteed. Also Levy flight distribution does not exhibit ergodicity property. Hence, improvement to the method is made in order to improve convergence rate and precision. Following Gauss distribution equation can be used to determine new solution as [18]

$$\sigma_s = \sigma_0 * e^{-mu*g}. \tag{5}$$

where σ_0 and mu are constant and g is current generation. Equation (5) instead of (2) is used for producing new solution S (t + 1) for cuckoo i. A Gauss distribution is used [18].

$$S_i^{t+1} = S_i^t + s \odot \sigma_s. \tag{6}$$

where s (s > 0) represents step size. Its value is subjected to the type and extent of the problem. Figure 1 represents the flowchart of proposed method.

Proposed algorithm works as follows:

1. First total population of nests is initialized, and then each nest using these four constraints p, q, r, k of transformation function is initialized that needs to be optimized.
2. Each nest corresponds to an enhanced image generated using Eq. (3) and satisfies the fitness function.
3. Only eggs (set of parameters) satisfying the fitness/objective function will be taken to the subsequent iteration.
4. New nests are created at new position using Gauss distribution as defined by Eqs. (5) and (6)
5. Fitness of all nests is calculated using fitness function defined in Eq. (4), and best nest and best fitness is recorded as best nest and Fbest.
6. Bad nests are discard by probability of discard p_d, and new nests are build at new location using Eq. (5).
7. Fitness of new nests is computed and compared with Fbest, and fitness with maximum value is retained as best value.
8. Termination criterion for the algorithm is set as maximum number of iterations.

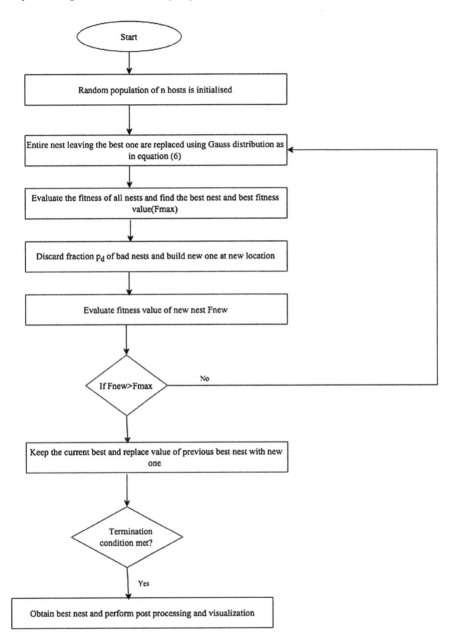

Fig. 1 Flowchart of proposed algorithm for image enhancement

9. And finally, enhanced image is obtained using best set of parameters which correspond to the best solution.

10. Post-processing and visualization is preformed to obtain the final enhanced image.

4 Results and Discussion

The proposed method was implemented on some images using MATLAB [19]. The main idea of the proposed method is to improve image's visual quality of the image in accordance with edgels and entropy which constitutes the fitness function. Number of nests = 25, discard probability p_d is 0.25, maximum number of iterations = 50. Window size is 3. Increased number of edgels put more emphasis on the edges in the image, and good measure of entropy leads to the uniform histogram distribution corresponding to the image. Results obtained from the proposed method are judged against original algorithm. Original grayscale images and corresponding enhanced images using original and proposed method are shown in Figs. 2, 3 and 4. Table 1 shows the summary of results obtained using original and proposed algorithm.

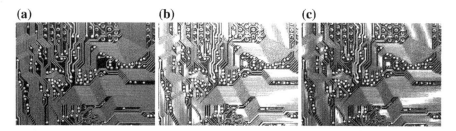

(a)   **(b)**   **(c)**

Fig. 2 **a** Circuit.jpg (original image) **b** image enhanced using CS **c** image enhanced using improved CS

(a)   **(b)**   **(c)**

Fig. 3 **a** Building.jpg (original image) **b** image enhanced using CS **c** image enhanced using improved CS

(a)   (b)   (c)

Fig. 4 **a** Forest.jpg (original image) **b** image enhanced using CS **c** image enhanced using improved CS

Table 1 Parameterized comparison of original and proposed algorithm

Image	Parameters	Original method	Proposed method
Circuit.jpg	Original fitness	4.1325	4.1325
	Fitness of enhanced image	7.1120	10.0856
	Edgels	14833	15027
	Computation time (in seconds)	22.14	20.46
Building.jpg	Original fitness	3.7524	3.7524
	Fitness of enhanced image	7.7081	11.8142
	Edgels	23613	25100
	Computation time (in seconds)	32.68	31.32
Forest.jpg	Original fitness	4.4965	4.4965
	Fitness of enhanced image	11.0856	13.7958
	Edgels	20543	27270
	Computation time (in seconds)	21.58	21.66

5   Conclusion and Future Work

In this paper, an improvement of original cuckoo search algorithm is proposed that use gauss distribution to find new solution and optimize the required parameters. Core purpose of this paper is to facilitate automatic grayscale image enhancement using meta-heuristic approach that eliminates requirement of human interpreter to assess the quality of the image. Results have proved that proposed method is better than the previous algorithm.

Work presented in this paper finds large applications in various spheres of computing and processing. The algorithm can be realized in other areas of digital image processing such as segmentation, feature selection or morphology. Same algorithm can be employed on color images as well to obtain better results than already existing algorithms for color image processing.

References

1. Gonzalez, R.C., Woods, R.E.: Digital image processing. ed: Prentice Hall Press, ISBN 0-201-18075-8 (2002).
2. Pratt, William K.: Digital Image Processing. PIKS Scientific Inside, Fourth Edition. p. 651–678, (1991).
3. Gonzalez, R.C., Woods, R.E., Eddins, S.L.: Digital image processing using MATLAB. 2nd ed: Prentice Hall Press.
4. Yao, H., Wang, S., Zhang, X.: Detect piecewise linear contrast enhancement and estimate parameters using spectral analysis of image histogram. p. 94–97, (2009).
5. Chi-Chia, S., Shanq-Jang, R., Mon-Chau, S., Tun-Wen, P.: Dynamic contrast enhancement based on histogram specification. IEEE Trans.Consum. Electron 51(4), p. 1300–1305, (2005).
6. Munteanu, C., Rosa, A.: Towards automatic image enhancement using genetic algorithms. In: Evolutionary Computation, Proceedings of the 2000 Congress on, vol. 2, p. 1535–1542, IEEE, (2000).
7. Saitoh, F.: Image contrast enhancement using genetic algorithm. In: Systems, Man, and Cybernetics, 1999. IEEE SMC'99 Conference Proceedings. 1999 IEEE International Conference on, vol. 4, p. 899–904, IEEE, (1999).
8. Munteanu, C., Rosa, A.: Gray-scale image enhancement as an automatic process driven by evolution. Systems, Man, and Cybernetics, Part B: Cybernetics, IEEE Transactions on 34, no. 2, p. 1292–1298, (2004).
9. Kennedy, J., Eberhart R.C.: Particle swarm optimization. In: Proceedings of IEEE International Conference on Neural Network, p. 1948–1995, (1995).
10. Ahmed, M.M., Zain J.M.: A study on the validation of histogram equalization as a contrast enhancement technique. In Advanced Computer Science Applications and Technologies (ACSAT), 2012 International Conference on, p. 253–256. IEEE, 2012.
11. Yang, X.S., Deb, S.: Engineering optimisation by cuckoo search. International Journal of Mathematical Modelling and Numerical Optimisation 1, no. 4, p. 330–343, (2010).
12. Clerc, M., Kennedy, J.: *The particle swarm-explosion, stability, and convergence in a multidimensional complex space.* Evolutionary Computation, IEEE Transactions on 6, no. 1, p. 58–73, (2002).
13. Hassanzadeh, T., Vojodi, H., Mahmoudi, F.: Non-linear grayscale image enhancement based on firefly algorithm. In: Evolutionary and Memetic Computing, vol 7077, p. 174–181. Springer, Berlin Heidelberg, (2011).
14. Sarangi, P.P., Mishra, B., Dehuri, S.: Gray-level image enhancement using differential evolution optimization algorithm. In: Signal Processing and Integrated Networks (SPIN), (2014).
15. Yang, X.S., Deb, S.: Cuckoo search via Lévy flights. In: Nature & Biologically Inspired Computing, 2009. NaBIC 2009. World Congress on, p. 210–214, IEEE, (2009).
16. Ghosh, S., Roy, S., Kumar, U., Mallick, A.: Gray Level Image Enhancement Using Cuckoo Search Algorithm. In: Advances in Signal Processing and Intelligent Recognition Systems, p. 275–286, Springer International Publishing, (2014).
17. Bouaziz, A., Draa, A., Chikhi, S.: A Cuckoo search algorithm for fingerprint image contrast enhancement. In: Complex Systems (WCCS), 2014 Second World Conference on, p. 678–685, IEEE, (2014).
18. Zheng, H., Zhou, Y.: A Novel Cuckoo Search Optimization Algorithm Base on Gauss Distribution. Journal of Computational Information Systems 8:10, p. 4193–4200, (2012).
19. http://www.mathworks.com/matlabcentral/fileexchange/29809-cuckoo-search-cs-algorithm.
20. Bunzuloiu, V., Ciuc, M., Rangayyan, R.M., Vertan, C.:Adaptive neighbourhood histogram equalization of color images. J. Electron Imaging 10(2), p. 445–449, (2001).
21. Yang X.S.: Nature-Inspired Metaheuristic Algorithms, Luniver Press, (2008).
22. Senthilnath, J.: Clustering Using Levy Flight Cuckoo Search. In: Proceedings of Seventh International Conference on Bio-Inspired Computing, vol. 202, p. 65–75, (2013).

Segmentation of Tomato Plant Leaf

S. Aparna and R. Aarthi

Abstract In today's world there is a huge need for constant monitoring of crops, plants in agricultural fields to avoid diseases in plants. Since we cannot rely on the ability and accuracy of the human eye, it is only natural to depend on electronic equipment to detect diseases in crops and plants. Use of electronics for monitoring crops will help prevent plants from infection since this is the need of the hour, hence, making this an essential research paper. Most crops such as tomato, chilli, paddy etc. are attacked by bacteria, fungus or viruses leading to change in color, texture or function of a plant as it responds to pathogens. Common fungal infections include leaf rust, stem rust or white mold formation on the plant. Bacterial infections such as leaf spot with yellow halo, fruit spot, canker and crown gall all affect crops severely. In plants that are effected by viruses, one can find ring spots, pale green color in leaves and the plant stops growing and becomes distorted. We take these visual changes into account and process these images to identify whether a plant is healthy or not. There are three main steps for segmentation and identification of this disease.

1. Acquiring the RGB image and converting it into a suitable color domain such as HSV, YCbCr etc.
2. Mask the green pixels using a suitable threshold.
3. Choose a particular component in the chosen color domain after analyzing which component gives the most feasible result.

S. Aparna (✉)
Department of Electronics and Communication Engineering,
Amrita School of Engineering, Amrita Vishwa Vidyapeetham,
Amrita University, Coimbatore, Tamil Nadu, India
e-mail: aparna_95@rediffmail.com

R. Aarthi
Department of Computer Science and Engineering, Amrita School of Engineering,
Amrita Vishwa Vidyapeetham, Amrita University, Coimbatore, Tamil Nadu, India
e-mail: aarthi.r4@gmail.com

© Springer Nature Singapore Pte Ltd. 2018   149
P.K. Sa et al. (eds.), *Progress in Intelligent Computing Techniques: Theory,*
Practice, and Applications, Advances in Intelligent Systems and Computing 518,
DOI 10.1007/978-981-10-3373-5_14

Keywords HSV · Masking · Plant diseases · Histogram equalization

1 Introduction

Healthy food and lifestyle has become a major issue in the recent years. Agriculturists work to provide crops that give maximum productivity with lower cost. Continuous assessment of crops and timely addressing of diseases is needed to ensure profit and quality. The general approach for monitoring of crops is by detecting the changes with our eyes. Since in most cases farmers are either not aware of all the diseases present in plants or lack the required skills to detect them, agriculture experts are generally employed to aid the farmers. Such help is generally provided to the farmer based on the request to the agriculture department or the surveys happen with the large time gap. However, this involves more cost and man power with a compromise in quality of crops. The lack of availability of such experts and unawareness of non-native diseases of crops makes the farmer's job tedious [1]. The other way to identify diseases is molecular designs like polymerase chain reaction. It is used to identify diseases in crops but these require detailed processing and laborious processing procedure [2]. Another way is to use electronic devices to take an image of the plant and it can be captured with the help of image processing tools, diseases can be detected with accuracy and much more efficiency.

Tomato is one of the most popular vegetables in TamilNadu state of India that gets affected by diseases easily [3]. Since tomato plants are in high demand, farmers try to find new ways to increase the productivity. Dry farming, amending soil with salt and crop rotation are some methods used by farmers to meet these demands [4]. However, in most cases, people are left with no choice but to compromise in the quality of this vegetable. Eating such vegetables will cause problems such as food poisoning, liver infection etc. Healthy generation and production of tomato plants will ensure satisfied and healthy consumers. In general, tomato plants that grow in wet, humid climate are susceptible to diseases. Dark brown or black spots appear on the surface of leaves first. It is then followed by yellowing or browning usually to those leaves at the lower branches. Removal of these infected leaves as soon as they appear may help reduce its spread [5]. Figure 1a shows a picture of a healthy leaf.

(a)   (b)   (c)   (d)

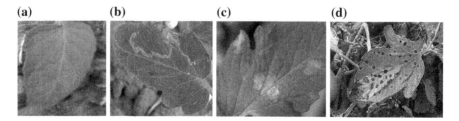

Fig. 1 **a** Healthy tomato leaf. **b** Leaf infected by 'leaf miner'. **c** Symptoms of bacterial cancer. **d** Leaf infected by 'early blight'

In (b), (c) and (d) the infected leaf is attacked by a disease called leaf miner, bacterial cancer and fungus alternaria solani. Leaf miner is a very common disease and it can be easily avoided if detected in its initial stages by careful monitoring. If detected in the later stages, the entire crop will have to be removed.

One of the major cues is the color change of the leaf from normal surface. Other way is to recognize patterns of the leaf. The major features that support to categorize the pattern of diseases is color and texture. The main challenge is to extract the tomato leaf regions in the screen so it can be classified [6]. So our main focus in this work is to design an algorithm that detects tomato leaves in various scenes so that the segmented part can be can be further processed to detect diseases. Segmentation is used to separate the leaf regions from the background so that the infected leaf can be classified. Watershed image segmentation is based on the theory of mathematical morphology. We can say that the watershed is a set of crest lines radiating from saddle points [7]. Another method for segmenting an image is by using k-means algorithm. In this method, a few seed points are taken and the distance between the seed points and the nearby pixel is calculated. All pixels within a given distance from the seed point will be grouped into that particular cluster. The process will be repeated till no two seed points have the same pixels [8]. Our procedure takes advantage of color cue to segregate the background and leaf region. The color cues are used for detection of various like skin regions [9], vehicles in roads [10].

1.1 The Proposed Approach

Here is the step-by-step procedure of the proposed approach:

1. Acquisition of RGB image.
2. Converting image to HSV color domain.
3. Masking the green pixels by choosing a suitable threshold.
4. Performing pre-processing techniques.
5. Segmentation of image.

1.1.1 Acquisition of RGB Image

The image is taken from a relative distance from the crop to avoid loss of data and clarity. Since the height of an average tomato plant will be around three feet, the camera height can be fitted from three and a half feet to capture the image with a top view [11]. To obtain an image with a horizontal view, an image can be captured from around one foot away from the plant. Illumination variation due to sun light is also a major problem. The shadow of adjacent leaves on a particular leaf may make it darker.

1.1.2   Converting to HSV Color Domain

HSV stands for Hue, Saturation and Value where hue is a color attribute that describes pure color as perceived by an observer. Saturation refers to the relative purity or the amount of white added to hue and value refers to amplitude of light. We use HSV over RGB because it can separate color from intensity. On doing so, we can remove shadow effect or create changes in lighting conditions [12]. It is also easier to implement with a direct conversion of RGB to this domain. We do not choose YCbCr because some of its values may be out of range of RGB. Also when we analyzed a sample image using RGB domain and masked the green, red and blue component separately by setting specific threshold values, we noticed that it is difficult to get a fixed value. There were colors which were present in all the three color components, thereby making the segmentation process difficult. Figure 2 shows the sample image analyzed and the results obtained.

Due to these inconsistencies and the fact that color and intensities can be separated in HSV, we have chosen it over our RGB color space. Surveys on color segmentation shows that HSV is the most suitable color domain in image processing [9].

1.1.3   Masking of the Green Pixels

Masking means setting the pixel value in an image to zero or some other value. Here average of the image containing healthy green is computed in hue, saturation and value [13]. For this, the image is initially cropped after converting it to HSV. The average of hue, saturation and value are computed separately and a range of

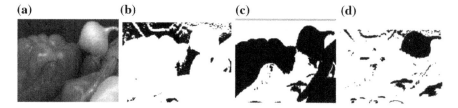

Fig. 2   a Original RGB image. **b** Masked red image. **c** Masked green image. **d** Masked blue image

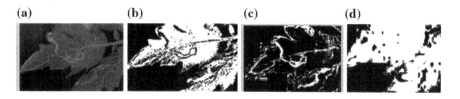

Fig. 3   a Acquired RGB image. **b** Hue component. **c** Saturation component. **d** Value component

Table 1 HSV values of tomato leaf before normalization

Name of image	Hue average	Saturation average	Value average	Hue range	Saturation range	Value range	Size of image
Image1	110.66	120.5	161	100–130	110–140	150–180	1222 × 617 × 3 uint8
Image2	55	122.18	147.95	50–80	120–140	140–180	194 × 260 × 3 uint8
Image3	82	112	87	80–110	120–150	85–110	184 × 274 × 3 uint8
Image4	70	134	174	60–90	130–180	165–200	1836 × 3264 × 3 uint8
Image5	89.1	80	120	80–120	130–160	115–140	1836 × 3264 × 3 uint8
Image6	83	95	147	80–120	90–130	150–185	1836 × 3264 × 3 uint8
Image7	99	107	143	90–120	90–130	150–185	760 × 994 × 3 uint8
Image8	107	117	143	105–130	115–140	150–180	1836 × 3264 × 3 uint8
Image9	54	173	105	50–90	175–210	100–140	194 × 259 × 3 uint8
Image10	57	170	106.11	50–90	150–190	100–140	1836 × 3264 × 3 uint8
Image11	51	201	128	50–90	190–230	120–150	1836 × 3264 × 3 uint8
Image12	76	96	157	90–120	100–130	150–180	1836 × 3264 × 3 uint8
Image13	74	103	155	80–110	100–140	150–190	1836 × 3264 × 3 uint8

Table 2 HSV values of tomato leaf after normalization

Name of image	Hue average	Saturation average	Value average	Hue range	Saturation range	Value range
Image10	87.7	77.6	134	60–90	60–100	100–140
Image11	76	145	142	60–90	130–180	130–180
Image12	125	95	175	120–150	130–180	160–200
Image13	93	83	144	70–120	70–120	140–180
Image14	100	134	137	70–110	130–160	130–160
Image15	56	31	142	40–80	20–50	140–180
Image7	110	132	134	100–130	120–150	140–180
Image6	96	99	114	90–130	120–150	110–140
Image4	97	60	174	80–130	50–90	150–190
Image1	120	162	162	80–130	150–190	150–190
Image8	98.7	143	136	100–130	150–190	130–160
Image17	77	143	140	60–100	130–160	130–160
Image16	75	65	163	70–100	60–90	160–180

values are given to them separately. Here any pixel falling within this range is set to white or 255 in grey scale. Any pixel outside this range is set to black or 0. In this way it is very easy to detect any color change apart from the normal. Figure 3 shows an RGB image and hue, saturation and value extracted separately and the green areas masked.

Table 1 gives the names of images of different tomato leaves analyzed and their respective hue, saturation and value averages and the most accurate ranges available.

These were the results before pre-processing techniques were done on the image.

1.1.4 Applying Pre-processing Techniques

Since the range of values for hue, saturation and value are so vast it is difficult to come to a narrow range. None of the images have a common average or range of values. This can be due to the effect of shadows on the leaf or even due to the lighting conditions. If the image was captured in the morning or the afternoon, the color of the image will also change due to the position of the sun. To avoid this problem we have to apply uniform lighting to all the images. This can be done by performing histogram equalization on the images by taking one image as reference. Since images have same content, we introduce histogram equalization to bring a contrast to the image. This will be enough to render any grey scale differences in our resulting image [11]. Hence, all of our images will have the same lighting conditions and one of the images will be kept as the reference image.

Table 2 gives us the range of values for hue, saturation and value using an image by name 'image5' as reference. Now, we can see a clear range of values for hue, saturation and value averages and their ranges.

1.1.5 Segmentation of Image

Taking the range of hue from 80 to 120, saturation range from 60 to 150 and value range from 135 to 170, we get the following results. Figure 4 shows a picture of the results obtained on setting the images to the ranges specified above. As seen in these sample images, most of the pictures used for analysis are close up images of the tomato leaf.

(a)   (b)   (c)   (d)

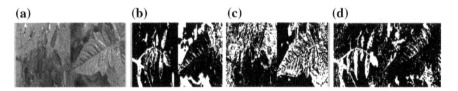

Fig. 4 **a** Original RGB image with reference image after pre-processing. **b** Saturation component of image. **c** Value component of image. **d** Hue component of image

2 Conclusions

We can clearly see that saturation component of image gives better results than the other two components. The leaf is clearly segmented too. Hence we can conclude by saying that saturation component in the HSV domain gives best results for detection of diseases in crops. With the information we have gathered we can use detect diseases in crops by acquiring the image of the crop and using saturation component in the HSV image. This will make the job of farmers easier thereby making our lives safer and healthier.

Acknowledgements I would like to thank my university for giving me the opportunity to present this project. Sincere thanks to my parents, family members and friends who provided me help and moral support through the course.

References

1. Hiary, H. A., Ahmad, S. B., Reyalat, M., Braik, M., & Alrahamneh, Z. (2011). Fast and Accurate Detection and Classification of Plant Diseases. International Journal of Computer Applications IJCA, 17(1), 31–38.
2. Sankaran, S., Mishra, A., Ehsani, R., & Davis, C. (2010). A review of advanced techniques for detecting plant diseases. Computers and Electronics in Agriculture, 72(1), 1–13.
3. TamilNadu Agriculture University. (n.d).Tomato. Retrieved on March 7,2016 from http://agritech.tnau.ac.in/banking/PDF/Tomato.pdf
4. New strategies for great-tasting tomatoes. (n.d.). Retrieved April 08, 2016, from http://www.growingformarket.com/articles/Improve-tomato-flavor
5. What are these lines on tomato leaves? (n.d.). Retrieved May 06, 2016, from http://gardening.stackexchange.com/questions/13706/what-are-these-lines-on-tomato-leaves
6. Sabrol, H., & Kumar, S. (2015). Recent Studies of Image and Soft Computing Techniques for Plant Disease Recognition and Classification. International Journal of Computer Applications IJCA, 126(1), 44–55.
7. Aly, A. A., Deris, S. B., & Zaki, N. (2011). Research Review for Digital Image Segmentation Techniques. International Journal of Computer Science and Information Technology IJCSIT, 3(5), 99–106.
8.] Kanungo, T., Mount, D., Netanyahu, N., Piatko, C., Silverman, R., & Wu, A. (2002). An efficient k-means clustering algorithm: Analysis and implementation. IEEE Transactions on Pattern Analysis and Machine Intelligence IEEE Trans. Pattern Anal. Machine Intell., 24(7), 881–892.
9. Padmavathi.S, Nivine. (2014). Survey on Skin Technology. International Journal of Engineering Research & Technology vol 3(issue 2).
10. Aarthi, R., Padmavathi, S., & Amudha, J. (2010). Vehicle Detection in Static Images Using Color and Corner Map. 2010 International Conference on Recent Trends in Information, Telecommunication and Computing. doi:10.1109/itc.2010.13
11. The Average Height for Tomato Plants. (n.d.). Retrieved March 28, 2016, from http://homeguides.sfgate.com/average-height-tomato-plants-50929.html
12. Why do we use the HSV colour space so often in vision and image processing? (n.d.). Retrieved May06, 2016, from http://dsp.stackexchange.com/questions/2687/why-do-we-use-the-hsv-colour-space-so-often-in-vision-and-image-processing
13. Gonalez, R. C. (2008). Digital signal processing. (3 ed).New Delhi: PHI Private Limited.

A Novel Image Steganography Methodology Based on Adaptive PMS Technique

Srilekha Mukherjee and Goutam Sanyal

Abstract The rapid sprout in the usage of secured information exchange through the Internet causes a major security concern today. In this paper, we have proposed a secured approach in the lexicon of modern image steganography frame. The adaptive form of the power modulus scrambling (PMS) technique has been used so as to scramble the pixels of the cover, thereby putting in a secured shield right from the beginning. Next, a comparative key-based permutation combination methodology adeptly carries out the embedding procedure. This ardently caters the concerning security issue while communication. The commendable results of the proposed approach evaluated with respect to substantial performance metrics maintain the foothold of imperceptibility.

Keywords Steganography · Adaptive power modulus scrambling · Peak signal to noise ratio · Cross-correlation coefficient

1 Introduction

The need of security in communication has become an evident issue these days. This paper consists of a proposed methodology that serves to meet all the needs of a secured system [1]. The chosen cover image is scuffled in the beginning of the procedure with the aid of adaptive PMS [2] algorithm. This scrambles the residing pixel distribution of the whole image, thereby generating an illegible form that is

S. Mukherjee (✉) · G. Sanyal
Computer Science and Engineering Department, National Institute
of Technology Durgapur, Durgapur, West Bengal, India
e-mail: srilekha.mukherjee3@gmail.com

G. Sanyal
e-mail: nitgsanyal@gmail.com

© Springer Nature Singapore Pte Ltd. 2018   157
P.K. Sa et al. (eds.), *Progress in Intelligent Computing Techniques: Theory,*
Practice, and Applications, Advances in Intelligent Systems and Computing 518,
DOI 10.1007/978-981-10-3373-5_15

difficult to trace. This is followed by the procedure of embedding [3] the secret bits into the chaotic distribution of the cover. The new methodology proposed incorporates the technique of both permutation and combination to decide upon the fashion of insertion of pixel bits. Depending upon the value of the keys, a comparative scheme comes into play. Finally, the execution of the inverse adaptive PMS consummates the spawning of the imperceptible stego-image [4]. The phase of extraction works by proficiently retrieving each inserted bit, thereby regenerating the hidden image. The quality of the inserted and extracted images has no differences at all. Also, the imperceptibility of the hatched stego-image is maintained.

2 Related Work

The simplest of all methods is data hiding by LSB [5] substitution. Though this method achieves commendable results, it is quite sensitive to delicate compression or cropping of images. Wu & Tsai proposed the PVD, i.e., pixel value differencing method, where the contrariety between each pair of pixel is considered to arbitrate the number of embedding pixels. Based on the PVD [6] concept, various other methodologies have been proposed. One among them is the one proposed by Chang et al. [7], which proffers a unique three-way pixel value differencing technique. Another methodology, namely gray-level modification, i.e., GLM proposed by Potder et al. [8], maps the secret data bits by comparing with the gray-level bits of the chosen pixel values. It employs the odd and even conception of numbers for the purpose of data mapping. Safarpour et al. [9] proposed a hybrid approach combining the PVD and GLM techniques. This basically increases the embedding capacity of the PVD method by adopting the GLM approach.

3 Proposed Approach

The host is scrambled using the adaptive power modulus scrambling (PMS) method. A chaotic representation of the original pixel correlation results. This adds a security stratum to the approach. Next, the embedding (shown in Fig. 1) of the secret pixel bits within the scrambled form of the host image is effectuated. Here, a comparative strategy of permutation and combination is put into effect. The reverse technique of the adaptive power modulus scrambling (PMS) wraps up the embedding stage. The entire procedure is efficient enough to sustain and preserve the imperceptibility of the output stego-image (depicted in Fig. 2).

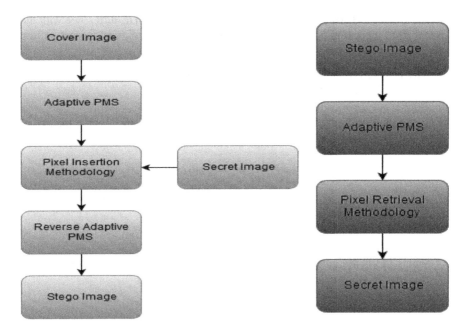

Fig. 1 Flow diagrams representing the embedding procedure (*left*) and the extraction procedure (*right*)

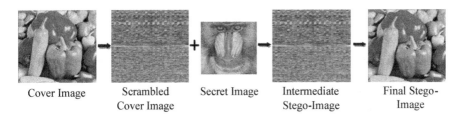

| Cover Image | Scrambled Cover Image | Secret Image | Intermediate Stego-Image | Final Stego-Image |

Fig. 2 Formation of the Stego-image

4 Algorithm

4.1 The Embedding Procedure

i. Take a cover image as input.
ii. Apply the adaptive PMS transformation on it.
iii. Reinforce the pixel insertion methodology to embed the secret bits.
iv. Apply the inverse adaptive PMS to generate the output stego.

Adaptive Power Modulus Scrambling (PMS) method. The difference with the original PMS algorithm lies in the step iv, where the odd and even property of the individual locations comes into effect.

   i. The host or cover image is the selected input.
  ii. Inspect the size of the input image (say m * n)
 iii. Figure out the location values of all the resident pixels.
 iv. For a pixel with the location L(i, j), trace out 'a,' 'b' and 'c' such that:
     If L(i, j) is odd, then

$$a = ((T - L(i,j) + 1) + FourLSB(L(i,j)))^2 \tag{1}$$

If L(i, j) is even, then

$$a = ((T - L(i,j) + 1) - FourLSB(L(i,j)))^2 \tag{2}$$

where 'T' is the total no. of pixels and 'Four LSB' is a function which takes decimal converted value of the four LSBs of a number.

$$b = (a\%((m/2)*n)) + 1 \tag{3}$$

where '(m/2) * n' is the last position value of the (m/2)th row

$$c = b + ((m/2)*n) \tag{4}$$

Here, 'c' is the location of a pixel that is decisively switched with the pixel at 'L(i, j)'

  v. If the pixel at location 'c' is found to be heretofore swapped, then linear probing is triggered.
 vi. If the collected value of 'c' exceeds the terminal value of the image matrix, 'L (m, n),' then the conditional checking restarts from the position value 'L((m/2), n) + 1' and the process of linear probing continues.

Pixel Insertion Methodology.

   i. Consider the entire rows (say m) and columns (say n) of the scrambled host.
  ii. Next, we get the gross total of pixels (say t = m * n).
 iii. For a pixel residing at location (i, j) position, compute the position values as:

$$L(i,j) = (i - 1)*m + j \tag{5}$$

 iv. From each pixel, generate four decimal values by taking into account the bits

Table 1 All the operation cases of conditional comparisons

Position Value	Operation for Key generation	Value of key	Operation cases
Even	Permutation	Even	1
		Odd	2
Odd	Combination	Even	3
		Odd	4

ii. 3rd 4th 5th 6th; 4th 5th 6th 7th; 5th 6th 7th 8th; 3rd 4th 7th 8th.
v. Compute the biggest (say 'high') and smallest (say 'low') element from the four estimated decimal values.
vi. Perform permutation or combination of the two obtained decimal values (i.e., 'high' and 'low') according to the even or odd values of position (displayed in Table 1).

4.2 The Extraction Procedure

i. Choose the input stego-image.
ii. Apply adaptive PMS transformation on it.
iii. Execute the pixel retrieval methodology to retrieve the secret bits.
iv. Finally the inserted secret image is retrieved.

Pixel Retrieval Methodology

i. Consider the entire rows (say m) and columns (say n) of the scrambled stego.
ii. Next, we get the gross total of pixels (say t = m * n) for the same.
iii. For a pixel residing at location (i, j) position, compute the position values as:

$$L(i,j) = (i-1) * m + j \qquad (6)$$

iv. From each pixel, generate four decimal values by taking into account the bits
v. 3rd 4th 5th 6th; 4th 5th 6th 7th; 5th 6th 7th 8th; 3rd 4th 7th 8th.
vi. Compute the biggest (say 'high') and smallest (say 'low') element from iv.
vii. Perform permutation or combination of the two obtained decimal values (i.e., 'high' and 'low') according to the even or odd values of position.
viii. Referring to the Table 1, the following four cases of data retrieval occur.
ix. The first two retrieved pixel values will occupy the foremost bit positions of an eight-bit pixel array and so on.
x. Likewise all the secret pixels are extracted by considering the filled-up eight-bit arrays, which individually accredits the hidden image.

5   Experimental Results

5.1   Payload and Cross-Correlation Coefficient

The maximum bit of information carried by a cover without being identified is referred to as payload or embedding capacity [10].

The normalized cross-correlation coefficient ('r') [11] is figured out as:

$$r = \frac{\sum (H(i,j) - m_H)(S(i,j) - m_S)}{\sqrt{\sum (H(i,j) - m_H)^2}\sqrt{\sum (S(i,j) - m_S)^2}} \tag{7}$$

where 'H' is the host/cover, 'S' is the obtained stego, 'm_H' is the mean value of the pixels of host/cover, and 'm_S' is the mean value of the pixels of stego (Fig. 3).

The above figure (on left) voices that the measured capacity of our proposed approach is better than the specified others. The quality of the images is not degraded as well, since the approach has its correlation scores close to 1 (from the figure on the right).

5.2   MSE and PSNR

MSE [12] can be estimated from the following equation (Eq. 8)

$$MSE = \frac{1}{(m*n)} \sum_{i=1}^{m} \sum_{j=1}^{n} [H(ij) - S(ij)]^2 \tag{8}$$

where 'H' is a cover or host image (with m x n pixels) and 'S' is the stego.

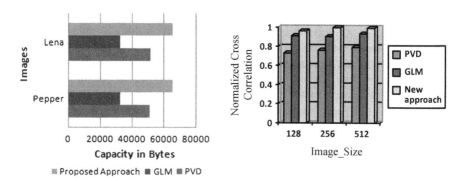

Fig. 3 Exemplification based on the embedding capacity (*left*) and similarity measure (*right*)

Table 2 Evaluation of an exclusive comparison of PSNR

Image_size	Without adaptive PMS		With adaptive PMS	
	MSE	PSNR	MSE	PSNR
Pepper_512 × 512	16.39	44.52	16.44	44.51
Pepper_256 × 256	16.44	44.64	16.38	44.65
Pepper_128 × 128	16.47	44.78	19.76	43.99
Lena_512 × 512	17.00	44.84	17.01	44.83
Lena_256 × 256	17.01	45.16	16.97	45.18
Lena_128 × 128	17.03	45.04	16.94	45.07

Therefore, PSNR [13] is assessed as (Eq. 9):

$$PSNR = 10 \log_{10} 255^2 / MSE \ db \tag{9}$$

Table 2 portrays the insignificant changes that occur for two specified cases. Thus, the adaptive PMS adds a security layer without forfeiting any kind of quality issue.

6   Conclusion

The proposed approach is highly proficient in its own information hiding [14] dexterity. The quantitative as well as qualitative metrics assure that all the obligations are met with efficacy and potency. The PSNR and cross-correlation values clearly prove that there is no visible degradation in the image quality [15].

References

1. Jaheel, H., Beiji, Z.: A Novel Approach of Combining Steganography Algorithms. International Journal on Smart Sensing and Intelligent Systems. Vol. 8 (2015).
2. Mukherjee, S., Ash, S., Sanyal, G.: A Novel Image Steganographic Methodology by Power Modulus Scrambling with logistic Mapping. TENCON, IEEE Region 10 Conference (2015).
3. Singh, K.: A Survey on Image Steganography Techniques. International Journal of Computer Applications. Vol 97 (2014).
4. Luo, X., et al.: On F5 Steganography in Images. The Computer Journal. Vol. 55 (2012).
5. Joshi, R., Gagnani, L., Pandey, S.: Image Steganography with LSB. International Journal of Advanced Research in Computer Engineering & Technology. Vol. 2 (2013).
6. Sanchetti, A.: Pixel Value Differencing Image Steganography Using Secret Key. International Journal of Innovative Technology and Exploring Engineering. Vol. 2 (2012).
7. Huang, P., Chang, K., Chang, C., et al.: A Novel Image Steganography Method using Tri-way Pixel Value Differencing. Journal of Multimedia (2008).
8. Potdar, V., Chang, E.: Gray Level Modification Steganography for Secret Communication. IEEE International Conference on Industrial Informatics, Berlin, Germany (2004) 355–368.

9. Safarpour, M., Charmi, M.: Capacity Enlargement of the PVD Steganography Method Using the GLM Technique. CoRRabs (2016) 1601.00299.
10. Lan, T., Mansour, M., Tewfik, A.: Robust High Capacity Data Embedding. ICIP (2000).
11. Reddy, V., Subramanyam, A., Reddy, P.: A Novel Technique for JPEG Image Steganography and its Performance Evaluation. IJAMC, Vol. 5 (2014) 211–224.
12. Subhedar, M., Mankar, V.: Current Status and Key Issues in Image Steganography: A Survey. Journal of Computer Science Review, Elsevier (2014) 95–113.
13. Almohammad, A., Ghinea, G.: Stego-Image Quality and the Reliability of PSNR. Image Processing Theory, Tools and Applications, IEEE (2010).
14. Roy, R., Samima, S., Changder, S.: A Map-Based Image Steganography Scheme for RGB Images. IJICS Vol. 7 (2015) 196–215.
15. Xue, W.: Study on Digital Image Scrambling Algorithm. JNW. Vol. 8 (2013) 1673–1679.

Defect Identification for Simple Fleshy Fruits by Statistical Image Feature Detection

Smita and Varsha Degaonkar

Abstract Maintaining the product quality of the fleshy fruits is the important criterion in the market. Quality assessment with computer vision techniques is possible with the proper selection of classifier which will give an optimal classification. Feature extraction is done in two steps: (1) Fruit image features were extracted using the 2-level discrete wavelet transform. (2) Statistical parameters like Mean and Variance of discrete wavelet transform features were calculated. A Feed-Forward back propagation neural classifier performed superior than the Support Vector Machine Linear classifier for identifying into three classes (Best, Average, and Poor) by achieving overall good accuracy.

Keywords Discrete wavelet transform · Mean · Variance · Feed-forward neural network · Support vector machine

1 Introduction

Pomegranate is the richest fruit in terms of its powerful medicinal properties and nutrients. As the saying goes prevention is better than cure, it is believed that the consumption of Pomegranate fruit is a preventive cure for many diseases including cancer and heart disease. It belongs to family Lythraceae and is a small tree or shrub. Pomegranate seeds may sometimes be sweet or some time sour. It is also widely consumed as juice.

Due to its rich, healthy benefits, it has become a popular fruit among the masses and is available in almost all markets. It is hence the fruit has a high export value which may still increase in the coming years. India also exports these fruits. Due to

Smita (✉)
MIT Academy of Engineering, Alandi (D), Pune, Maharashtra, India
e-mail: sskulkarni@mitaoe.ac.in

V. Degaonkar
International Institute of Information Technology, Hinjawadi, Pune, Maharashtra, India
e-mail: varshad@isquareit.edu.in

© Springer Nature Singapore Pte Ltd. 2018
P.K. Sa et al. (eds.), *Progress in Intelligent Computing Techniques: Theory,*
Practice, and Applications, Advances in Intelligent Systems and Computing 518,
DOI 10.1007/978-981-10-3373-5_16

its high produce, its essential to maintain the export quality grade and the common man should also be able to identify the quality of this fruit [1].

Following are the categories of identification:

1. Best—The pomegranate in this category is the superior class that is they must be free of defects in terms of shape and color, and these qualities are highly recommended for export.
2. Average—The pomegranate in this category should be of average quality. In this category, there may be a slight defect in the appearance of fruits which may include skin defect and defect in shape appearance. This category is not suitable for export.
3. Poor—This class does not qualify even the minimum requirements. This quality is absolutely poor in terms of shape coloring and skin disease. This quality is not advisable even for consumption.

The external appearance of the fruit gives the idea of the internal quality directly. Due to this, the purchasing of the fruits is affected a lot. Due to use of color grading in the system, the processing directly affects the fruit income as quality of fruit directly linked with color of fruit. In the existing systems, the quality of the fruit is given by color parameters [2].

Many rating systems are designed for fruit identification like a tomato is used as the product that to be tested for food superiority. The system was carried out to calculate the fruit ripeness based on their color. Evolutionary methodologies, by using several image processing techniques including image capture, image improvement, and feature extraction were implemented. To recover image superiority, the collected images were converted to color space format. A Back propagation neural network was used to perform classification of tomato ripeness based on color [3].

Through this research, we have used the following algorithm which will help to identify and distinguish between the various qualities of pomegranate fruit.

2 Feature Extraction

2.1 Database

We have created our own database. Color images of the pomegranate fruit samples are captured by using a regular digital camera. All images are resized to 256×256. The captured color image of fruit is converted into a gray color image. Preprocessing is done by the Gaussian filter for image database for removal of noise.

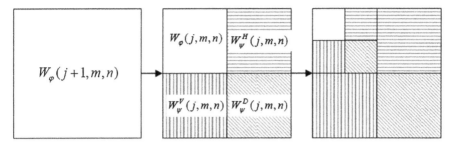

Fig. 1 A two-level decomposition

2.2 Discrete Wavelet Transform

Wavelets represent the scale of features in an image, as well as their position. DWT uses multi-resolution property. Wavelet extracts image information in terms of the frequency domain. In low frequency components subband, image energy is spread and preserved as features of the image. High frequency components subband contains image edge information which may degrade image quality, so high frequency components are rejected [4]. 2-level DWT x(t) is signified by translations and dilations of a fixed function called mother wavelet function. The representation of the DWT can be written as (1):

$$x(t) = \sum_{k=z} u_{j0,k}\phi_{jo,k}(t) + \sum_{j=-\infty}^{j_0} \sum_{k=z} w_{j,k}\psi_{j,k} \tag{1}$$

At the first level of decomposition, image is divided into four parts, as shown in Fig. 1 [5]. At the second level of decomposition, the low frequency component is again divided into four parts. These are considered as the features of the input image.

3 Classifiers

To classify pomegranate fruit into three classes (Best, Average and poor), two classifiers are developed, Linear SVM and ANN.

3.1 SVM

To classify pomegranate images, linear Support Vector Machine is used. SVM uses a main separating line and two other lines called hyperplane to classify the fruit images into three classes (Best, Average, and Poor). The best separating line is a line that located in the middle of classes. This best line obtained by maximizing margin between the hyperplane and main separating line [6]. Transposition of W and b compared to the margin width M. The margin function for an input is statistical DWT features of fruit image is given by Eq. (2) [7],

$$M(x) = W \cdot x - b \tag{2}$$

where,

$$x \epsilon \begin{cases} Class\ A\ (Best) & if\ M(x) > 0 \\ Class\ B\ (Average) & if\ M(x) < 0 \\ Class\ C\ (Poor) & if\ M(x) = 0 \end{cases} \tag{3}$$

The margin from x to the hyperplane is set by the Eq. (4)

$$\frac{M(x)}{||W||} \tag{4}$$

Input to Liner SVM classifier is features extracted from DWT. For each fruit image, 8 features are extracted. Let training data set for each fruit image of 8 points,

$$(\overrightarrow{x1}, y1), \dots, (\overrightarrow{x8}, y8) \tag{5}$$

Let input features of the image represented by $x_1, \dots, x_8$ and their individual group classification be represented by $y_1, \dots, y_8$ Where [7]

$$f(x) = \begin{cases} +1, & x_i \epsilon A \\ -1, & x_i \epsilon B \\ 0, & x_i \epsilon C \end{cases} \tag{6}$$

To maximize $\frac{M(x)}{||W||}$, it implies to minimize W and, in order to prevent data points falling into the margin M, add the limit to each equation [7]:

$$W \cdot x_i - b \geq +1,\ y_i = +1 \tag{7}$$
$$W \cdot x_i - b \leq -1,\ y_i = -1 \tag{8}$$
$$W \cdot x_i - b = 0,\ y_i = 0 \tag{9}$$

Thus, using the linear SVM classifier, the quality of pomegranate test is done by creating hyperplanes with margin for best fruit class and average or poor fruit class.

3.2 ANN

To classify pomegranate, according to good and defective quality, back propagation Feed-Forward multilayered network is used as a classifier. Statistical parameters are calculated from discrete wavelet transform features of fruit image, and these statistical parameters correspond to input layer neurons and the output layer neurons corresponds to three classes (Best, Average, and Poor) according to quality. The Feed-Forward neural network is developed and tested using statistical binary parameters (accuracy, sensitivity and specificity). The back propagation algorithm uses supervised learning; the algorithm includes the following steps [8]:

3.2.1 Training

1. Each input neuron from input layer receives image features x_i from feature extraction and apply it to the hidden layer neuron ($i = 1$ to n).
2. Input of the hidden layer neuron calculated by:

$$z_{inj} = v_{oj} + \sum_{i=1}^{n} x_i \cdot v_{ij} \tag{10}$$

By applying activation functions, output of the hidden layer is calculated by:

$$z_j = f(z_{inj}) \tag{11}$$

3. Input of the output layer neuron is:

$$y_{ink} = w_{ok} + \sum_{j=1}^{p} z_j w_{jk} \tag{12}$$

By applying activation functions, output of the outputlayer is calculated by:

$$z_k = f(y_{ink}) \tag{13}$$

4. To achieve the targeted class of fruits, the error function is calculated and according to that weights are updated in the training phase.

$$\delta_k = (t_k - y_k)f'(y_ink)$$
$$\Delta w_{jk} = \alpha \delta_k z_j \tag{14}$$
$$w_{jk}(new) = w_{jk}(old) + \Delta w_{jk}$$
$$w_{ok}(new) = w_{ik}(old) + \Delta w_{ok}$$
$$v_{ij}(new) = v_{ij}(old) + \Delta v_{ij}$$

5. The training will stop when the target output is achieved according to fruit class (Best, Average and Poor).

3.2.2 Testing

1. Weights are taken from the training algorithm.
2. Input of the hidden layer neuron calculated by:

$$z_{inj} = v_{oj} + \sum_{i=1}^{n} x_i.v_{ij} \tag{15}$$

By applying activation functions, output of the hidden layer is calculated by:

$$z_j = f(z_{inj}) \tag{16}$$

3. Now, compute the output of the output layer unit. For $k = 1$ to me

$$y_{ink} = v_{ok} + \sum_{j=1}^{p} z_i.w_{jk} \tag{17}$$

$$y_k = f(y_{ink}) \tag{18}$$

Here, the sigmoid activation function is used to calculate the output.

4 Methodology

Methodology for the proposed work is as follows:

Step 1: Fruit images are captured and stored in database which includes Grade A (Best), Grade B (Better), Grade C (Poor), and Grade D (Worst) images.

Step 2: To get the precise features, preprocessing is done. Basically, the images which are obtained during image acquisition may not be directly suitable for identification and classification purposes because of some factors, such as noise, weather conditions, and poor resolution of an image and unwanted background.

Step 3: 2-level discrete wavelet transform is used to extract the features of these images.

Step 4: From these features, statistical parameters such as Mean and Variance is calculated.

Step 5: For defect identification of fleshy fruits, two classifiers (SVM and ANN) are trained with these statistical features.

Step 6: Performance of the classifiers is tested using statistical measures such as Accuracy, Sensitivity, and Specificity.

5 Result and Discussion

For defect identification, fruit database has been created, including 200 image samples of each class (Best, Average, Poor). For identifying the quality of pomegranate fruit images, ANN and SVM classifiers have been used. The 75% of pomegranate images have been used to train the system and remaining to test it. To measure the performance of classifiers (ANN and SVM), Binary classification and statistical parameters such as Sensitivity, Specificity, and Accuracy have been used. Sensitivity specifies the test prediction level of one category, and Specificity specifies the test prediction level of another category. Whereas Accuracy specifies the test prediction level of both categories [9].

5.1 Sensitivity [10]

Sensitivity specifies the test's ability to appropriately identify fruits category. Mathematically, this can be expressed as follows: Sensitivity = Correctly Selected/Correctly Selected + Mistakenly Rejected.

5.2 Specificity [10]

Specificity relates to the test's ability to appropriately identify fruits without any condition. Mathematically, this can also be written as follows: Specificity = Correctly Rejected/Correctly Rejected + Mistakenly Selected.

5.3 Accuracy [10]

The accuracy is defined as the ratio of correctly recognized image samples to the total number of test image samples. Accuracy = (Correctly Selected + Correctly Rejected)/(Correctly Selected + Mistakenly Selected + Correctly Rejected + Mistakenly Rejected). Table 1 shows the Percentage Accuracy, Percentage Sensitivity, and Percentage Specificity.

Table 1 Percentage accuracy, percentage sensitivity, and percentage specificity

Fruit type	% Accuracy		% Sensitivity		% Specificity	
	ANN	SVM	ANN	SVM	ANN	SVM
Grad A Best	80.08	61.86	78.14	60.66	84.75	68.64
Grade B Good	86.54	75.64	75.38	64.18	84.62	56.41
Grade C Poor	88.75	76.25	85.86	75.86	81.75	62.5

6 Conclusion

For defect identification of fleshy fruits Pomegranate, linear SVM and ANN (Feed-Forward back propagation) classifiers have been implemented. As pomegranate fruit defect identification is done among three classes (Best, Average and Poor). In our fruit classification, multiple outputs are expected; here, SVM needs to be trained for each class, one by one where as ANN can be trained at a time for all fruit classes. ANN makes more sense than linear SVM, so it was difficult to use linear classifier.

The numerical values of sensitivity represent the probability of defective fruit taste. In the ANN, sensitivity of a classification is greater than SVM means test on a Pomegranate fruit with certain defect will be identified as average or poor class fruit. This test with high sensitivity is often used to identify defective fruit.

The numerical values of specificity represent the probability of defect-free fruit taste. In the ANN, specificity of a classification is greater than SVM means that the test on a Pomegranate fruit identifies as good class fruit. This is a test with high specificity is often used to identify for defect-free fruit.

Accuracy in ANN is more than SVM regardless of test results are accurate for both defective and defect-free fruit. Hence, for all these statistical parameters, ANN (Feed-Forward back propagation) classifier has given good results.

References

1. A. K. Bhatt, D. Pant: Automatic apple grading model development based on back propagation neural network and machine vision, and its performance evaluation, AI & Soc. 30, 45–56 (2015)
2. Dah-Jye Lee, James K. Archibald, and Guangming Xiong: Rapid Color Grading for Fruit Quality Evaluation Using Direct Color Mapping, IEEE Transaction on Automation Science and Engineering, Vol. 8, No. 2, 292–302 (2011)
3. Navnee S. Ukirade: Color Grading System for Evaluating Tomato Maturity, International Journal of Research in Management, Science & Technology, Vol. 2, No. 1, 41–45 (2011)
4. Mallat, S.: A theory for multiresolution signal decomposition: the wavelet representation, IEEE Pattern Anal And Machine Intell., Vol. 11, No. 7, 674–693 (1989)
5. Lee, Tzu-Heng Henry: Citeseer. Wavelet Analysis for Image Processing, Institute of Communication Engineering, National Taiwan University, Taipei, Taiwan, ROC. http://disp.ee.ntu.edu.tw/henry/wavelet_analysis.pdf
6. Muhammad Athoillah, M. Isa Irawan, Elly Matual Imah: Support Vector Machine with Multiple Kernal Learning for Image Retrieval, IEEE International Conference on Information, Communication Technology and System, 17–22 (2015)
7. Thome, A.C.G.: SVM Classifiers-Concepts and Applications to Character Recognition, In Advances in Character Recognition, Ding, X., Ed., InTech Rijeka, Croatia, 25–50 (2012)
8. S. N. Sivanandam, S. N. Deepa: Principles of Soft Computing, 2nd Edn, Wiely India (2012)
9. Jagadeesh Devdas Pujari, Rajesh Yakkundimath, Abdulmunaf Syedhusain Byadgi:Grading and Classification of Anthracnose Fungal Disease of Fruits based on Statistical Texture Features, International Journal of Advanced Science and Technology, Vol. 52, 121–132 (2013)
10. Dr. Achuthsankar, S. Nair, Aswathi B. L: Sensitivity, Specificity, Accuracy and the Relationship between them, a Creative Commons Attribution-India License.Based on a work at, http://www.lifenscience.com

Extraction of FECG Signal Based on Blind Source Separation Using Principal Component Analysis

Mahesh B. Dembrani, K.B. Khanchandani and Anita Zurani

Abstract Fetal electrocardiogram (FECG) gives faithful medical information of heartbeat rate of the fetal living. Extraction of FECG from abdomen of maternal woman consists of interferences and motion artifacts and noises. Maternal electrocardiogram (MECG) is a main source of interference signal present in FECG. This paper focuses on FECG extraction from blind adaptive filtering using principal component analysis (PCA). The abdominal ECG (AECG) is obtained by blind adaptive algorithm which consists of MECG and FECG QRS complex. Principal component analysis separates the two MECG and FECG. The experiments show that it can simultaneously accomplish maternal ECG and fetal QRS complexes enhancement for their detection. The simulation results show that FECG extracted from the peaks of R-R interval is noise-free signal, and extract FHR.

Keywords FECG extraction · MECG · Blind source separation (BSS) · Principal component analysis (PCA)

1 Introduction

Blind source separation is referred as BSS, and the term blind says that the signal obtained is the linear mixture source that are independence sources. The signal consisting of mixture sources can be analyzed by number of different techniques. Blind source separation based on principal components analysis (PCA) and independent component analysis (ICA) can be used to analyze such mixture signals [1].

M.B. Dembrani (✉) · K.B. Khanchandani · A. Zurani
Shri Sant Gajanan Maharaj College of Engineering, Shegaon 444203,
Maharashtra, India
e-mail: mahesh.dembrani@gmail.com

K.B. Khanchandani
e-mail: kbkhanchandni@rediffmail.com

A. Zurani
e-mail: anita.zurani@gmail.com

© Springer Nature Singapore Pte Ltd. 2018
P.K. Sa et al. (eds.), *Progress in Intelligent Computing Techniques: Theory,*
Practice, and Applications, Advances in Intelligent Systems and Computing 518,
DOI 10.1007/978-981-10-3373-5_17

These techniques are suitable for representing data in statistical domain instead of frequency or time domain. The statistical-based techniques use the Fourier components of a data segments as fixed, and PCA-based transform of data structure is analyzed. The method of projection on another set of axes is required to separate the data components or sources which will clearly show the signal analyzed in that projection.

(i) **Principal component analysis**

PCA uses the variance measure for obtaining the new axes known as orthogonal axes to analyze the signals. The data are de-correlated in second-order sense, and dot product of any set of pair of the newly discovered axes is zero. The PCA is used to perform both on lossy and on lossless transformation by multiplying recorded data and separation or de-mixing of matrix. PCA involves the projection of data on orthogonal axes to determine the nature of data based on blind source separation BSS. This paper shows that the accuracy is obtained without using any feature extraction method by using PCA to the signal source or input. PCA is used to reduce the size of input vectors of data.

(ii) **Database signal for FECG extraction**

The ECG recording database of a pregnant woman is taken from the physionet database in European database format. This database consists of five woman labor, cardiologist-verified annotations of fetal heartbeats, five-minute multichannel fetal recordings from university of Silesia, Poland. This recording of woman in labor consists of signals taken from four abdomens and one signal taken from the scalp of fetal known as direst signal. The five women in labor signal are taken at 1-kHz frequency with 16-bit resolution. The signals are taken during the 38 and 41 weeks of pregnant woman.

2   Literature Survey

M.A. Hasan et al. [2] proposed new algorithms on FECG signal extraction and different methods for nature and monitoring of fetal ECG. The performance measure based on accuracy and signal quality detection was presented. G.D. Clifford [3] presented the blind source separation (BSS) based on the principal component analysis and independent component analysis methods to obtain the signal separation of independent source in statistical domain. K. V. K. Ananthanag et al. [4] proposed a BSS method at higher-order statistics which does not get affected due to the electrode placement. All these algorithms were able to extract ECG considerably if the amount of the input SNR is high. R. Sameni et al. [5] proposed JADE ICA algorithm which obtained the data from the 4 who clearly correspond to mother heart, 2 from fetal heart, 2 from noise and 8 from extracted components. S. Sargolzaei et al. [6] proposed adaptive filtering methods for extraction of FECG real and synthetic signals based on wavelet transform. P. P. Kanjilal et al. [7] proposed a technique based on PCA and SVD single-channel maternal ECG signal to extract fetal ECG signal with low SNR.

3 Implementation and Results

In this paper, the BSS-based PCA is used to extract the FECG signal and Fig. 1 shows the flowchart of the proposed method. The raw ECG signal is taken from the physionet ATM database in European database format. The filtering process is carried out to remove the interference or noise signal present in the signal. BSS-based PCA is used for the extraction of FECG signal. BSS-based PCA is used for the suppression of MECG. Thus, we can get the extracted FECG and measure the FHR, accuracy and positive predictivity.

Figure 2 shows the signal obtained from the database which consists of direct signal from the scalp of the fetal and 4 signals which are taken from the abdomen of the pregnant woman. Figure 2 shows initially direct signal and remaining 4 signals.

Fig. 1 Flowchart for proposed FECG extraction

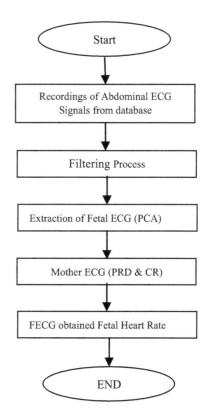

Fig. 2 Raw ECG signal
obtained from database

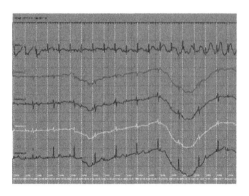

Fig. 3 Measured signal after
filtering

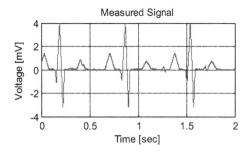

Fig. 4 Fetal heartbeat signal

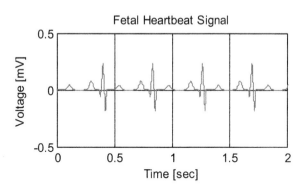

Figure 3 shows the measured signal obtained by using blind adaptive filtering
which consists of mixed signal of FECG and MECG along with interferences.

Figure 4 shows the fetal heartbeat signal, and Fig. 5 shows the maternal heartbeat signal obtained after applying PCA which separates the FECG and MECG for
further processing. Since by applying the DBE to R wave, the QRS signal complex
of FECG can be monitored.

Figure 6 shows the filtered FECG signal which consists of no noise signal and
detection of R-R interval for accurate detection of fetal heartbeat rate.

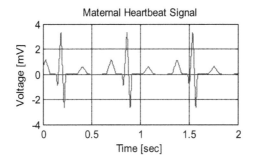

Fig. 5 Maternal heartbeat signal

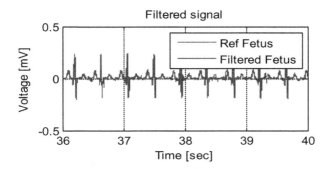

Fig. 6 Filtered FECG

Table 1 Performance measure for FECG

Raw ECG signal	Sensitivity	Positive predictivity	Accuracy	FHR (BPM)
1	49	59	60	103.3
2	73	85	78	112
3	68	72	67	135.3
4	53	69	65	105
5	58	36	57	140.5

Table 1 shows the response for the 5 ECG signals obtained from database. The performance measures for detection of FHR and parameters such as sensitivity, positive predictivity and accuracy are also calculated.

4 Performance Measures

(i) Percent root mean square difference (PRD)

The percent root mean square difference for an ECG signal can be calculated using the equation given by

$$PRD = \sqrt{\frac{\sum\limits_{n=1}^{N} (x[n] - x^\wedge[n])^2}{\sum\limits_{i=1}^{N} x^2[n]}}$$

where N is the number of ECG samples, where x[n] is original ECG signal, xˆ[n] is reconstructed enhanced ECG signal.

(ii) Compression ratio (CR)

The compression ratio for an ECG signal can be calculated using the equation given by

$$CR = \frac{x[n] - x^\wedge[n]}{x[n]}$$

where x[n] is data before compression of ECG signal and xˆ[n] is data after compression ECG signal.

The data compression techniques can be applied to ECG signals by using Tompkins method such as turning point algorithm, AZTEC algorithm and FAN algorithm.

Table 2 shows the PRD for the ECG signal, and compression ratio is obtained as 50% by using the turning point method for obtaining. The PRD is obtained for AZTEC method and FAN method, and threshold is considered as 0.05 while calculating the PRD using AZTEC and FAN method. PRD is measure for both considering and without consideration of 7-point filter.

Table 2 PRD measurement using TP, AZTEC and FAN method

Raw ECG signal	Turning point method (%)	AZTEC method		FAN method	
		PRD (no filter) (%)	PRD (with filter) (%)	PRD (no filter) (%)	PRD (with filter)
1	4.566	11.692	10.05	23.472	23.408
2	5.151	11.692	10.405	15.301	15.388%
3	2.597	5.358	5.118	5.487	5.801%
4	2.051	9.287	8.245	17.067	17.062%
5	3.633	11.797	10.480	15.827	15.871%

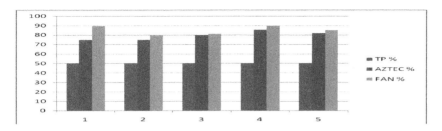

Fig. 7 Graphical representation of calculating of compression ratio (CR)

Figure 7 shows the graphical representation of compression ratio for five ECG signals based on the compression methods. From graphical representation, it is clear that turning point method provides constant compression ratio and FAN method provides efficient compression ratio as compared to AZTEC method.

5 Conclusion

This paper is presented on primary focus to extract FECG. FECG is extracted based on blind source separation adaptive filtering to extract MECG, and PCA is used for suppression of MECG and extraction of the FECG. The performance measures are normally used to estimate the sensitivity, positive predictivity and accuracy of the proposed method. Along with the ECG compression techniques such as turning point algorithm, AZTEC algorithm and FAN algorithm are used to obtain compression ratio and PRD. The future works can be applied to proposed algorithm by using ICA and Fast ICA for improving the performance measures and for exact detection of FECG.

References

1. G. D. Clifford, "Biomedical Signal and Image Processing", Course materials for HST.582J / 6.555J / 16.456J, MIT Open Course Ware (http://ocw.mit.edu), 2007.
2. M. A. Hassan, M. B .I. Reaz, M. I. Ibrahimy, M .S. Hussain and J. Uddin, "Detection and processing techniques of FECG signal for fetal monitoring", Biological procedures online, vol. 11, issue 1, pp. 263–295, 2009.
3. G. D. Clifford, Course materials for HST.582J / 6.555J / 16.456J, "Biomedical Signal and Image Processing", MIT Open Course Ware (http://ocw.mit.edu), 2007.
4. K. V. K. Ananthanag and J. S. Sahambi, "Investigation of blind source separation methods for extraction of fetal ECG", Proceedings of Canadian Conference on Electrical and Computer Engineering (IEEE CCECE), vol. 3, pp. 2021–2024, May 2003.
5. R. Sameni, C. Jutten and M. B. Shamsollahi, "What ICA provides for ECG processing: application to noninvasive fetal ECG extraction", IEEE International Symposium on Signal Processing & Information Technology, pp. 656–661, Vancouver, BC, Aug. 2006.

6. S. Sargolzaei, K. Faez and A. Sargolzaei, "Signal processing based techniques for fetal electrocardiogram extraction", International Conference on Bio-Medical Engineering and Informatics (BMEI), vol. 2, pp. 492–496, sanya, May 2008.
7. P. P. Kanjilal, S. Palit and G. Saha, "Fetal ECG extraction from single channel maternal ECG using SVD", IEEE Transactions on Bio-medical Engineering, vol. 44, issue 1, pp. 51–59, Jan. 1997.

An Improved Method for Text Segmentation and Skew Normalization of Handwriting Image

Abhishek Bal and Rajib Saha

Abstract This paper proposed an off-line cursive handwriting segmentation method and an efficient skew normalization process of the handwritten document. The proposed segmentation method based on horizontal and vertical projection, which has been already used for different purposes in handwriting analysis. But to tolerate the text lines overlapping and multi-skewed text lines, present work implements modified version of horizontal and vertical projection, which can segment the text lines and words even if text lines are overlapped. Present work also proposed a skew normalization method which is based on orthogonal projection toward the x-axis. The proposed method was tested on more than 550 text images of IAM database and sample handwriting image which are written by the different writer on the different background. The experimental result shows that proposed algorithm achieves more than 96% accuracy.

Keywords Document image · Feature extraction · Line segmentation · Word segmentation · Skew normalization · Orthogonal projection

1 Introduction

Throughout the previous few years, handwriting analysis has been a demanding research topic. Generally, handwriting analysis for document image has four parts that are preprocessing, segmentation, feature extraction and classification. Image preprocessing procedure [1, 2] is used to enhance the excellence of image quality

A. Bal (✉) · R. Saha
Department of Computer Science & Engineering, RCC Institute of Information Technology,
Kolkata, West Bengal, India
e-mail: abhisheknew1991@gmail.com

R. Saha
e-mail: rajibsaha2009@gmail.com

© Springer Nature Singapore Pte Ltd. 2018   181
P.K. Sa et al. (eds.), *Progress in Intelligent Computing Techniques: Theory,*
Practice, and Applications, Advances in Intelligent Systems and Computing 518,
DOI 10.1007/978-981-10-3373-5_18

for easy and efficient processing in the next steps. The steps of the preprocessing are thresholding [3, 4] and noise removal [5–10]. Types of the noises that can appear in the scanned handwriting document are marginal noise [7], shadow noise [6, 7], stroke-like pattern noise [7], salt-and-pepper noise [6, 7] and background noise [6, 7]. Salt-and-pepper noise can appear during conversion processes and also may occur by dust on the document image [6, 7]. Salt-and-pepper noise can be removed by median [8] and k-fill [9] filters. Background noise [9] can occur due to the bad contrast, spots in the background, moisture in different areas and rough backgrounds. The next step after preprocessing is segmentation [2, 11–22] which partitions the binary image into multiple segments depending on requirement. The present segmentation methods are Hough transformation [23, 24] and projection profile [23], etc. The next step after segmentation is feature extraction [25] which is the most crucial task that extracts the feature from the segmented image using various techniques [25]. The last step is classification [25–27] method which is the heart of the handwriting recognition technique. There are previous classification approaches are available which are referred in [25–27].

Due to the high inconsistency of handwriting styles, handwriting analysis techniques should be more robust. Toward this goal, this paper proposed an off-line cursive handwriting text document segmentation technique and an efficient skew normalization process of the handwritten text document. The proposed segmentation method based on horizontal and vertical projection, which have been already used for different purposes in handwriting analysis. To tolerate text lines overlapping and different skew in text lines in the single handwriting image, this paper implements modified version of horizontal and vertical projection, which can segment the text lines and words even if text lines are overlapped. In order to verify the stability of proposed word segmentation technique, the threshold which is used to differentiate between intra-word and inter-word gaps from each other is not fixed. If the threshold value is small, then it may cause over-segmentation, whereas large threshold may cause under-segmentation.

Present work also proposed an efficient method for skew detection and normalization on handwriting document. The proposed method is based on orthogonal projection [28] length toward the x-axis. Generally, machine-printed document skew occurs at the time of scanning process due to the incorrect arrangement of the pages, whereas skew [23, 28–31] in handwriting document can occur due to the human's behavior as well as by the scanner during the scanning process. Generally, skew recognition in handwriting document is more difficult than the skew in printed document due to the variation of present mind condition of the writer and the difficulties at the time of scanning. In the case of the correct orientation of the pages, handwriting document still consists of smaller and larger skew [23, 28–31] due to writer variation.

Technically, skew [23, 28–33] is defined as an alignment of the text lines and words with respect to the horizontal direction. There are four types of skew are available that are negative or ascending skew, positive or descending skew, normal or straight skew and wavy skew, which are shown in Fig. 1.

Fig. 1 Orientation of skew in handwriting text document image

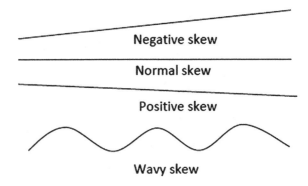

Negative skew

Normal skew

Positive skew

Wavy skew

2 Related Work

In 2016, Abhishek Bal and Rajib Saha [32] proposed a method, skew detection and normalization method which uses the orthogonal projection. The proposed method was tested on IAM database. The proposed method can detect exact skew angle and also able to normalize the skew angle. The experimental result shows that proposed algorithm achieves higher accuracy for all type skew angles.

In 1973, N. Otsu [3] proposed a threshold selection technique that can automatically measure the threshold for image segmentation. The proposed method is simple and also useful for multi-threshold problems. The experimental result shows the accuracy of the proposed method for grayscale image.

In 2012, Subhash Panwar and Neeta Nain [30] proposed a skew normalization technique which uses the orthogonal projection of lines. The proposed method is tested on different handwriting document images and achieves more than 98% accuracy.

In 2012, Jija Das Gupta and Bhabatosh Chanda [31] proposed a method for slope and slant detection and correction for handwriting document images. The algorithm was applied on IAM database and archive most promising experimental result. Comparison of result showed that proposed method achieved the better result than the method proposed by B. Gatos et al and Moises pastor et al.

In 2007, Florence Luthy et al. [14] proposed a method for segmentation of off-line handwriting text document. Hidden Markov models are used in this proposed method to distinguish between inter-word and intra-word gaps. The proposed method was tested on off-line handwriting document images and was appropriate for writer-dependent and writer-independent handwriting document.

In 2011, Fotini Simistira, Vassilis Papavassiliou and Themos Stafylakis proposed an enhancement method of their previously word segmentation method [34] by exploiting local spatial features. The proposed method has been tested on ICDAR07, ICDAR09, ICFHR10 and IAM handwriting databases and performs the better result than winning algorithm.

3 Proposed Method

At first, proposed work collects color and grayscale handwriting document from the IAM database [13, 35] and scanned sample handwriting images which are written by the different writer on the different background. Proposed approach assumes that given handwriting document perfectly scanned so only the skew which is introduced by the writer considers. This present work considers that scanned handwriting document may consist of salt-and-pepper noise [6, 7] and background noise [6, 7] which may be occurred before or after scanning process. The steps of the proposed work are as follows.

3.1 Preprocessing

Image preprocessing technique [1, 2] is used to increase the excellence of image quality for easy and efficient processing in the next steps. Handwriting analysis needs to perform preprocessing steps such as binarization [3, 4] and noise removal [5–9] for better recognition.

3.1.1 Noise Removal

A major issue that can occur during scanning is noise. The most common noises that can occur in handwriting document are salt-and-pepper noise and background noise. The median filtering technique is the most popular approach for removing salt-and-pepper noise. Sliding window technique was used for median filtering over a grayscale image where pixel value was replaced by the gray value of its neighborhood pixels. Figure 2 shows the sample handwriting image from IAM database

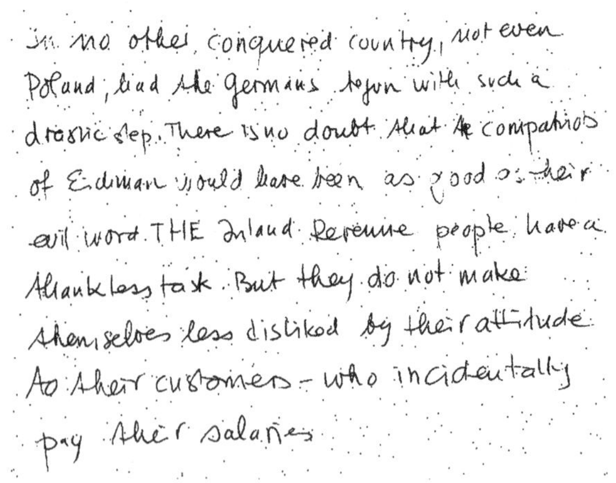

Fig. 2 Image before noise removal

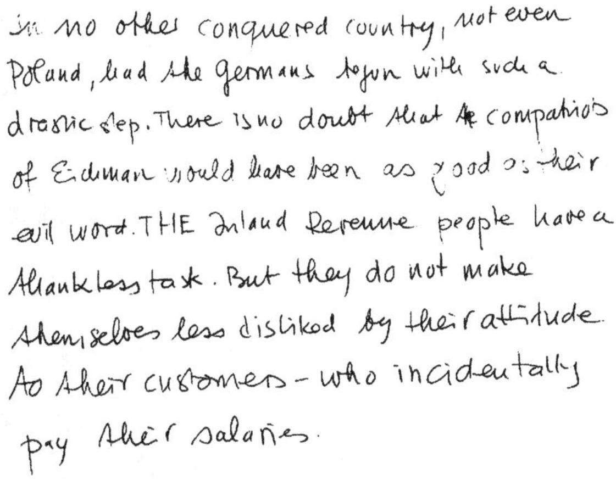

Fig. 3 Image after removal
of salt-and-pepper noise

in no other conquered country, not even Poland, had the germans begun with such a drastic step. There is no doubt that the compations of Eichmann would have been as good as their evil word. THE Inland Revenue people have a thankless task. But they do not make themselves less disliked by their attitude. As their customers - who incidentally pay their salaries.

which contains salt-and-pepper noise and background noise. The image after removal of salt-and-pepper noise from the original document by the median filter is shown in Fig. 3. But still image contain the background noise which can be removed with the help of binarization.

3.1.2 Thresholding

Thresholding is a technique which converts pixel intensity of a grayscale image to black if the pixel intensity is less than the constant value T; otherwise, pixel intensity is converted into white. Thresholding [3, 4] is applied to grayscale handwriting image to convert the grayscale image into binary image for increasing better processing and decrease storage space in the next steps. In this proposed method, Otsu thresholding [3] technique is used for converting the grayscale image into binary image.

3.2 Segmentation

3.2.1 Line Segmentation

After binarization, text lines are segmented from the binary document image. The present work implements a modified horizontal projection method of an image that can segment individual text line from the previous and following text lines based on rising section of the horizontal projection histogram of document image shown in Fig. 4.

In english handwriting document image, most of the time no gaps are present between two lines, which may create incorrect line segmentation due to overlapping

Fig. 4 Horizontal projection
of binary image

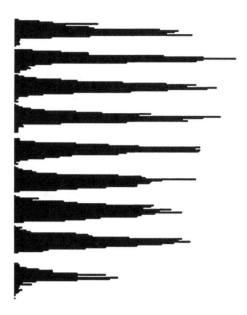

between two lines if simple horizontal projection histogram is concern. In the proposed method, after creating the horizontal projection histogram of a binary document image, count the number of rising section and height of each rising section. The average height of the rising sections is treated as the threshold. Then consider each and every rising section and check the height of that rising section is greater than or equals to the threshold or not. If yes, then based on that rising section of the horizontal histogram, the line was individually segmented from the actual binary document image; otherwise, neglect that rising section as a false line segment. These types of the false rising section may occur due to overlapping between two lines or the presence of a bar in an upper letter. The most important are that when a rising section is treated as a false line segment and next rising section is treated as a true line segment then the portion of the false line segment is added to true line segment for the segmentation of next line from the actual document image; otherwise, some features are removed. Figure 7 shows the segmented line sample.

After line segmentation, it may happen different lines may have the different skew angle. The proposed skew normalization process is based on orthogonal projection length which is shown in Fig. 5. To normalize the skew angle, orthogonal projection method that applied on the segmented lines and words to normalize the skew angle is shown in Fig. 7. The normalized line corresponding to the skewed line is shown in Fig. 8. This method efficiently deals with higher as well as smaller skew of handwriting document.

Figure 5 shows the orthogonal projection of the handwriting image. The handwriting is considered to be written within the rectangle box. The skew angle is Θ, and the first and last black pixels of the word are pointed by O_{start} and O_{last}. The actual length of the word is denoted by A_l, and projected length is denoted by O_l. The orthogonal projection of the text line is calculated as

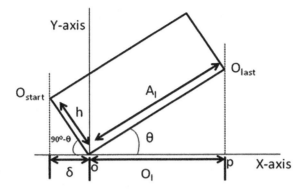

Fig. 5 Orthogonal projection technique for line and word in handwriting text document image

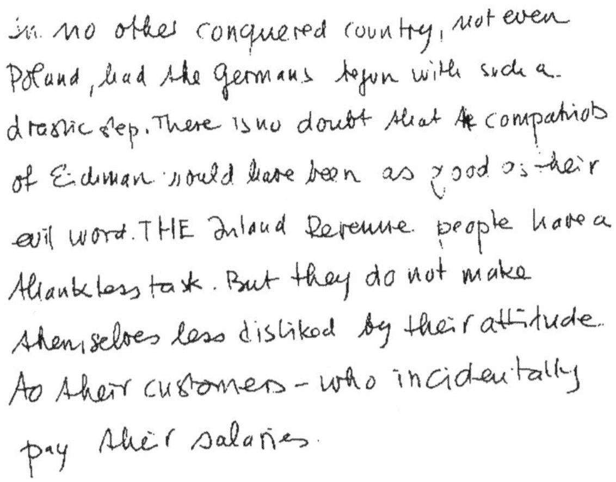

in no other conquered country, not even
Poland, had the germans begun with such a
drastic step. There is no doubt that the compatriots
of Eidman would have been as good as their
evil word. THE Inland Revenue people have a
thankless task. But they do not make
themselves less disliked by their attitude.
As their customers - who incidentally
pay their salaries.

Fig. 6 Binary image after noise removal

themselves less disliked by their attitude.

Fig. 7 Line before skew correction

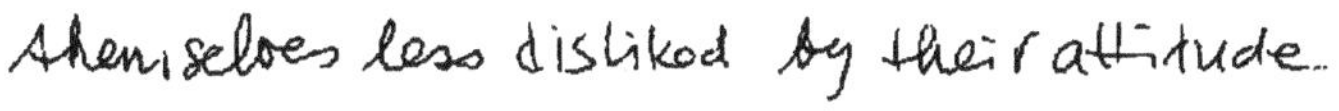

themselves less disliked by their attitude.

Fig. 8 Line after skew correction

$$O_l = O_{last} - O_{start} - \delta \tag{1}$$

Here δ is the projection and h denotes the actual height of the text line. The relationship between text line and the projection is

$$\delta = h \times \cos\left(90^0 - \Theta\right) \tag{2}$$

$$\Theta = \tan^{-1}\left((y2 - y1)/(x2 - x1)\right) \tag{3}$$

In this case, value of δ belongs 0 to h. If $\Theta = 0$, then value of δ is 0, and if the value of $\Theta = 90^{\circ}$, then the value of δ is h. The proposed method normalizes the skew angle using the orthogonal projection length which is clearly explained in the experimental part. Here, (x_1, y_1) is the first left down most black pixel of the word that is starting of the baseline coordinate and (x_2, y_2) is the last right down most black pixel of the word that is ending of the baseline coordinate.

3.2.2 Word Segmentation

In the case of word segmentation, to segment the words from the line, firstly inter-word and intra-word gaps are measured. Inter-word gaps denote the gaps between two words and intra-word gaps denote gaps within a word. Generally, gaps between the words are larger than the gaps within a word. These proposed methods construct the vertical projection histogram to measure the width of each inter-word and intra-word gaps; then, it measures the threshold value to differentiate between inter-word and intra-word gaps. If the width of gaps is greater than or equals to threshold, then gaps are treated as inter-word gaps and words are segmented individually from the line depending on the threshold. Figure 9 shows the intra-word and inter-word gaps with the help of vertical black lines.

If a line has global skew, then it may be possible that several words within a line may have a different skew. So it may require normalizing the skew of the words for a single line. For that reason again proposed skew normalization method is applied to the each segmented word separately.

Fig. 9 Inter-word and intra-word gaps within a line

3.3 Algorithm

3.3.1 Algorithm for Line Segmentation

Line segmentation algorithm is carried out by horizontal projection as follows

Step 1: Read a handwriting document image as a multi-dimensional array.
Step 2: Check the image is a binary image or not. If binary image then stores it into a 2-d array IMG[ ][ ] with size $M \times N$ and go to Step 4, otherwise go to Step 3.
Step 3: Convert the image into binary image and store in a 2-d array IMG[ ][ ].
Step 4: Construct the horizontal projection histogram of the image IMG[ ][ ] and store in a 2-d array HPH[ ][ ].
Step 5: Measure the height, starting row position and ending row position of each horizontally rising section of horizontal projection histogram image and store in 3-d array LH[ ][ ][ ] sequentially.
Step 6: Count the number of rising section by counting the rows of the 3-d array LH[ ][ ][ ]. Then measure the threshold (T_i) value by calculating average height of rising sections from the 3-d array LH[ ][ ][ ].
Step 7: Select one by one rising section from 3-d array LH[ ][ ][ ] and check the height of that rising section is less than the threshold or not. If yes then this rising section is not considered as a line and go to Step 9, otherwise rising section is treated as a line and go to Step 8.
Step 8: Find the rising section's starting and ending rows number from the array LH [ ][ ][ ]. Let starting row and ending row are r_1 and r_2, respectively. Extract the line segment between r_1 and r_2 from the original binary image denoted by IMG[ ][ ].
Step 9: Go to Step 7 for next rising sections till all rising sections are not under consideration, otherwise go to next step.
Step 10: End.

3.3.2 Algorithm for Word Segmentation

Word segmentation algorithm is carried out by vertical projection as follows

Step 1: Read a segmented binary line as 2-d binary image LN[ ][ ].
Step 2: Construct the vertical projection histogram of the line LN[ ][ ] and store in a 2-d array LVP[ ][ ].
Step 3: From the vertical projection histogram (LVP[ ][ ]), measures width of each gaps and store the width into 1-d array GAPSW[ ].
Step 4: Count total number gaps as TGP by calculating the size of GAPSW[ ]. Add width of all gaps by adding the elements of GAPSW[ ] and store in TWD.
Step 6: Calculate the threshold (T_i) as follows

$$T_i = TWD/TGP \qquad (4)$$

In Eq. (4), T_i is the threshold value denoting average width of inter-word gaps, TWD denotes total width of all gaps and TGP denotes the total number of gaps.
Step 7: For each i, if GAPSW[i] $\geq$ T_i then this gap is treated as inter-word gaps, otherwise gap is treated as an intra-word gaps. Depending on inter-word gaps width, words are segmented from the line.
Step 8: End.

3.3.3   Algorithm for Skew Normalization

Skew Normalization algorithm is carried out by orthogonal projection as follows

Step 1: Read a segmented line or word as 2-d binary image BW[ ][ ].
Step 2: Calculate the skew angle (Θ) and orthogonal projection length (O_l) of the image. Rotate the image (BW[ ][ ]) anticlockwise and clockwise direction with 1 degree and store in two 2-d arrays I1[ ][ ] and I2[ ][ ], respectively.
Step 3: For each image I1[ ][ ] and I2[ ][ ], calculates the orthogonal projection length and store in O_{l+} and O_{l-}, respectively. Initialize a variable X with 1.
Step 4: If $\Theta < 0$ AND $O_+ > O_l$ then the actually segmented line or word (BW[ ][ ]) contain the positive skew and go to step 5, otherwise go to step 7.
Step 5: Rotate the original image anticlockwise direction with X degree and store in a 2-d array I[ ][ ]. Calculate the skew angle and orthogonal projection length of the rotated image (I[ ][ ]) and store in Θ (theta) and O_l, respectively.
Step 6: Increment X with 0.25 degree and continue the step 5 till calculated projection length O_+ is larger than the projection length O_l and Θ (theta) is less than zero, otherwise go to step 9.
Step 7: If segmented line or word contain negative skew, then rotate the original image clockwise direction with X degree and store in a 2-d array I[ ][ ]. Calculate the skew angle and orthogonal projection length of the rotated image (I[ ][ ]) and store in Θ (theta) and O_l, respectively.
Step 8: Increment X with 0.25 degree and continue the step 7 till calculated projection length O_- is larger than the projection length O_l and Θ (theta) is greater than zero, otherwise go to step 9.
Step 9: At the end of the above processing, the balanced line or word denoted by 2-d array I[ ][ ].
Step 10: End.

3.4 Experimental Result

The proposed work implemented in MATLAB on IAM database [13] over 550 text images containing 3800 words and some sample handwriting image which are written by the different writer on the different background.

An example of text lines of IAM database is shown in Fig. 2. At first, noise removal techniques are applied on the handwriting text document if noise is present. Image quality is improved by removing noise from the handwriting text document. Sample text document after removal of salt-and-pepper noise is shown in Fig. 3. After that thresholding [3] is applied on noiseless grayscale handwriting image to convert the grayscale image to binary image which is shown in Fig. 6. Then proposed line segmentation method is applied on binary document image that segments individual text line from the previous and following text lines based on rising section of the horizontal projection histogram of the document image. After line segmentation, different segmented lines may contain a different skew. During recognition process, handwriting document should free from the unbalanced skew angle for better recognition. Proposed skew normalization method is applied on each segmented line to normalize the segmented line with respect to skew angle. Sample outputs for text line normalization process are shown in Figs. 7 and 8. After lines skew normalization, proposed method measures width of each inter-word and intra-word gaps from the vertical projection histogram and segments the words from the line depending on the threshold value. Here proposed method calculates the threshold value to differentiate between inter-word and intra-word gaps. Inter-word and intra-word gaps details are shown in Fig. 9 with the help of vertical black lines, and segmented words are shown in Fig. 10. If a line has global skew, then it may be possible that after word segmentation several words within a line may have a different skew. So it is required to normalize the skew of the segmented words for each line. For that reason, again proposed method was applied to the each segmented word separately. Sample outputs of skewed and skewed less word are shown in Figs. 11 and 12.

In order to enhance the performance of the algorithms, proposed method has been applied on subset [a-f] of the IAM database and sample handwriting image which was written by the different writer on the different background. Test datasets consist of more than 550 handwriting document images of IAM database. Further details for the IAM datasets are included in the related paper [13]. The evolution results of proposed segmentation and skew normalization methods are shown in Tables 1 and 2, respectively. The percentage of over- and under-segmented lines and words is given in Table 1.

Fig. 10 Words after segmentation

Fig. 11 a Positive skewed image, **b** normalized image after skew correction

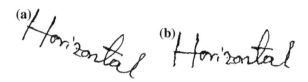

Fig. 12 a Negative skewed image, **b** normalized image after skew correction

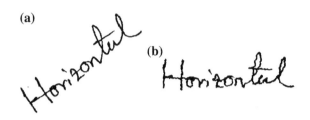

Table 1 Segmentation accuracy on form a–f of the IAM database and percentage of over- and under-segmentation results for lines and words

Segmentation type	Accuracy (%)	Over (%)	Under (%)
Line	95.65	1.45	2.9
Word	92.56	2.85	4.59

Table 2 Skew normalization result produced by proposed method

Sl. no.	Actual skew	Normalized skew	Accuracy (%)
1	+29.4275	+0.1916	99.4
2	+16.5838	−0.8654	95
3	+29.2170	−1.5608	94.7
4	+18.1757	+1.4730	92
5	+24.9967	−0.2906	98.9
6	+3.01285	+0.1609	95
7	+0.85513	+0.8551	100
8	−0.71627	−0.7162	100
9	−16.7861	+1.2825	93
10	−14.8455	−0.4308	97
11	−20.2931	+1.6211	93
12	−12.5860	−0.5439	96
13	−13.3768	−0.1758	98.7

Figures 11a and 12a show positive and negative skewed words, respectively. After applying the proposed method on above-skewed words, resultant normalized skewed free words are shown in Figs. 11b and 12b. Experiment results for skew normalization which were tested on the dataset are shown in Table 2. The significant improvements in proposed skew normalization method over existing methods are shown in Fig. 12. Experimental part considers that the words with the skew

Fig. 13 Skew comparison chart between proposed method and existing methods

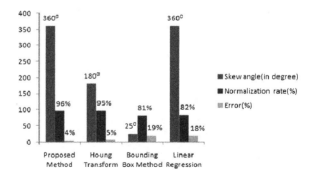

range between −1° and +1° are normalized words because these small amounts of skew angle do not make so much effect on words characteristic.

From Fig. 13, it is observed that proposed method is much efficient for skew normalization and can deal with any types of skew angle up to 360°. The comparative evolution results of four different algorithms with proposed algorithm are shown here. The success rate of proposed skew normalization method is 96% which is better than the existing methods. Further evolution results are shown in Table 2. The failure cases of the proposed method are mainly caused by misclassification of lower case letters.

4   Conclusion and Future Work

This current work proposed a segmentation method of off-line cursive, writer irrespective handwriting document and an efficient skew normalization process of the handwritten document. The proposed method has been applied to more than 550 text images of IAM database and sample handwriting image which are collected from surroundings and written by the different writer on the different background. Using the proposed method, 95.65% lines and 92.56% words are correctly segmented from the IAM datasets. Proposed work also normalizes 96% lines and words perfectly with very small error rate. Proposed skew normalization method deals with the exact skew angle and extremely efficient with compare to on-hand technique and can be normalized the skew up to 360°. Proposed method could be applied to various languages writing style with more or less same accuracy.

As a future work, we want to overcome the weakness of proposed methods which are mainly caused due to the misclassification of lower case letters and obtaining more appropriate system. We also want to construct a tool for analysis of human behavior which can predict personality pattern trail with computer-aided automated interaction. For analysis of human behavioral, it should required to analyze all 26 alphabets in English for English handwriting and most important features are letter size, word space, slant, writing pressure, letter connection, signature, etc.

Acknowledgements The authors sincerely express their gratitude to TEQIP-II & Dr. Arup Kumar Bhaumik, Principal of RCC Institute of Information Technology College, for giving regular encouragement in doing research in the field of image processing.

References

1. Stéphane Nicolas, Thierry Paquet, Laurent Heutte: Text Line Segmentation in Handwritten Document Using a Production System. In: Proceedings of the 9th Int'l Workshop on Frontiers in Handwriting Recognition (IWFHR-9 2004), IEEE, 2004.
2. Khaled Mohammed bin Abdl, Siti Zaiton Mohd Hashim: Handwriting Identification: a Direction Review. In: IEEE International Conference on Signal and Image Processing Applications (2009), 978-1-4244-5561-4/09.
3. N. Otsu: A threshold selection method from Gray level histogram. In: IEEE Transaction on system, Man, Cybernetics (1979), VOL. SMC-9, pp. 62–66.
4. W. Niblack: An Introduction to Digital Image Processing. In: Englewood Cliffs, New Jersey Prentice-Hall (1986).
5. Rejean Plamondon, Sargur N. Srihari: On-Line and Off-Line Handwriting Recognition. In: A Comprehensive Survey. IEEE transactions on pattern analysis and machine intelligence (2000), vol. 22, no. 1.
6. Yan Solihin, C.G. Leedham: Noise and Background Removal from Handwriting Images, IEEE (1997).
7. Atena Farahmand, Abdolhossein Sarrafzadeh, Jamshid Shanbehzadeh: Document Image Noise and Removal Methods. In: Proceedings of the International MultiConference of Engineers and Computer Scientists, Hong Kong (2013), Vol I.
8. G. Story, L. O'Gorman, D. Fox, L. Schaper, H. Jagadish: The rightpages image-based electronic library for alerting and browsing. In: IEEE Computer Society Press Los Alamitos, CA, USA (1991), vol. 25, no. 9, pp. 17– 26.
9. N. Premchaiswadi, S. Yimgnagm and W. Premchaiswadi: A scheme for salt and pepper noise reduction and its application for ocr systems. In: Wseas Transactions On Computers (2010), vol. 9, pp. 351–360.
10. Andria, G. Savino, M. Trotta, A.: Application of Wigner-Ville distribution to measurements on transient signals. In: Instrumentation and Measurement, IEEE Transactions (1994), vol. 43, no. 2, pp. 187–193.
11. U.-V. Marti, H. Bunke: Text Line Segmentation and Word Recognition in a System for General Writer Independent Handwriting Recognition. In: IEEE, 0-7695-1263-1/01 (2001).
12. Partha Pratim Roy, Prasenjit Dey, Sangheeta Roy, Umapada Pal, Fumitaka Kimura: A Novel Approach of Bangla Handwritten Text Recognition using HMM. In: 14th International Conference on Frontiers in Handwriting Recognition (2014), IEEE, 2167-6445/14.
13. Matthias Zimmermann, Horst Bunke: Automatic Segmentation of the IAM Off-line Database for Handwritten English Text, IEEE, 1051:465:l/02 (2002).
14. Florence Luthy, Tamas Varga, Horst Bunke: Using Hidden Markov Models as a Tool for Handwritten Text Line Segmentation. In: Ninth International Conference on Document Analysis and Recognition (ICDAR 2007), 0-7695-2822-8/07.
15. Themos Stafylakis, Vassilis Papavassiliou, Vassilis Katsouros, George Carayannis: Robust Text-Line and Word Segmentation for Handwritten Documents Images. In: Greek Secretariat for Research and Technology under the program (2008), IEEE.
16. A. Sánchez, P.D. Suárez, C.A.B. Mello, A.L.I. Oliveira, V.M.O. Alves: Text Line Segmentation in Images of Handwritten Historical Documents. In: Image Processing Theory, Tools & Applications, IEEE, 978-1-4244-3322-3/08 (2008).

17. Jija Das Gupta, Bhabatosh Chanda: A Model Based Text Line Segmentation Method for Off-line Handwritten Documents. In: 12th International Conference on Frontiers in Handwriting Recognition, IEEE, 978-0-7695-4221-8/10 (2010).
18. Mohammed Javed, P. Nagabhushan, B.B. Chaudhuri: Extraction of Line-Word-Character Segments Directly from Run-Length Compressed Printed Text-Documents, Jodhpur (2013), IEEE, ISBN: 978-1-4799-1586-6, INSPEC: 14181850, 18–21.
19. Xi Zhang, Chew Lim Tan: Text Line Segmentation for Handwritten Documents Using Constrained Seam Carving. In: 2014 14th International Conference on Frontiers in Handwriting Recognition, IEEE, 2167-6445/14 (2014).
20. Matthias Zimmermann, Horst Bunke: Hidden Markov Model Length Optimization for Handwriting Recognition Systems. In: Proceedings of the Eighth International Workshop on Frontiers in Handwriting Recognition (IWFHR 2002), IEEE, 0-7695-1692-0/02.
21. Joan Pastor-Pellicer, Salvador Espana-Boquera, Francisco Zamora-Martınez, Marıa Jose Castro-Bleda: Handwriting Normalization by Zone Estimation using HMM/ANNs. In: 14th International Conference on Frontiers in Handwriting Recognition (2014), IEEE, 2167-6445/14.
22. Fotini Simistira, Vassilis Papavassiliou, Themos Stafylakis: Enhancing Handwritten Word Segmentation by Employing Local Spatial Features. In: 2011 International Conference on Document Analysis and Recognition (2011), IEEE, 1520-5363/11.
23. Wesley Chin, Man Harvey, Andrew Jennings: Skew Detection in Handwritten Scripts. In: Speech and Image Technologies for Computing and Telecommunications, IEEE TENCON (1997).
24. Srihari, S.N., Govindraju: Analysis of textual image using the Hough transform. In: Machine Vision Applications (1989) Vol. 2 141–153.
25. Champa H N, K R AnandaKumar: Automated Human Behavior Prediction through Handwriting Analysis. In: 2010 First International Conference on Integrated Intelligent Computing, IEEE, 978-0-7695-4152-5/10 (2010).
26. Abdul Rahiman M, Diana Varghese, Manoj Kumar G: Handwritten Analysis Based Individualistic Traits Prediction. In: International Journal of Image Processing (2013), Volume 7 Issue 2.
27. Abhishek Bal, Rajib Saha: An Improved Method for Handwritten Document Analysis using Segmentation, Baseline Recognition and Writing Pressure Detection. In: 6th IEEE International Conference on Advances in Computing and Communications (ICACC-2016), Sept 6–8, 2016, Volume 93, Pages 403–415, doi:http://dx.doi.org/10.1016/j.procs.2016.07.227
28. Subhash Panwar, Neeta Nain: Handwritten Text Documents Binarization and Skew Normalization Approaches. In: IEEE Proceedings of 4th International Conference on Intelligent Human Computer Interaction, Kharagpur, India (2012).
29. A. Roy, T.K. Bhowmik, S.K. Parui, U. Roy: A Novel Approach to Skew Detection and Character Segmentation for Handwritten Bangla Words. In: Proceedings of the Digital Imaging Computing: Techniques and Applications, IEEE, 0-7695-2467-2/05 (2005).
30. Subhash Panwar, Neeta Nain: A Novel Approach of Skew Normalization for Handwritten Text Lines and Words. In: Eighth International Conference on Signal Image Technology and Internet Based Systems, IEEE, 978-0-7695-4911-8/12 (2012).
31. Jija Das Gupta, Bhabatosh Chanda: Novel Methods for Slope and Slant Correction of Off-line Handwritten Text Word. In: Third International Conference on Emerging Applications of Information Technology (2012), IEEE, 978-1-4673-1827-3/12.
32. Abhishek Bal, Rajib Saha: An Efficient Method for Skew Normalization of Handwriting Image. In: 6th IEEE International Conference on Communication Systems and Network Technologies, Chandigarh (2016), pp. 222–228, ISBN: 978-1-4673-9950-0.
33. M Sarfraj, Z Rasheed: Skew Estimation and Correction of Text using Bounding Box. In: Fifth IEEE conference on Computer Graphics, Imaging and Visualization (2008), pp. 259–264.

34. V. Papavassiliou, T. Stafylakis, V. Katsouros, G. Carayannis: Handwritten document image segmentation into text lines and words. In: Pattern Recognition, vol. 43, pp. 369-377, doi:10. 1016/j.patcog (2009).
35. Marcus Liwicki and Horst Bunke: IAM-OnDB - an On-Line English Sentence Database Acquired from Handwritten Text on a Whiteboard. In: Proceedings of the 2005 Eight International Conference on Document Analysis and Recognition (ICDAR 2005), IEEE, 1520-5263/05.

Enhanced Mutual Information-based Multimodal Brain MR Image Registration Using Phase Congruency

Smita Pradhan, Ajay Singh and Dipti Patra

Abstract In intensity-based image registration methods, similarity measure plays a vital role. Recently, mutual information and the variations of MI have gained popularity for the registration of multimodal images. As multimodal images have contrast changes, it is difficult to map them properly. To overcome this issue, phase congruency of the images that gives the significant features of illumination changed images. Also, the soft tissues present in the brain images have same intensity value in different regions. Hence, another assumption is that different pixels have unique distribution present in different regions for their proper characterization. For this challenge, utility measure is incorporated into enhanced mutual information as a weighted information to the joint histogram of the images. In this paper, spatial information along with features of phase congruency is combined to enhance the registration accuracy with less computational complexity. The proposed technique is validated with 6 sets of simulated brain images with different sets of transformed parameters. Evaluation parameters show the improvement of the proposed technique as compared to the other existing state of the arts.

Keywords Phase congruency · Enhanced mutual information · Target registration error

S. Pradhan (✉) · A. Singh · D. Patra
IPCV Lab, Department of Electrical Engineering, National Institute of Technology,
Rourkela, Odisha, India
e-mail: ssmita.pradhan@gmail.com

A. Singh
e-mail: ajayniya.singh@gmail.com

D. Patra
e-mail: dpatra@nitrkl.ac.in

© Springer Nature Singapore Pte Ltd. 2018
P.K. Sa et al. (eds.), *Progress in Intelligent Computing Techniques: Theory,*
Practice, and Applications, Advances in Intelligent Systems and Computing 518,
DOI 10.1007/978-981-10-3373-5_19

1 Introduction

Nowadays, in medical imaging applications, high spatial and spectral information from a single image is required to monitor and diagnose during treatment process. These information can be achieved by multimodal image registration. Different modalities of imaging techniques give several information about the tissues and organ of human body. According to their application range, the imaging techniques are CT, MRI, fMRI, SPECT and PET. A computed tomography (CT) image detects the bone injuries, whereas MRI defines the soft tissues of an organ such as brain and lungs. CT and MRI provide high-resolution image with biological information. The functional imaging techniques such as PET, SPECT and fMRI give low spatial resolution with basic information. To get the complete and detailed information from single modality is challenging task, which necessitates the registration task to combine multimodal images. The registered or fused image is more suitable for radiologist for further image analysis task.

Image registration has numerous applications such as remote sensing and machine vision. Several researchers discussed and proposed different image registration techniques in the literature [1, 2]. Image registration technique can be divided into intensity-based technique and feature-based technique [3]. Feature-based techniques consider different features such as line, point and textures whereas intensity-based techniques calculate the similarity metric between the images according to pixel correspondences of the spatial information from neighbouring pixels. Different similarity metrics generally used are maximum likelihood, cross-correlation and mutual information. Recently, mutual information (MI) gains more interest, which matches the data points based on mutual dependence among the images. MI is a powerful approach for multimodal medical image registration. A tremendous survey of this approach and their variations is discussed in [4, 5].

Several variations of MI have been proposed. Regional MI (RMI) technique has been adopted with spline-based interpolation for nonrigid registration in [6]. Another variation of MI, quantitative-qualitative mutual information (QMI), is proposed by Zhang et. al. They experimented for deformed multimodal medical image registration combining phase congruency and QMI [7]. Ye et al. used minimum moment of phase congruency for the selection of feature points of the floating images, and determined the correspondences in the reference image using normalized cross-correlation. They achieved the final set of correspondences of the images by projective transformation [8]. For phase congruency representations of multimodal images, Xia et al. used scale-invariant feature transform. Afterwards, they obtain the putative nearest neighbour matching on the SIFT descriptor space [9]. Also, SIFT has been used as feature extraction for score-level fusion in [10].

In this paper, we proposed a novel registration technique for multimodal brain MR image by incorporating the phase congruency into a new similarity measure-enhanced mutual information. The detailed phase congruency and enhanced mutual information are depicted in Sect. 2. In Sect. 3, proposed method is described. Performance evaluation with experimental results is discussed in Sect. 4 with a conclusion in Sect. 5.

2 Materials and Methods

Let I_r and I_f be the input images such as reference and floating images, where the floating image is transformed with affine transformation. For the alignment of the transformed floating image I_f^* relating to reference image I_r, the objective is to get the transformation parameter p that optimizes the cost function $SM(I_r, I_f)$.

$$P^* = \arg \max_p SM(p; I_r, I_f) \tag{1}$$

2.1 Phase Congruency

According to physiological evidences, human visual system takes strong responses towards the important points for feature detection with high phase congruency. According to Peter, phase congruency (PC) is based on local energy model [11]. PC is a dimensionless quantity that acquires the vital aspects of texture, i.e. contrast, scale and orientation, and has advantages over gradient-based techniques. It is normalised by dividing the sum over all orientations and scales of the amplitudes of the individual wavelet responses at the location α in the image, it can be characterized as

$$PC(\alpha) = \frac{\sum_\theta \sum_s \omega_\theta(\alpha)[\Lambda_{\theta s}(\alpha)\Delta\Phi_{\theta s}(\alpha) - \eta_\theta]}{\sum_\theta \sum_s \Lambda_{\theta s}(\alpha) + \kappa} \tag{2}$$

where ω defines the frequency spread weighting parameter, $\Lambda_{\theta s}$ is the amplitude at θ orientation and sth scale, η is the noise compensation parameter achieved independently in each orientation. κ is a constant factor.

$$\Delta\Phi_{\theta s}(\alpha, \theta) = \cos(\phi_s(\alpha, \theta)) - \overline{\phi(\alpha, \theta)} - \left|\sin(\phi_s(\alpha, \theta) - \overline{\phi(\alpha, \theta)})\right| \tag{3}$$

2.2 Enhanced Mutual Information

In an image, salient regions are visually pre-attentive distinct portions. Saliency can be measured by entropy of local segments within an image. Some of salient key points are detected and represented using speeded up robust feature (SURF) descriptor [12]. Luan et al. measured the saliency or utility by entropy of the image and proposed a new similarity measure, known as quantitative-qualitative mutual information (QMI). QMI is calculated as

$$QMI(I_r, I_f) = \sum_{r \in I_r} \sum_{f \in I_f} W(I_r, I_f) P(I_r, I_f) log \frac{P(I_r, I_f)}{P(I_r) P(I_f)} \tag{4}$$

Similarly, Pradhan et al. have adopted the weighted information as the utility to mutual information measure and proposed enhanced mutual information (EMI) [13]. EMI can be defined as

$$EMI(I_r, I_f) = \sum_{r \in I_r} \sum_{f \in I_f} P(I_r, I_f) log \frac{P(I_r, I_f)}{P(I_r) P(I_f)}$$
$$+ W(I_r, I_f) P(I_r, I_f) \tag{5}$$

where $W(I_r, I_f)$ is joint weighted information of I_r and I_f, which can be calculated by saliency measure of both reference and floating images. Here, the self-similarity measure, i.e. maximum entropy, is considered and multiplied with a dissimilarity measure to get the saliency of each pixel. Then, the joint utility of each intensity pair is calculated as

$$W_n(r, f) = \sum_{j, k \varepsilon \Omega} \delta_{I_r}(j) . \delta_{I_f}(k) \tag{6}$$

where Ω is the overlap area of both images and $\delta_{I_r}(j)$ and $\delta_{I_f}(k)$ are the weighted information of I_r and I_f, respectively. In the registration procedure, the joint utility is upgraded progressively. The utility varies with the transformation step 1 with a condition

$$W_n(r, f) + \alpha(l) . (1 - W_n(r, f)) \tag{7}$$

where $\alpha(l)0 : 1$.

3   Proposed Methodology

To incorporate the feature information into spatial information through higher saliency value, a hybrid technique known as PC-EMI has been proposed. Figure 1 shows the block diagram of proposed algorithm.

In the first step, phase congruency features of both the reference image and transformed floating image are extracted. Then, the marginal and joint entropies are calculated with the extracted phase congruency of the images. Those entropies are used for the calculation of enhanced mutual information. Saliency or utility measure is another important factor for the calculation of enhanced mutual information. The utility or saliency of the images is evaluated from original images instead of phase congruency mappings. Simultaneously, the joint utility is defined for every intensity pair of the images. Then, it is utilized as the weight to the joint histogram from PC features of the corresponding intensity pair. Then, for a set of transformation parameters, the cost function is evaluated and optimized to get the higher value. For optimal geometric transformation, quantum-behaved particle swarm optimization (QPSO)

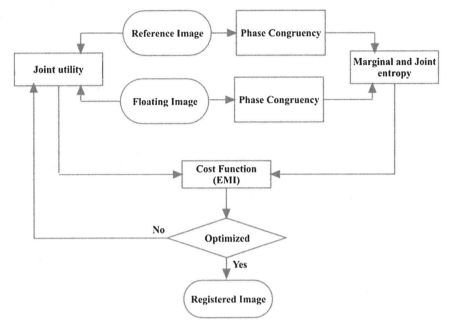

Fig. 1 Block diagram of proposed technique

technique is adopted for maximization of the similarity measure, i.e. enhanced mutual information. The registration approach is summarized in Algorithm 1.

Algorithm 1: PC-EMI Algorithm

    Input: $I_r, T(I_f)$

    Output: P^* *with high costfunction*

1 Initialize geometric transformations

2 **for** *each transformation* $-$ *step* $= 1$ **to** n **do**

3   Calculate the cost function as given by Eq. 5

4   **repeat**

5   Use QPSO technique to solve the optimization problem as in Eq. 1

6   Update the joint utility using Eq. 6

7   Recalculate the cost function

8   **until** ;

9   The difference of cost function in three consecutive transformation steps is ≤ 0.01

10 **end**

---

4 Simulation and Results

To know the performance, the proposed PC-EMI algorithm is tested on 6 set of multimodal brain image data sets. Here, only MR T1 and T2 weighted with transformed images are shown for simulation and analysis. The images are translated, rotated and scaled manually and validated with proposed technique. The proposed algorithm is validated with each data set and compared using different performance measure indices to proof the robustness and accuracy. The performance indices calculated are normalized cross-correlation (NCC), peak signal-to-noise ratio (PSNR), mean squared error (MSE) and normalized absolute error (NAE). The simulation is done in MATLAB R 2013a with a system specification Intel (R) Core(TM) i5$-$2400 CPU @ 3.10GHz. The simulated brain MR images were taken from the database (http://brainweb.bic.mni.mcgill.ca/brainweb/).

The images must be aligned with a particular transformation parameters that have higher cost function. This is the only condition taken for the registration procedure with minimization of MSE. The input images shown in Fig. 2a, b show MR T2 weighted and T1 weighted, respectively. The size of images is 182×182 with slice thickness of 3 mm. The cost function, i.e. similarity measure value, for EMI is calculated and compared with existing similarity measures, i.e. qualitative-quantitative mutual information (QMI) and mutual information (MI).

MR T2-T2 translated along y-axis image: The registration procedure is done with a set of brain MR T2 weighted and translated T2 weighted along y-axis. The cost function, i.e. the similarity measure value, for PC-EMI is 0.76, which is higher than PC-QMI and PC-MI. To verify the performance of proposed algorithm visually, checker board image (CBI) of registered image and the reference image is shown in Fig. 2c. Figure 1d and e shows the CBI of PC-QMI and PC-MI, respectively. The different performance measures are organized in Table 1. The computational time is additionally arranged in Table 1.

MR T1-T2 translated along x-axis image: In this test, the T2 image is transformed along x-axis manually. Then, the translated T2 image is registered with respect to T1 image using PC-EMI algorithm. The final EMI value is higher whereas the MSE value is lower than the other two methods. The checker board image of reference image and registered image is shown in Fig. 2f–h for PC-EMI, PC-QMI and PC-MI, respectively. Also, the NCC value of proposed technique is higher than that of other methods. All other parameters are presented in Table 1.

MR T1-T2 rotated image: Finally, one set of rotated T2 image and T1 image are registered. The alignment of rotated T2 image is tested within a range $[-20, 20]$. The fourth row of Fig. 2i–k shows the CBI using PC-EMI, PC-QMI and PC-MI, respectively. The convergence plot for the three techniques is presented in Fig. 3. From the plot, it is observed that EMI has higher value than that of QMI and MI. The computational time with other performance measures is tabulated in Table 1. It is also observed that the computational time for proposed technique is lower as compared to PC-QMI technique. The lower MSE proves the efficiency of PC-EMI algorithm.

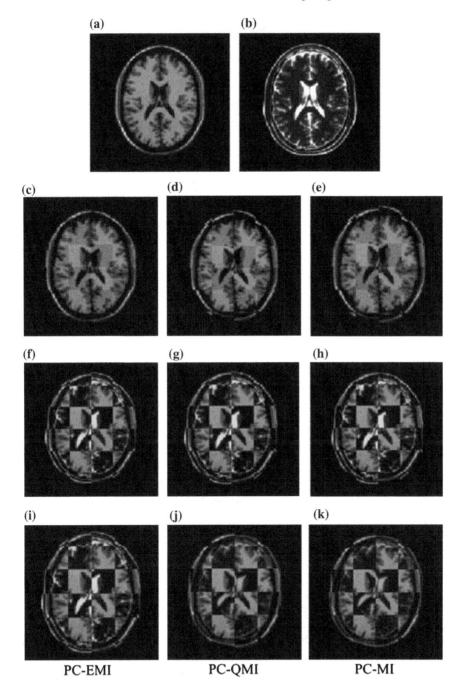

Fig. 2 Checker board image for all techniques

Table 1 Calculated cost value with different performance measures

Data Set	Methods	SM value	MSE	PSNR	NCC	NAE	Time (s)
T2-T2 translated along y-axis Brain MR image	PC-EMI	0.88	1.42×10^3	16.58	0.83	0.34	644.18
	PC-QMI	0.73	3.63×10^3	12.59	0.58	0.73	766.63
	PC-MI	0.62	3.89×10^3	12.09	0.54	0.78	28.83
T1-T2 translated along x-axis Brain MR image	PC- EMI	1.67	3.54×10^3	12.63	0.58	0.84	636.38
	PC-QMI	1.55	4.34×10^3	11.75	0.50	0.93	780.86
	PC-MI	0.51	4.61×10^3	11.49	0.47	0.96	29.53
T1-T2 rotated Brain MR image	PC-EMI	0.92	3.15×10^3	14.97	0.36	0.77	436.38
	PC-QMI	0.60	3.26×10^3	12.95	0.34	0.79	519.33
	PC-MI	0.15	3.30×10^3	12.84	0.33	0.81	19.53

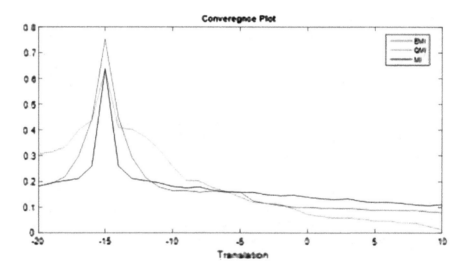

Fig. 3 Convergence plot for T1-T2 rotated image data set

5 Conclusion

The paper proposed a new registration technique, based on phase congruency into enhanced mutual information with the ability of taking feature and structural information. The proposed method is robust towards contrast and scale changes. The method is also rotation invariant. The utility factor gives the weighted information, and phase congruency gives feature information that gives an improvement in the calculation of EMI-based similarity measure for higher registration accuracy than that of other state of the arts. This work can be extended towards nonrigid registration.

References

1. J.B.A. Maintz, M.A. Viergever, A survey of medical image registration, Med. Image Anal, 2 (1), pp. 1–36, (1998).
2. Mani, V.R.S and Rivazhagan, S. Survey of Medical Image Registration, Journal of Biomedical Engineering and Technology, 1 (2), 8–25 (2013).
3. Oliveira, F. P., Tavares, J.M.R. Medical image registration: a review, Computer methods in biomechanics and biomedical engineering, 17 (2), 73–93 (2014).
4. A. Studholme, D.L.G. Hill, D.L. Hawkes, An overlap invariant entropy measure of 3D medical image alignment, Pattern Recogn, 32, pp. 71–86 (1999).
5. J.P.W. Pluim, A. Maintz, M.A. Viergever, Mutual information based registration of medical images: a survey, IEEE Trans. Med. Imaging, 22(8), pp. 986–1004 (2003).
6. Pradhan, S., Patra, D. RMI based nonrigid image registration using BF-QPSO optimization and P-spline, AEU-International Journal of Electronics and Communications, 69 (3), 609–621 (2015).
7. Zhang, Juan, Zhentai Lu, Qianjin Feng, and Wufan Chen.: Medical image registration based on phase congruency and RMI. IEEE International Conference on Medical Image Analysis and Clinical Applications (MIACA), pp. 103–105, (2010).
8. Ye, Yuanxin, Lian Xiong and Jie Shan.: Automated Multi-Source Remote Sensing Image Registration Based on Phase Congruency. ISPRS-International Archives of the Photogrammetry, Remote Sensing and Spatial Information Sciences, 1, pp. 189–194, (2012).
9. Xia, R., Zhao, J. and Liu, Y., 2013, October. A robust feature-based registration method of multimodal image using phase congruency and coherent point drift. In Eighth International Symposium on Multispectral Image Processing and Pattern Recognition, International Society for Optics and Photonics, (2013).
10. Bakshi, S., Das, S., Mehrotra, H., Sa, P. K. Score level fusion of SIFT and SURF for iris. IEEE International Conference on In Devices, Circuits and Systems (ICDCS), pp. 527–531, (2012).
11. Kovesi, Peter.: Image features from phase congruency. Videre: Journal of computer vision research, 1.3, pp. 1–26, (1999)
12. Sahu, B., Sa, P. K., Majhi, B. Salient Keypoint Detection using Entropy Map for Iris Biometric. In Proceedings of 2nd International Conference on Perception and Machine Intelligence, pp. 104–109, (2015).
13. Pradhan, S., Patra, D. Enhanced mutual information based medical image registration, IET Image Process., 10 (5) , pp. 418–427, (2015).

Enhanced Super-Resolution Image Reconstruction Using MRF Model

Rajashree Nayak, L.V. Sai Krishna and Dipti Patra

Abstract In this paper, an enhanced super-resolution image reconstruction method based on Markov random field (MRF) is proposed. This work incorporates an efficient training set. The most appropriate training set is utilized to find the similar patches based on the associated local activity of the image patch to improve the quality of reconstruction and to reduce the computational overload. Secondly, locality-sensitive hashing (LSH) method is used to search the similar patches from the training set. Finally, a robust and transformation invariant similarity measure named Image Euclidean distance (IMED) is incorporated to measure the patch similarity. Unlike traditional Euclidean distance, IMED measure considers the spatial relationships of pixels and provides an instinctively reasonable result. Experimental results demonstrate the outperforming nature of the proposed method in terms of reconstruction quality and computational time than several existing state-of-the-art methods.

Keywords Super resolution · Image reconstruction · MRF · IMED · LSH

1 Introduction

In modern-day applications, learning-based or example-based super-resolution reconstruction (SRR) methods outperform in terms of reconstruction quality at a cost of computational complexity. Among a variety of learning-based SRR methods, SRR methods using Markov random field [1–4] are prominent. SRR based on

R. Nayak (✉) · L.V. Sai Krishna · D. Patra
Department of Electrical Engineering, National Institute of Technology,
Rourkela 769 008, Odisha, India
e-mail: rajashreenayak17@gmail.com

L.V. Sai Krishna
e-mail: saikrishna_lv@yahoo.in

D. Patra
e-mail: dpatra@nitrkl.ac.in

© Springer Nature Singapore Pte Ltd. 2018
P.K. Sa et al. (eds.), *Progress in Intelligent Computing Techniques: Theory,
Practice, and Applications*, Advances in Intelligent Systems and Computing 518,
DOI 10.1007/978-981-10-3373-5_20

Markov network consists of two stages, one called as learning stage and the other as inference stage [3]. In learning stage, the relations in the network are obtained from the training set, whereas in inference stage, high resolution (HR) patch related to the test patch is estimated. However, some of the bottlenecks such as choice of an optimal training data set which should show sufficient interrelationship between the corresponding image patch pairs, efficient search algorithms to search for similar patches in the high-dimensional data space, and difficulty in finding an efficient distance measure to find the similarity between the similar patches, limit the performance of SRR methods based on MRF model.

To alleviate these above bottlenecks, the present paper utilizes an improved training data set which provides an excellent co-occurrence relationship among the LR and HR image pairs. Aiding to this, instead of K-NN search, the proposed method incorporates the locality-sensitive hashing (LSH) technique [5] to search the optimal patches. In order to find out the amount of similarity among the similar patches, image Euclidean distance is used. The MRF model followed by the belief propagation algorithm is used to approximate the MAP estimation to reconstruct the final HR patch.

The rest of the paper is organized as follows: A brief idea about MRF-based SRR method is presented in Sect. 2. The proposed method is discussed in Sect. 3. The result analysis of the proposed method compared with some of the existing methods is presented in Sect. 4. Section 5 describes the concluding remarks of the proposed work.

2   MRF-based SRR

In the MRF-based SRR methods, the main aim is to estimate a HR image I_{HR} provided its low-resolution counterpart image I_{LR} by utilizing the probabilistic method. The estimation process tries to optimize the posterior probability $P(I_{HR}/I_{LR})$. Equation (1) provides the formulation for estimating the HR patch $\hat{i}_{HRj}$ by utilizing th MRF model [4].

$$\hat{i}_{HRj} = \prod_{(i,j)} \psi(i_{HRi}, i_{HRj}) \prod_{k} \varphi(i_{HRk}, i_{LRk}) \tag{1}$$

However, Eq. (1) can be efficiently solved by implementing the Belief propagation algorithm and the formulation is given in Eq. (2).

$$\hat{i}_{HRj} = \arg \max_{i_{HRj}} \varphi(i_{HRj}, i_{LRj}) \prod_{k} M_j^k \tag{2}$$

where K is the index neighboring to jth patch. jth LR patch is represented as i_{LRj} and its corresponding HR image patch is given by i_{HRj}. The message transition probability from the HR patch i_{HRj} to its neighbor patch i_{HRk} is defined by M_j^k and the detailed description is given in [3].

3 Proposed Method

The proposed work implements the SRR of image using MRF model solved by BP algorithm as given in [3, 4] but some modifications are introduced. The contributions of the proposed work are summarized as follows, and the block diagram of the proposed method is presented in Fig. 1.

(1) Generation of efficient training set T_{eff} using the mutual information criterion and clustering of T_{eff} into $T_{eff}(smooth)$ and $T_{eff}(non-smooth)$ depending upon the type of patch.
(2) Selection of optimal similar training patches by utilizing the LSH [5] method.
(3) Implementation of MRF model for the SRR of image by utilizing the image Euclidean distance [6] for measuring the similarity between image patches.

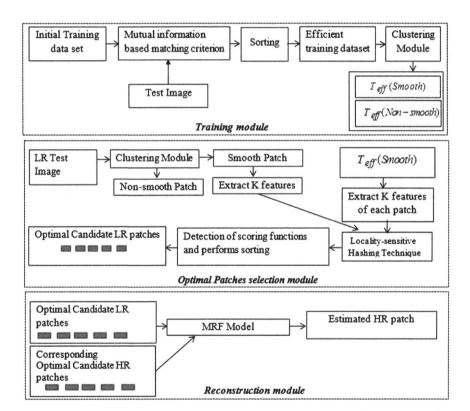

Fig. 1 Block diagram of the proposed SRR method

3.1 Generation of Training Set

In the proposed work, we have used a mutual information-based similarity measure to chose a training set. The idea behind using this type of criterion is to generate the training set containing images which are much similar to the test image. The whole process to select the effective training set is described in Algorithm 1 and is graphically presented as training module of Fig. 1.

Algorithm 1

Input: Randomly selected training set T, Test LR image I_{LR}
Output: Efficient training set T_{eff}
1: **for** each image I in T **do**
2:    Calculate mutual information (MI), with respect to I_{LR} image
3: **end for**
4: Sort the MI values
5: Take N number of images from T having best MI values
6: Construct T_{eff} with the selected N number of images

After the successful generation of T_{eff}, the training images in T_{eff} are divided into different patches and are clustered as smooth or non-smooth patches and the whole T_{eff} training set is divided two different training sets such as $T_{eff}(smooth)$ and $T_{eff}(non-smooth)$. Depending upon the type of test patch, the corresponding training set is chosen to find out the similar patches. The whole mechanism is provided in Algorithm 2. This mechanism helps to reduce the computational complexity of the algorithm. After that, the $T_{eff}(smooth)$ and $T_{eff}(non-smooth)$ training sets are up-sampled to get their corresponding HR training sets such as $\left[T_{eff}(smooth)\right]_{HR}$ and $\left[T_{eff}(non-smooth)\right]_{HR}$.

Algorithm 2

Input: $T_{eff}, C1, C2,$ *sampling parameter s*
Output: $T_{eff}(Smooth),\ T_{eff}(Non-smooth)$
1: **for** each patch t_i in T_{eff} **do**
2:   **if** var(t_i) $\leq$ *threshold* **then**
3:     **return** t_i *is a smooth patch*, $C1 = C1 + 1$
4:   **else**
5:     **return** t_i *is a non-smooth patch* $C2 = C2 + 1$
6:   **end if**
7:   Continue until all patches of T_{eff} are not tested
8: **end for**
9: Construct training set $T_{eff}(Smooth)$ by taking the smooth patches
10: Construct training set $T_{eff}(Non-smooth)$ by taking the non-smooth patches
11: Up-sample $T_{eff}(Smooth)$ and $T_{eff}(Non-smooth)$ by a factor of S to find $\left[T_{eff}(Smooth)\right]_{HR}$ and $\left[T_{eff}(Non-smooth)\right]_{HR}$

3.2   Selection of Optimal Similar Training Patches

This subsection describes to find out D number of optimal similar patches for a LR test image patch i_{LRi} from the training set. In the literature, K-nearest neighborhood is the popular choice for finding the optimal similar patches. In this paper, instead of K-nearest neighborhood, LSH method is used to find the similar patches by taking the first and second-order gradient features of the considered patches. As compared to K-nearest neighborhood method, LSH method needs less parameters to adjust and hence provides a faster searching process. The complete process for the choice of optimal similar patches for a smooth low-resolution patch is described in optimal patches selection module of Fig. 1.

3.3   Implementation of MRF Model for the SRR of Image

After finding the optimal LR patches, their corresponding HR patches are picked from either $\left[T_{eff}(Smooth)\right]_{HR}$ or $\left[T_{eff}(Non-smooth)\right]_{HR}$ for each test LR image and then Eq. (1) is solved to estimate the HR patch corresponding to the LR patch. The detailed process is defined in reconstruction module of Fig. 1. In this paper instead of Euclidean distance, image Euclidean distance (IMED) is used to find out the similarity among optimal patches to calculate the compatibility functions. Equation (3) provides the formulation for compatibility function $\varphi(i_{HRk}, i_{LRk})$ between the HR patch i_{HRk} and test LR patch i_{LRk}. Equation (4) provides the formulation for compatibility function $\psi(i_{HRj}, i_{HRk})$ between the overlap areas of jth and kth HR patch.

$$\varphi(i_{HRk}, i_{LRk}) = \exp\left(-\frac{IMED(i_{HRk}, i_{LRk})}{2\sigma^2}\right) \tag{3}$$

$$\psi(i_{HRj}, i_{HRk}) = \exp\left(-\frac{IMED(i_{HRj}, i_{HRk})}{2\sigma^2}\right) \tag{4}$$

where, $IMED(i_{HRk}, i_{LRk})$ defines the image Euclidean distance between i_{LRk} and downsampled patch of i_{HRk}. $IMED(i_{HRj}, i_{HRk})$ measures the image Euclidean distance between the overlap areas of jth and kth HR patch.

Euclidean distance between two patches is performed by the summation of pixel-wise intensity differences. There is no provision to consider the spatial relationship among the pixels. As a result, small perturbation results in a significant change in the distance measure. On the other hand, in IMED measure, the spatial relationship among the pixels is taken into consideration and hence provides intuitively improved results than Euclidean distance. Equation (5) provides the mathematical formulation of the IMED measurement between two image patches $x(i,j)$ and $y(i,j)$ for $i,j = 1, 2.............MN$.

$$IMED^2(x, y) = \frac{1}{2\pi} \sum_{i,j=1}^{MN} \exp\left\{ -\left(\left|P_i - P_j\right|^2 \middle/ 2\right) \right\} \left(x^i - y^i\right)\left(x^j - y^j\right) \quad (5)$$

where, P_i, P_j are pixels for $i, j = 1, 2............MN$. $\left|P_i - P_j\right|$ defines the pixel distance.

4 Result and Analysis

In this section, we describe the SR results obtained by our proposed method on various test images compared with other existing state-of-the-art methods in the literature. We have considered a training set T_{eff} of 41 images from Berkely database [7, 8]. The LR test image is divided into patches with a patch size of 5×5 with overlapping of 3 pixels. We use $D = 30$, $sigma = 15$. The robustness and effectiveness of the proposed method is compared with some existing methods such as ExSR [2], NLKR [9], and document super-resolution image using structural similarity and Markov random field (SR+SSMRF) [4]. Several image quality assessment parameters such as FSIM [10], SSIM [11], and PSNR are calculated to measure the quality of reconstructed SR images.

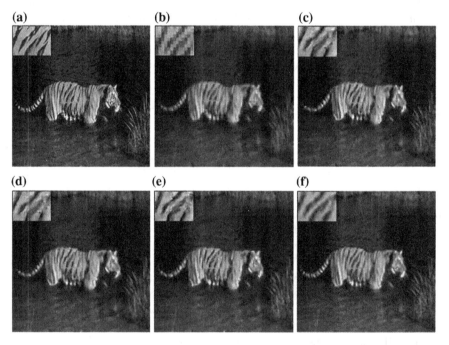

Fig. 2 Visual representation of reconstructed HR solutions for image "Tiger" using proposed method and different state-of-the-art SRR methods. **a** Original HR image **b** LR image **c** ExSR **d** NLKR **e** SR+SSMRF **f** Proposed

Fig. 3 Visual representation of reconstructed HR solutions for image "Flower" using proposed method and different state-of-the-art SRR methods. **a** Original HR image **b** LR image **c** ExSR **d** NLKR **e** SR+SSMRF **f** Proposed

The proposed work is validated for a sequence of test images from Berkely database [8]. Experimental results for three images are presented in this paper for the comparison purpose. Visual representation of the reconstructed results based on proposed method compared with the other state-of-the-art methods are presented in Figs. 2, 3, and 4. From the visual perception, it is observed that the SR results using NLKR provide better results than the ExSR. However, the implementation of inherently smooth kernel in NLKR method limits the performance of the method. As compared to NLKR method, SR+SSMRF method provides clearer result. But in the presence of blurring and for more resolution enhancement factor, reconstruction process degrades. Among all methods, the proposed method shows sharp SR images. To quantify the visual presentation, PSNR, SSIM, and FSIM values are presented in Table 1. Processing time for the reconstructed results is also given in Table 1. From the tabular data analysis, it is observed that the proposed method provides better PSNR and SSIM values than the other compared methods, and our proposed method takes less time.

(a)   (b)   (c)

(d)   (e)   (f)

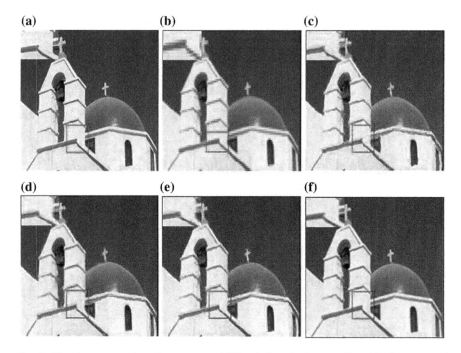

Fig. 4 Visual representation of reconstructed HR solutions for image "Church" using proposed method and different state-of-the-art SRR methods. **a** Original HR image **b** LR image **c** ExSR **d** NLKR **e** SR+SSMRF **f** Proposed

Table 1 Comparison of different subjective quality measures for SR results using different SRR methods

Images	Methods	PSNR	SSIM	FSIM	Avg. run time (seconds)
Tiger	ExSR	21.02	0.7212	0.9211	321
	NLKR	21.60	0.7314	0.9272	192
	SR+SSMRF	21.84	0.7408	0.9310	100
	Proposed	22.37	0.7878	0.9478	79
Flower	ExSR	23.13	0.9172	0.9712	343
	NLKR	24.47	0.9213	0.9769	246
	SR+SSMRF	24.98	0.9432	0.9783	176
	Proposed	25.22	0.9442	0.9809	132
Church	ExSR	23.01	0.8812	0.9513	303
	NLKR	23.54	0.8873	0.9619	198
	SR+SSMRF	23.79	0.8912	0.9732	134
	Proposed	24.05	0.9096	0.9843	86

5   Conclusion

This paper presents a patch-wise super-resolution image reconstruction method using Markov random field model. In this paper, by considering the local characteristics of the image, an efficient training set is proposed to take care of smooth image patch and non-smooth image patches separately. Afterward, we used a locality-sensitive hashing (LSH) method based searching technique to obtain the similar patches in the training sets. We reduced the computational complexity in terms of processing time by choosing efficient training set along with LSH technique. An efficient distance measure, i.e., image Euclidean distance measure which can be applicable to images of any size and resolution, is used in this paper to measure the similarity between the optimal patches. The proposed approach provides higher PSNR, SSIM, and FSIM values when compared to several other approaches in the literature.

References

1. Rajan, Deepu and Chaudhuri, Subhasis.: An MRF-based approach to generation of super-resolution images from blurred observations. Journal of Mathematical Imaging and Vision. 16, 5–15 (2002)
2. Freeman, William T and Jones, Thouis R and Pasztor, Egon C.: Example-based super-resolution. J. Computer Graphics and Applications, IEEE. 22, 56–65 (2002)
3. Freeman, William T and Pasztor, Egon C and Carmichael, Owen T.: Learning low-level vision. J. International journal of computer vision. 40, 25–47 (2000)
4. Chen, Xiaoxuan and Qi, Chun.: Document image super-resolution using structural similarity and Markov random field. J. Image Processing, IET. 8, 687–698 (2014)
5. Gionis, Aristides and Indyk, Piotr and Motwani, Rajeev and others.: Similarity search in high dimensions via hashing: VLDB. 99, 518–529 (1999)
6. Wang, Liwei and Zhang, Yan and Feng, Jufu.: On the Euclidean distance of images. J. Pattern Analysis and Machine Intelligence, IEEE Transactions on. 27, 1334–1339 (2005)
7. The Berkeley Segmentation Dataset and Benchmark, https://www.eecs.berkeley.edu/Research/Projects/CS/vision/bsds/
8. Computer Vision Group, http://www.eecs.berkeley.edu/Research/Projects/CS/vision/grouping/
9. Zhang, Haichao and Yang, Jianchao and Zhang, Yanning and Huang, Thomas S.: Non-local kernel regression for image and video restoration:Computer Vision–ECCV 2010, 566–579 (2010)
10. Zhang, Lin and Zhang, Lei and Mou, Xuanqin and Zhang, David.: FSIM: a feature similarity index for image quality assessment. J. Image Processing, IEEE Transactions on. 20, 2378–2386 (2011)
11. Channappayya, Sumohana S and Bovik, Alan Conrad and Heath Jr, Robert W.: Rate bounds on SSIM index of quantized images. J. Image Processing, IEEE Transactions on. 17, 1624–1639 (2008)

Part III
Biometrics Security Artifacts for Identity Management

An Exploration of V-HOG on W-Quartette Space for Multi Face Recognition Issues

Bhaskar Belavadi and K.V. Mahendra Prashanth

Abstract Illumination, Pose, and Expression variations are the major factors that put down the performance of face recognition system. This paper presents a dewy-eyed withal efficient hypothetical description for complex face recognition based on combination of w-Quartette Colorspace with Variant Hoglets (v-HOG) in sequel of our earlier work of fSVD [1]. Firstly, face image is mapped onto w-Quartette Colorspace to effectively interpret information existing in the image. Further variant Hoglet is employed to extract substantive features exhibiting linear properties to map line singularities and at the same time to derive tender features of face contortions. To foster the extracted features, the features are projected to a lower dimensional space for efficient face recognition. Five distinct distance measures are adopted as classifier to obtain appreciable recognition rate.

Keywords fSVD · w-Quartette Colorspace · v-HOG · r-HOG · Similarity measures

1 Introduction

Faces are complex ophthalmic inputs which cannot be depicted by elementary shapes and patterns. Face Recognition is a domain wherein dynamic research and assorted methods have been purported to execute this chore. Image gradients with schur decomposition are invariant to changes in illumination and pose variations [2]. Face images can be represented in terms of Eigen faces [3], which facilitates gradual changes in face and also simple to compute. A better and unquestionable 2DPCA algorithm proposed in [4], used statistical intercellular data using basic image inter-cellular data. In [5], volume measure was used as classifier with 2DPCA, which

B. Belavadi (✉) · K.V. Mahendra Prashanth
SJBIT, BGS Health & Education City, Bengaluru 560060, Karnataka, India
e-mail: bhaskar.brv@gmail.com

K.V. Mahendra Prashanth
e-mail: kvmprashanth@sjbit.edu.in

© Springer Nature Singapore Pte Ltd. 2018   219
P.K. Sa et al. (eds.), *Progress in Intelligent Computing Techniques: Theory,*
Practice, and Applications, Advances in Intelligent Systems and Computing 518,
DOI 10.1007/978-981-10-3373-5_21

outperformed classical distance measure classifiers. For effectual face representation [6], (2D) 2PCA was coined which works on row and column directions of face images with much diluted coefficients for face representation. Multiple Kernel Local Fisher Discriminant analysis (MKLFDA) was proposed [7] to delineate the nonlinearity of face images. Independent component neighborhood preserving analysis [8] (IC-NPA) was proposed to retain the inviolable incisive potential of LDA and at the same time upholding the intrinsic geometry of the within-class information of face images. As stated earlier, combining the features always aids in improved efficiency, in [9] Eigen and Fisher face are combined using wavelet fusion, and 2FNN (Two-Feature Neural Network) is used for classification to improve the recognition rate. To surmount the problem of small sample size, an unsupervised nonlinear Spectral Feature Analysis (SFA) [10] was minted, which extracts discriminative features out of a small sample-sized face image. In [11], the author proposed a method for the recognition of facial expressions in face images using landmark points in linear subspace domain.

Nonlinear kernel sparse representations [12] aids in robustness of face recognition against illumination and occlusion by developing pixel-level and region-level kernels hiring local features. Immix of DWT, DFT, and DCT, expeditiously educe pose, translation, and illumination invariant features [13]. Decimated redundant DWT was proposed in [14] to cauterize the effect of translation of face image. Gabor jets [15] feature extractor improves accuracy rate and also reduces the computational time with the facilitation of Borda count classifier. Gray scale Arranging Pairs (GAP) [16] delineates the holistic data in the whole face image and demonstrates mellowed raciness against illumination changes. Combination of two effectual local descriptors Gabor wavelet and enhanced local binary pattern with generalized neural network [17] is insensitive to small changes in input data and is robust-to-slight variation of imaging conditions and pose variations. A more versatile improved Quaternion LPP [18] was purported to exlploit the color information in face images.

2 Proposed Methodology

In this section, we key out the proposed face image depiction and classification technique. We take into account the primal issues of face recognition and reduce the effect of these in improving the accuracy of recognition. Firstly, face image is mapped onto w-Quartette Colorspace to effectively interpret information existing in the image. Further variant Hoglet extracts discriminative features exhibiting linear properties to map line singularities and at the same time to derive tender features of face contortions. To foster the extracted features, the features are projected to a lower dimensional space for efficient face recognition. Finally, for classification, five different similarity measures are used to obtain an average correctness rate.

2.1 W-Quartette Colorspace

Different Color models possess different discriminative index. Compared to greyscale images, a color image embeds in it much more utile information for improving face recognition performance [19]. Making use of these properties, we derive an image from the original image by mapping it on Quartette Colorspace. The color face image $F(x, y, z)$ of size $(p \times q \times 3)$ is mapped as given below,

$$F_{ci}(x, y, z) = \{F(x, y, z) \rightarrow \{a * Lab, b * LUV, c * YCgCr, d * YUV\}\} \quad (1)$$

$$\text{where } i = 1, 2, 3, 4 \text{ and a, b, c, d are weights for colorspace}$$

The weights for the Quartette Colorspace are set based on the behavior of the Colorspace with fSVD.

$$I_{fi}(x, y, z) = \frac{(F_{ci}(x, y, z) + \xi_1 * D_i(x, y, z))}{(1 + (\xi_1, \xi_2, \xi_3, \xi_4))} \quad (2)$$

$$\text{where } D_i(x, y, z) = U * \Sigma^{\gamma_i} * V^T$$

$$[U, \Sigma, V] = SingularValuedecomp(Inorm(x, y))$$

$$\xi_1 = 0.07, \xi_2 = 0.05, \xi_3 = 0.05, \xi_4 = 0.5, \gamma_1 = 0.695, \gamma_{2,3,4} = 0.895$$

2.2 Variant Hoglets

HOG features were first introduced by Navneed Dalal and Bill Triggs [20]. The essence of the histogram of oriented gradient descriptor or signifier is that local

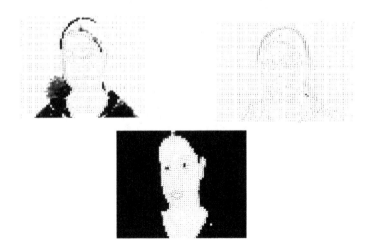

Fig. 1 v-HOG image in LAB space

object appearance and shape within an image can be described by the distribution of local intensity gradients or adjoin directions. The effectuation of these signifiers can be achieved by dividing the image into small colligated regions called cells, and for each cell amassing a histogram of gradient directions or adjoin orientations for the pixels within the cell. The combination of these histograms then represents the signifiers. The HOG signifiers has few crucial advantages like invariance to geometric and photometric transformation, which are helpful in face recognition applications. Gradient orientation preserves the important structure of the image and is much useful in face recognition [21]. To improve the face recognition accuracy, in our work we propose variant Hoglets by adopting HOG with various gradient filters which is adopted based on the colorspace. In an attempt to imbibe the behavior of wavelets with HOGs under different colorspace, we have incubated daubechies as filter kernels

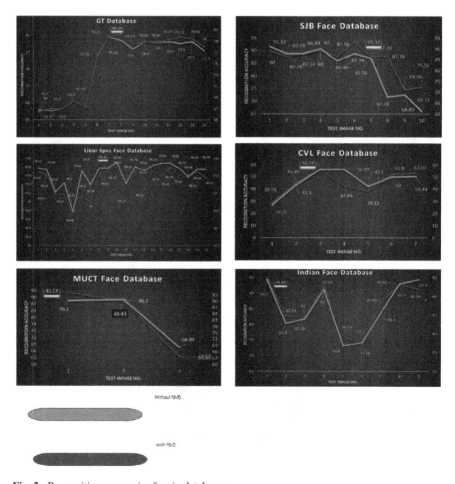

Fig. 2 Recognition accuracies for six databases

for HOG which are defined below.

$$dbn = Coeff1, Coeff2, \dots, Coeffn,$$

where $n = 7$ and 9, are used in combination with YCgCr and LAB colorspace.
Furthermore, the below filter kernels are also adopted with LUV and YUV colorspace, respectively.

$$[-1, 0, 1] \, and \, [-1, 0, 1]^T, [0.707, 0, -0.707] \, and \, [0.707, 0, -0.707]^T$$

Below is an example of v-HOGs under different planes of LAB colorspace:
The features obtained from the above proposed method is further compressed by projecting them on to the lower dimensional subspace and classified using an aggregate of five different distance measures (Figs. 1 and 2).

3   Result

This section acquaints the results primly on six different widely used databases. We embark into the recognition system by mapping face image onto w-Quartette Colorspace, which prepares the image for recognition in the best possible way. fSVD is then applied to normalize the mapped image. Further to extract the shape descriptors, v-HOG is applied to the derived face image and is projected onto the reduced dimensionality eigen subspace. We have used five different distance measures to formalize the inquiry face image.

To appraise our proposed technique, we have incorporated six different database viz., Georgia Tech, Dr. Libor Spacek Faces94, CVL, MUCT, Indian face and SJB face database. GTech consists of 50 persons × 15 images/person. This database shows frontal face with variations in expressions, lighting conditions, cluttered background, and scale. Libor consists of 152×20 images with tilt of head and considerable expression changes. CVL consists of 114×7 with pose and expression variations. MUCT consists of 3755 out of which we have used 199×12 images, and it has diversity of lighting, age, and ethnicity. Indian face database has 671 images out of which we have used 61×7 images, and it has considerable variations in pose and expression. SJB face database is our own face database which consists of 10 persons × 10 images/person. This also has substantial variations in expression, pose, and illumination.

To assess our algorithm, single image from every class is used for training and rest of the images are entailed for testing purpose. We performed experiments with different γ and ξ in both training and test phase. Finally, the value is set to $\xi_1 = 0.07, \xi_2 = 0.05, \xi_3 = 0.05, \xi_4 = 0.5, \gamma_1 = 0.695, \gamma_{2,3,4} = 0.895$, to achieve appreciable results.

3.1 Experiment-I—Effect of fSVD on the Extracted Features

In this experiment, we have found the correlation coefficients for features of images with and without fSVD. For experimentation purpose, only two database have been considered.

Below table suggests that the correlation coefficient is high for features extracted using fSVD than without fSVD. It also says that fSVD has better ability to discriminate within and between class features (Tables 1 and 2).

3.2 Experiment-II—Comparision of v-HOG and rHOG

In this experiment, we have compared the invidious ability of within and between class images by regular HOG (rHOG) and variant HOG (v-HOG). In this experiment, fSVD is retained in both the cases. The results suggests that v-HOG has better correlative feature extraction capacity than rHOG. v-HOG decorrelated the between-class images and at the same time upholds the correlative coefficient of within-class images.

Table 1 Average correlation coefficient

Database	With fSVD		Without fSVD	
	Within	Between	Within	Between
CVL	0.9719	0.9426	0.9616	0.9485
GT	0.9584	0.9213	0.9490	0.9229

Table 2 Comparison of v-HOG and rHOG

Database	rHOG		v-HOG	
	Within	Between	Within	Between
GT	0.9477	0.9374	0.9584	0.9213
SJB	0.9867	0.9228	0.9973	0.8702
Dr Libor	0.9965	0.9607	0.9977	0.9274
CVL	0.9697	0.9593	0.9719	0.9462
Muct	0.9898	0.9509	0.9856	0.9378
Indian	0.9837	0.9729	0.9911	0.8962

3.3   Experiment-III—Effect of v-HOG on Recognition Accuracy with and Without fSVD

In this section of experiments, we have found the recognition accuracy for with and without fSVD for all six face databases. In every database, only one image is trained per person and rest of the images are used as query images. The result suggests that the proposed method has the ability to correctly recognize the face image. The result also evoke that the proposed method has the ability to subdue the effect of single sample per person (SSPP), Illumination, expression, and pose in face recognition to a larger extent. Below figures shows the recognition accuracies of six different databases with fSVD and without fSVD.

4   Conclusion

This paper has developed an efficient algorithm for complex face recognition based on combination of w-Quartette Colorspace with Variant Hoglets (v-HOG) in coaction fSVD. w-Quartette Colorspace has the ability to effectively interpret information existing in the image. Further to it variant Hoglet extracts substantive features exhibiting linear properties to map line singularities and at the same time to derive tender features of face contortions. The simulation experiments suggest that the proposed technique yields overwhelming results under single sample per person, variation in pose, illumination, and expression.

References

1. Wang, F., Wang, J., Zhang, C., and Kwok, J. (2007). Face recognition using spectral features. Pattern Recognition, 40(10), 2786–2797.
2. Ghinea, Gheorghita, Ravindran Kannan, and Suresh Kannaiyan. "Gradient-orientation-based PCA subspace for novel face recognition." Access, IEEE2 (2014): 914–920.
3. Turk, Matthew, and Alex Pentland. "Eigenfaces for recognition." Journal of cognitive neuroscience 3.1 (1991): 71–86.
4. Yang, Jian, et al. "Two-dimensional PCA: a new approach to appearance-based face representation and recognition." Pattern Analysis and Machine Intelligence, IEEE Transactions on 26.1 (2004): 131–137.
5. Meng, Jicheng, and Wenbin Zhang. "Volume measure in 2DPCA-based face recognition." Pattern Recognition Letters 28.10 (2007): 1203–1208.
6. Zhang, Daoqiang, and Zhi-Hua Zhou. "(2D) 2PCA: Two-directional two-dimensional PCA for efficient face representation and recognition."Neurocomputing 69.1 (2005): 224–231.
7. Wang, Ziqiang, and Xia Sun. "Multiple kernel local Fisher discriminant analysis for face recognition." Signal processing 93.6 (2013): 1496–1509.
8. Hu, Haifeng. "ICA-based neighborhood preserving analysis for face recognition." Computer Vision and Image Understanding 112.3 (2008): 286–295.

9. Devi, B. Jyostna, N. Veeranjaneyulu, and K. V. K. Kishore. "A novel face recognition system based on combining eigenfaces with fisher faces using wavelets." Procedia Computer Science 2 (2010): 44–51.
10. Bhaskar, Belavadi, K. Mahantesh, and G. P. Geetha. "An Investigation of fSVD and Ridgelet Transform for Illumination and Expression Invariant Face Recognition." Advances in Intelligent Informatics. Springer International Publishing, (2015). 31–38.
11. Aifanti, Niki, and Anastasios Delopoulos. "Linear subspaces for facial expression recognition." Signal Processing: Image Communication 29.1 (2014): 177–188.
12. Kang, Cuicui, Shengcai Liao, Shiming Xiang, and Chunhong Pan. "Kernel sparse representation with pixel-level and region-level local feature kernels for face recognition." Neurocomputing 133 (2014): 141–152.
13. Krisshna, NL Ajit, V. Kadetotad Deepak, K. Manikantan, and S. Ramachandran. "Face recognition using transform domain feature extraction and PSO-based feature selection." Applied Soft Computing 22 (2014): 141–161.
14. Li, Deqiang, Xusheng Tang, and Witold Pedrycz. "Face recognition using decimated redundant discrete wavelet transforms." Machine Vision and Applications 23.2 (2012): 391–401.
15. Perez, Claudio A., Leonardo A. Cament, and Luis E. Castillo. "Methodological improvement on local Gabor face recognition based on feature selection and enhanced Borda count." Pattern Recognition 44, no. 4 (2011): 951–963.
16. Zhao, Xinyue, Zaixing He, Shuyou Zhang, Shunichi Kaneko, and Yutaka Satoh. "Robust face recognition using the GAP feature." Pattern Recognition 46, no. 10 (2013): 2647–2657.
17. Sharma, Poonam, K. V. Arya, and R. N. Yadav. "Efficient face recognition using wavelet-based generalized neural network." Signal Processing 93.6 (2013): 1557–1565.
18. Wu, Shuai. "Quaternion-based improved LPP method for color face recognition." Optik-International Journal for Light and Electron Optics 125.10 (2014): 2344–2349.
19. Torres, Luis, Jean-Yves Reutter, and Luis Lorente. "The importance of the color information in face recognition." Image Processing, 1999. ICIP 99. Proceedings. 1999 International Conference on. Vol. 3. IEEE, 1999.
20. Dalal, Navneet, and Bill Triggs. "Histograms of oriented gradients for human detection." In Computer Vision and Pattern Recognition, 2005. CVPR 2005. IEEE Computer Society Conference on, vol. 1, (2005) pp. 886–893.
21. Tzimiropoulos, G., Zafeiriou, S., and Pantic, M. (2011, March). Principal component analysis of image gradient orientations for face recognition. In Automatic Face & Gesture Recognition and Workshops (FG 2011), 2011 IEEE International Conference on (pp. 553–558). IEEE.

Comparative Analysis of 1-D HMM and 2-D HMM for Hand Motion Recognition Applications

K. Martin Sagayam and D. Jude Hemanth

Abstract Hand motion recognition is an interesting field in the development of virtual reality applications through the human–computer interface. The stochastic mathematical model hidden Markov model (HMM) is used in this work. There are numerous parametric efforts in HMM for temporal pattern recognition. To overcome the recursiveness in the forward and backward procedures, dimensionality and storage problem in Markov model, 2-D HMM has been used. The experimental results show the comparison of 2-D HMM with 1-D HMM in terms of performance measures.

Keywords Hand motion recognition · HCI · HMM · 1-D HMM · 2-D HMM

1 Introduction

In recent years, human–computer interaction (HCI) is very much significant in the development of artificial intelligence system. Among different gestures in human body, hand gestures plays major role in the non-verbal communication such as clapping, praying, counting etc. between the user and the system [1, 2]. The problems in gesture recognition system (GRS) are illumination, transformation and background subtraction [3, 4].

Figure 1 shows the proposed work of hand motion recognition system based on HMM model for temporal pattern analysis. The extended version of 1-D HMM has been experimented with various hand gestures for real-time applications. To decode the path of the state sequence model of HMM by using 2-D Viterbi algorithm. Here

K.M. Sagayam (✉) · D.J. Hemanth
Department of ECE, Karunya University, Coimbatore, Tamil Nadu, India
e-mail: martinsagayam.k@gmail.com

D.J. Hemanth
e-mail: jude_hemanth@rediffmail.com

© Springer Nature Singapore Pte Ltd. 2018
P.K. Sa et al. (eds.), *Progress in Intelligent Computing Techniques: Theory, Practice, and Applications*, Advances in Intelligent Systems and Computing 518, DOI 10.1007/978-981-10-3373-5_22

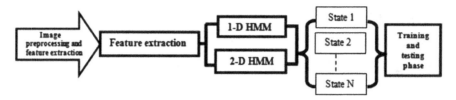

Fig. 1 Proposed methodology

comparison made between 1-D HMM and 2-D HMM in terms of performance is discussed in Sect. 4 and its simulation results in Sect. 5.

2 Related Work

The hand gestures were collected from 5 subjects with 60 variations in each set. The proposed method discussed about the hidden sates in the state sequence of the observation matrix and its variable states in Viterbi algorithm.

2.1 Architecture and Hidden States Inference Problem

Devijver [5] has encountered the problem in second- and third-order probabilities in 2-D HMM which represents the hidden states in the state sequences of the observation matrix. Li et al. [6] have discussed about the image classification based on causal 2-D HMM by using Viterbi algorithm. This tends to the decoded path of the variable state. Schonfeld and Bouaynaya [7] have proposed an idea for the best search in the state sequence of HMM classifier by using 2-D Viterbi algorithm. To find the most probable state sequence called state sequence analysis (SSA) is a problem that requires evaluation of probabilities of state sequence [8].

3 Image Preprocessing and Feature Extraction

Raman and Himanshu [9] have compared the different edge detection techniques such as Sobel, Prewitt, Canny and Robert operator. Neha, Deorankar [10] have proposed histogram of oriented gradient (HOG) feature descriptor to find the shape, appearance and angler position of an image by using edges in x and y directions.

$$
\begin{aligned}
|\text{Grad}| &= \left(\text{Grad in } x\ -\text{direction}^2 +\text{Grad in } y\ -\text{direction}^2\right)\\
\text{Angle } (\theta) &= \arctan(\text{Grad in } y\ -\text{direction} /\text{Grad in } x\ -\text{direction})
\end{aligned}
\tag{1}
$$

Neha et al. [11] have proposed real-time hand gesture by using HOG feature descriptor by fine motion of hand gesture in spatiotemporal pattern.

4 HMM-Based Recognition

Hidden Markov models (HMMs) are unique mechanism for data classifications in pattern recognition such as speech, face, character, texture analysis presented by Bobulski and Kubanek [12]. Bobulski and Kubanek [12] have presented the traditional Markov model which have complicated structure with respect to the model parameter. This process is quite complex with its feature vectors of the input patterns, due to the scalability issue. Janusz Bobulski [13] stated the problems faced in the work [12] can overcome by using the extended version of 1-D HMM for pattern recognition applications.

4.1 1-D HMM

A HMM involves with a stochastic mathematical modeling approach for pattern recognition applications [14]. Bobulski [12] have analyzed the temporal pattern sequence by using 1-D HMM. This model have their state sequence in linear passion. Figure 2 shows the structure of 1-D HMM which contains four states. Each states link to an adjacent state in forwarding direction either in horizontal or in vertical called as left–right model.

Statement 1 Determine $P(O|\lambda)$, by observing $O = \{o_1, o_2, \ldots, o_T\}$ and model $\lambda = \{A, B, \pi\}$

Statement 2 Determine model parameters $\lambda = \{A, B, \pi\}$ that increases $P(O|\lambda)$.

The statement 1 can be solved by forward-backward approach [15], whereas the statement 2 can be solved by Baum-Welch approach [12].

Fig. 2 Structure of 1-D HMM

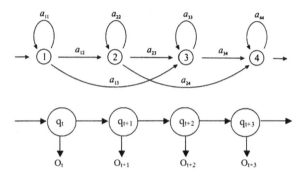

4.2   2-D HMM

Jai Li [14] have proposed the extended version of 1-D HMM called as 2-D HMM as shown in Fig. 3, to overcome the dimensionality problem of input patterns. This reduces the memory storage in the HMM model. Consider 'N' represents the number of states so it have N^2 possible state variable in the model. From each state in the HMM determine the observation state, transition state and self-looping state sequence. The transition state probability changes with respect to time with $N = 2$ which produces the classified output without any loss of information.

The parameter re-estimation algorithm solves the statement 2 [16]. At time 't' the initial state at ith and jth sequence in the HMM model increments the data sequence by 't + 1' as given in Eq. 2.

$$\xi(i,j,l) = \frac{\alpha_t(i,j,k)\, a_{ijl}b_{ij}(o_{t+1})\beta_{t+1}(i,j,k)}{P(O|\lambda)} \tag{2}$$

The observation sequence from the state probability at the state i, j, l, t is given by $\gamma_t(i,j)$

$$\gamma_t(i,j) = \sum_{l=1}^{T}(i,j,l) \tag{3}$$

Suppose the data $y = \{(y_1, y_2 \ldots y_n)^T\}$ have to be observed as to fit with a posterior model by maximizing log-likelihood decision rule.

$$l(\theta, y) = \log p(y|\theta) \tag{4}$$

Let us consider a hidden variable x is known and explicit form $p(y|\theta)$ is unknown. There is more difficulty in maximization of $l(\theta, y)$ because the log term cannot be further reduced. Alternatively, Expectation-Maximization (EM)

Fig. 3 Structure of 2-D HMM

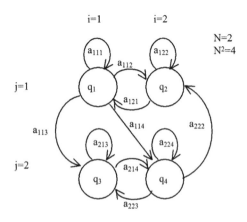

algorithm is used in the pattern recognition is to determine maximum likelihood in the model, where the model depends on the hidden state sequences [17]. If the joint probability $p(y, x|\theta)$ falls into exponential function, then it should be expressed as

$$p(y, x|\theta) = \exp\{\langle g(\theta), T(y, x)\rangle + d(\theta) + s(y, x)\} \tag{5}$$

5 Experimental Results and Discussion

The dataset consists of 300 hand images from 5 classes; each class consists of 60 images in different angle with 320×240 gray-level images. In this work, edge detection is used to identifying and locating sharp discontinuities in the image which in turn characterized boundaries of an object in the image. Figure 4 shows the difference between four type operators and the filters that have been tried to detect the edges of an object. The reason why Canny filter is not used is because of its high sensitivity which can detect even minute irregularities that lead to improper edge detection.

Both Prewitt Fig. 4b and Sobel operators Fig. 4a were efficient in detecting the objects. Instead of Prewitt, Sobel operator is preferred to avoid false edge detection.

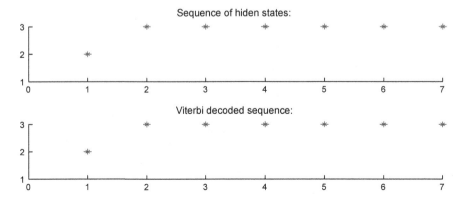

(a) Sobel filter   (b) Prewitt filter   (c) Canny filter   (d) Roberts filter

Fig. 4 Preprocessed hand gestures. **a** Sobel filter **b** Prewitt filter **c** Canny filter **d** Robert filter

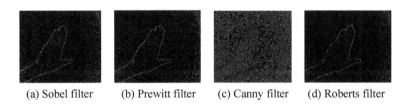

Fig. 5 Testing and evaluation of 1-D HMM

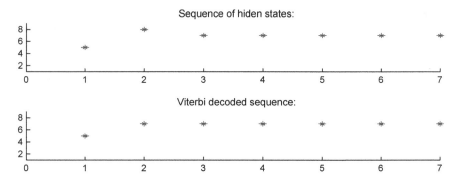

Fig. 6 Testing and evaluation of 2-D HMM

Table 1 Comparative analysis of 1-D HMM and 2-D HMM

Technique	No. of objects in dataset	No. of objects recognized	Recognition rate (%)
1-D HMM	300	255	85
2-D HMM	300	270	90

5.1 1-D HMM

The longest path is to travel from one state to another state by using Viterbi algorithm. This leads to the loss of information from two-dimensional quality of an image. At last it recognized the wrong part of the hand image from the dataset. This is due to the recursive operation in both forward and backward estimation analysis.

Figure 5 shows the result of testing and evaluation of 1-D HMM. All input data are passed to HMM classifier after feature extraction. The plot for sequence of hidden states and Viterbi decoded path is shown. It infers that the hidden state result perfectly matched with the Viterbi decoded path data of {2, 3, 3, 3, 3, 3, 3} at 30th iteration.

5.2 2-D HMM

The two-dimensional data are accounted for using 2-D model, to overcome the problem in 1-D HMM. This leads to improve its robustness of the detection against size variations. Figure 6 shows the result of testing and evaluation of 2-D HMM. All input data are passed to 2-D HMM classifier after feature extraction. The plot for sequence of hidden states and Viterbi decoded path is shown. It infers that the hidden state result perfectly matched with the Viterbi decoded path data of {5, 7, 7, 7, 7, 7, 7} at 30th iteration.

Table 1 shows that the comparative analysis of 1-D HMM and 2-D HMM of hand gestures in temporal pattern. Equation 6 represents the mathematical formula to calculate the recognition rate.

$$\text{Recognition rate } (\%) = \frac{\text{No. of objects recognized} \times 100}{\text{Total no. of objects present}} \tag{6}$$

6 Conclusions

This research work is an attempt to explore the advantage of HMM approach for hand gesture recognition system for virtual reality applications. The simulation result reveals 5% improvement for 2-D HMM over 1-D HMM in terms of recognition rate. The computational time of 2-D HMM is very much less comparing to 1-D HMM. As a future work, modification in the existing algorithms can be performed to enhance the performance of the hand gesture recognition system for gaming application in virtual environment. Hybrid HMM models can also be developed for such applications to improve the performance metrics of the system.

References

1. Mokhtar M. Hasan and Pramoud K. Misra, Brightness factor matching for Gesture Recognition system using Scaled Normalization, International Journal of Computer Science & Information Technology (IJCSIT), vol. 3, no. 2, April 2011.
2. Sushmita Mitra and Tinku Acharya, Gesture Recognition: A Survey, IEEE Trans. on Systems, Man, and Cybernetics, Part C: Applications and Review, vol. 37, no. 3, pp. 311–324, 2007.
3. Mokhtar M. Hasan and Pramoud K. Misra, Robust Gesture Recognition using Euclidian Distance, IEEE International Conference on Computer and Computational Intelligence, China, vol. 3, pp. 38–46, 2010.
4. Mokhtar M. Hasan and Pramoud K. Misra, HSV Brightness factor matching for Gesture Recognition system, International Journal of Image Processing, Malaysia, vol. 4, no. 5, pp. 456–467, 2010.
5. P. A. Devijver, Modeling of digital images using hidden Markov mesh random fields, Signal Processing IV: Theories and Applications, pp. 23–28, 1998.
6. J. Li, A. Najimi and R. M. Gray, Image classification by a two-dimensional hidden Markov model, IEEE Trans. on Signal Processing 48, pp. 517–533, 2000.
7. D. Schonfeld and N. Bouaynaya, A new method for multidimensional optimization and its application in image and video processing, IEEE Signal Processing Letters 13, pp. 485–488, 2006.
8. Yuri Grinberg and Theodore J. Perkins, State Sequence Analysis in Hidden Markov Model, grant from the Natural Sciences and Engineering Research Council of Canada (NSERC).
9. Raman, Himanshu, Study and Comparison of Various Image Edge Detection Techniques, International Journal of Image Processing (IJIP), vol. 3, Issue 1, 2009.
10. Arindam Misra, Abe Takashi, Takayuki Okatani, Koichiro Deguchi, Hand Gesture Recognition using Histogram of Oriented Gradients and Partial Least Squares Regression, IAPR Conference on Machine Vision Applications, Japan, pp. 479–482, 2011.

11. Neha V. Tavari, A. V. Deorankar, Indian Sign Language Recognition based on Histograms of Oriented Gradient, International Journal of Computer Science and Information Technologies, vol. 5(3), ISSN: 0975-9646, 2014.
12. Bobulski. J, Kubanek. M, Person identification system using sketch of the suspect, Optica Applicata, 4(42), pp. 865–873, 2012.
13. Janusz Bobulski, Comparison of the effectiveness of 1D and 2D HMM in the pattern recognition, Image Processing & Communication, vol. 19, no. 1, pp. 5–12, 2015.
14. Jai Li, Gray, Robert M., Image Segmentation and Compression using Hidden Markov Model, Springer International Series in Engineering and Computer Science, 2000.
15. Rabiner. L.R., A tutorial on hidden Markov models and selected application in speech recognition, Proceedings of the IEEE, 77(2), pp. 257–285, 1989.
16. Bobulski. J, Adrjanowicz. L, Part I. In Artificial Intelligence and Soft Computing, Springer publication, pp. 515–523, 2013.
17. E. de Souza e Silva, R. M. M. Leao and Richard R. Muntz, Performance Evaluation with Hidden Markov Models, International Federation for Information Processing (IFIP), LNCS 6821, pp. 112–128, 2011.

Signature Classification Using Image Moments

Akhilesh Kushwaha, Aruni Singh and Satyendra Kumar Shrivastav

Abstract Here, in this contribution we want to demonstrate the person's identity by their behavioral biometric characteristics, which is the signature of the person. The shape of the signature is used for the evaluation purpose to classify the signatures. For the feature extraction of signatures, we have explored the idea of Hu and Zernike moments. The experiments incorporates the classification using SVM and found accuracy 81–96% for online signature data and 59–72% for offline signature data.

Keywords Moments · Hu · Zernike · SVM · Biometrics

1 Introduction

The most powerful technique for the identification of the user is the third-level security which is called biometric. In the real-life scenario, biometric techniques are also acceptable and intrusive throughout the world. That is why so many researches are going on for the classification of real user with the imposter. Very often, signatures are used for authentication of a person. There are two kinds of signature verification techniques used online and offline. Stylus, touch screen, digitizer, etc. are generally used to obtain dynamic values of online signature, to extract their properties such as coordinate values, pressure, time, and speed. This technique achieves high accuracy and is difficult to forge due to its dynamic nature. In the offline signature verification techniques, image of the signatures done on the paper are scanned which are taken as snapshot. In the absence of stable dynamic characteristic of processing

A. Kushwaha (✉) · A. Singh · S.K. Shrivastav
K.N.I.T, Sultanpur, Uttar Pradesh, India
e-mail: akhilesh.16august@gmail.com

A. Singh
e-mail: arunisingh@rocketmail.com

S.K. Shrivastav
e-mail: ssknit1@gmail.com

© Springer Nature Singapore Pte Ltd. 2018 235
P.K. Sa et al. (eds.), *Progress in Intelligent Computing Techniques: Theory,*
Practice, and Applications, Advances in Intelligent Systems and Computing 518,
DOI 10.1007/978-981-10-3373-5_23

is complex due to containing a lot of noise in offline signature comparison to online signature. In this work, image moments (geometric moments and Zernike moments) use as a classification feature for online and offline signature database. In 1961, Hu et al. demonstrated to explode the properties of image moments for analysis and instance representation [1]. They used geometric moments to derive a set of invariants which results in automatic character recognition. During four decades from 1962, so many researchers have classified the acquired object from their images using image moments such as Smith et al. identified a ship in 1971 [2], Dudani et al. identify an aircraft [3], and Dirilten et al. a patten matching in 1977 under affine transformation [4]. Wong et al. classified the image seen in 1978 [5], and in 1980 Teague et al. introduced the orthogonal moments to explore the basic concept of Legendre and Zernike moments [6]. In 1986, Sheng et al. also demonstrated Fourier-million descriptor which provides the generalized model to derive the invariants [7]. In 2003, P. Yap et al. introduced a new group of orthogonal moments depending on the discrete classical Krawtchouk polynomial [8].

The number of researches have been incorporated for online and offline signature verification. Pippin et al. proposed various filters to find velocity-based stroke segmentation for encoding [9]. F.A. Afsar et al. demonstrate that the global features are derived from the spatial coordinates in the data acquisition stage. One-dimensional wavelet transform [10] is used here. C.F. LAM et al. acquired X and Y coordinates from the normalized signatures and transformed into the frequency domain using fast Fourier transformation technique [11]. Alisher et al. worked for online signature verification and described problems such as two-class pattern recognition problem [12]. The idea from the above literature has motivated our thought to work in this scope because it filters high-level global and local characteristics of the image of shape. In this work, we consider signature shape the pivot for feature extraction.

2 Methodology

A set of moments computed from a digital image keeps global characteristic of the image shape and provides a lot of information about different types of geometrical feature of image.

Moments Definition [13]

$$\phi_{pq} = \iint_{\zeta} \psi_{pq} I(x, y) dx dy \tag{1}$$

where p and q are +ve natural numbers, $I(x, y)$ is two-dimensional image density function, ζ denotes the image region of the x–y plane, ϕ_{pq} order of moments (p + q), and ψ_{pq} the moment weighting kernel or the basis set.

2.1 Geometry Moments

Geometric moments [14] are the simplest among moments functions. The two-dimensional geometric moment of order $P + Q$ of a function $I(x, y)$ is defined as

$$M_{PQ} = \int_{a_1}^{a_2} \int_{b_1}^{b_2} x^P y^Q I(x, y) dx \, dy \tag{2}$$

where $x^P y^Q$ is the kernel of geometric moments.

In 1962, six absolute and one skew orthogonal invariant equations were derived by Hu [1] based upon geometric invariants. These equations are independent of parallel projection as well as position, size, and orientation. Seven equations are given below

$$(\Psi_1) = v_{20} + v_{02}. \tag{3}$$

$$(\Psi_2) = (v_{20} - v_{02})^2 + 4v_{11}^2. \tag{4}$$

$$(\Psi_3) = (v_{30} - 3v_{12})^2 + (3v_{21} - v_{03})^2. \tag{5}$$

$$(\Psi_4) = (v_{30} - v_{12})^2 + (v_{21} - v_{03})^2. \tag{6}$$

$$(\Psi_5) = (v_{30} - 3v_{12})(v_{30} + v_{12})[(v_{30} + v_{12})^2 - 3(v_{21} + v_{03})^2] \\ + (3v_{21} - v_{03})(v_{21} + v_{03})[3(v_{30} + v_{12})^2 (v_{21} + v_{03})^2] \tag{7}$$

$$(\Psi_6) = (v_{20} - v_{02})[(v_{30} - v_{12})^2 - (v_{21} + v_{03})^2] + 4v_{11}(v_{30} - v_{12})(v_{21} + v_{03}) \tag{8}$$

$$(\Psi_7) = (3v_{21} - v_{30})(v_{30} + v_{12})[(v_{30} + v_{12})^2 - 3(v_{21} + v_{30})^2] \\ - (v_{30} - 3v12)(v_{21} + v_{03})[3(v_{30} + v_{12})^2 - (v_{21} + v_{30})^2 \tag{9}$$

where v is the central moment of an image defined as:

$$(v_{PQ}) = \int_{a_1}^{a_2} \int_{b_1}^{b_2} (x - \bar{X})^P (y - \bar{Y})^Q I(x, y) dx \, dy \quad \text{where } \bar{X} = \frac{M_{10}}{M_{00}} \text{ and } \bar{Y} = \frac{M_{01}}{M_{00}}. \tag{10}$$

2.2 Zernike Moments

In 1990, Zernike introduced special polynomial [14], which is a set of orthogonal functions with simple rotational properties. This polynomial is defined as:

$$\xi_{nm}(x, y) = \xi_{nm}(\rho \sin \theta, \rho \cos \theta) = \bar{R}_{nm}(\rho)e^{jm\theta} \tag{11}$$

where n is non-negative, and m is chosen such that $n - |m| = even$ and $|m| <= n$. The radial polynomial $\bar{R}_{nm}(\rho)$ is defined as:

$$\bar{R}_{nm}(\rho) = \sum_{s=0}^{(n-|x|)/2} (-1^s)\frac{(n - s)!}{s!(\frac{n+|m|}{2} - s)!(\frac{n-|m|}{2} - s)!}\rho^{n-2s} \tag{12}$$

with $\bar{R}_{n(-m)}(\rho) = \bar{R}_{nm}(\rho)$.

Zernike Moments Zernike moments of order of n with repetition of m for an image $I(x, y)$ are defined as

$$Z_{nm} = \frac{n + 1}{\pi} \int \int_{x^2+y^2 <=1} I(x, y)\xi_{nm}^*(\rho, \theta)dx\, dy \tag{13}$$

or, in polar coordinate

$$Z_{nm} = \frac{n + 1}{\pi} \int_0^{2\pi} \int_0^1 I(\rho, \theta)\bar{R}_{nm}(\rho)e^{-jm\theta} \rho d\rho d\theta \tag{14}$$

Pseudo-Zernike Moments In Zernike polynomial definition, if we eliminate the condition $n - |m| = even$, then ξ_{nm} becomes the set of pseudo-Zernike polynomials [14]. Radial polynomial $\bar{R}_{nm}(\rho)$ for pseudo-Zernike moments is defined as

$$\bar{R}_{nm}(\rho) = \sum_{s=0}^{n-|m|} (-1)^s \frac{(2n + 1 - s)!}{s!(n - |m| - s)!(n + |m| + 1 - s)}\rho^{n-s} \tag{15}$$

where n is +ve natural number and m is natural number with condition $|m| <= n$.

Fig. 1 Block diagram of experiment evolution step

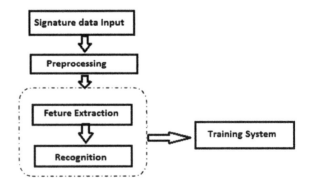

3   Experimental Evolution

We have carried out our experiment in four phases (which is shown in Fig. 1):

1. Preprocessing.
2. Feature Selection.
3. Feature Extraction.
4. Training and Classification.

3.1   Preprocessing

The obtained images are raw and not upto the label for experimental work. Therefore, we need to some preprocessing using factors such as alignment's, illumination and resizing for the benchmark experimental protocol. We used coordinates (x, y) for reconstructing a signature image for online signature database. We are also resizing the image of signature to minimize computational cost, and in offline signature preprocessing we used noise removal (morphological operation) and resizing.

3.2   Feature Selection

Choosing appropriate features for any classification technique is a very important task to design a good classifier. If we choose a wrong feature, then our result will not be according to our expectations. There are so many classification techniques and feature extraction methods for signature matching, but we use image moments for feature selection. In this paper, we are using SVM as a classifier for online and offline signature classification. Image moments can be classified into two types based on their kernel. Moments with simple kernel is called geometric moments, and moments with complex kernel is called complex moments. In this paper, we used both simple moments (Hu) and complex moments (Zernike) for our experimental work.

3.3   Feature Extraction

After selecting feature for classification, we need to extract those features from signature. Implementation of Hu equations is simple compared to Zernike moments because of its kernel. We store our result in the form of two-dimensional matrix $a \times b$. Here, a is the number of signature in database, and b is the number of extracted features whose values vary according their moments' technique.

3.4 SVM Training and Classification

After extracting features from the signature, the next step is to classify signature with the help of extracting features. We used the LibSVM package to train and classify our database. LibSVM packages are freely available at [15]. We used radial base function (RBF) as a kernel function for SVM. We are using 5-, 10-, 15-, and 20-fold cross validation for performing a grid search to find the optimal cost and Gamma parameters.

4 Database Profile and Result

4.1 Online Database

We used standard database that is publicly for "First International Signature Verification Competition" called SVC2004 in July 2004. This database is available at [16]. Here, select database consists of 100 sample signatures by 5 different users. Each signature has seven features, and each feature has a separate text file. We only choose coordinate information of the signature, and after this we construct an image with the help of these coordinates value and save it in JPG format. So now we have 100 images of signature, done by 5 different users 20 of each. Figure 2 shows the sample of signature images.

Fig. 2 Sample of offline signature

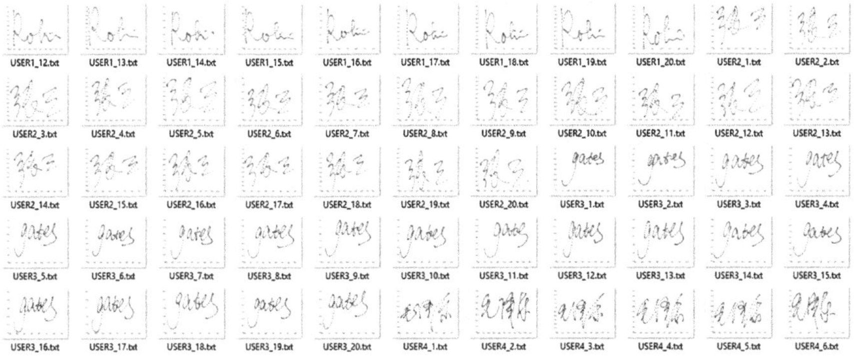

Fig. 3 Sample of online signature

4.2 Offline Database

Two hundred signatures from 10 different people have been collected for our experiment. For a collection of offline database, A4 size sheet of paper is given to the signers, and each signer was requested to give their signature (20 signature images from each signer). Therefore, 200 signatures were collected for our experiment. Figure 3 shows the sample of offline signature database.

5 Result

5.1 Online Database

Here, features were extracted by us in two ways from online database:

1. On original image size that is 1200×900
2. After resizing, image size that is 150×900

After feature extraction, we have got four extracted matrices $D_{H1}, D_{H2}, D_{Z1},$ and D_{Z2}.

D_{H1} matrix contains HU feature from original signature size, and D_{H2} matrix contains the features for the resized signatures database. D_{H1} and D_{H2} matrices have a size 100×7. Here are 100 signatures, and each has 7 features extracted by Hu moments.

Similarly, D_{Z1} matrix contains Zernike feature from original signature size and D_{Z2} matrix for the resized signature database. D_{Z1} and D_{Z2} matrices have a size 100×441, where 100 is the number of signatures in database and each has 441 features extracting by Zernike moments. The Libsvm package was used to train and classify signature. We are using radial base function (RBF) as a kernel function for SVM.

Table 1 Hu result online signature

HU result online signature				
Fold size use to train SVM	5	10	15	20
Accuracy of image size 1200×900	87	89	92	93
Accuracy of image size 150×900	81	84	85	86

Table 2 Zernike result online signature

Zernike result online signature				
Fold size use to train SVM	5	10	15	20
Accuracy of image size 1200×900	94	95	95	96
Accuracy of image size 150×900	87	87	88	89

Table 1 represents classifier accuracy of Hu moments with different fold training in online signature database. Similarly, Table 2 represents the accuracy of Zernike moments with different fold training in online signature database.

5.2 Offline Database

After feature extraction of offline signature database, we have two matrices: for Hu moments we have D_H, and for Zernike moments we have D_Z. Our database has 200 signatures. So D_H matrix has a size 200×7, where 200 is the number of signatures in the database, and each has 7 features extracted by Hu moments. Similarly, D_Z matrix has a size 200×441, where 200 is the number of signatures in the database, and each has 441 features extracting by Zernike moments. Table 3 represents the accuracy result of offline signature database a different fold training, and Table 4 represents the result for Zernike moments.

Table 3 Hu result offline signatures

HU result offline signature				
Fold size use to train SVM	5	10	15	20
Accuracy of image size 150×900	59	61	61	63

Table 4 Zernike result offline signatures

Zernike result offline signature				
Fold size use to train SVM	5	10	15	20
Accuracy of Image Size 150×900	67	69	71	72

6 Conclusion and Discussion

As we got the information about the efficiency for online signature verification which is better than the offline signature verification due to their dynamic properties. In the same way, when we considered the signature as a shape of object and use Hu and Zernike moments for our assertion and for signature verification, so many complex modules are generally used to evaluate the dynamic properties of the signature such as pressure azimuth, altitude, tangent, and slope. But in this work, those complex properties of the signatures are not considered, but the verification accuracy leads to maturity value which are 81–96% for online data and 59–72% for offline database.

In this paper since we have not incorporated the dynamic signature properties but the experimental results shows the consideration of these properties will be beneficial better in future.

References

1. Hu, M.K.: Visual pattern recognition by moment invariants. Information Theory, IRE Transactions on **8**(2) (1962) 179–187
2. Smith, F.W., Wright, M.H.: Automatic ship photo interpretation by the method of moments. Computers, IEEE Transactions on **100**(9) (1971) 1089–1095
3. Dudani, S.A., Breeding, K.J., McGhee, R.B.: Aircraft identification by moment invariants. Computers, IEEE Transactions on **100**(1) (1977) 39–46
4. Dirilten, H., Newman, T.G.: Pattern matching under affine transformations. Computers, IEEE Transactions on **100**(3) (1977) 314–317
5. Wong, R.Y., Hall, E.L.: Scene matching with invariant moments. Computer Graphics and Image Processing **8**(1) (1978) 16–24
6. Teague, M.R.: Image analysis via the general theory of moments. J. Opt. Soc. Am **70**(8) (1980) 920–930
7. Sheng, Y., Arsenault, H.H.: Experiments on pattern recognition using invariant fourier-mellin descriptors. JOSA A **3**(6) (1986) 771–776
8. Yap, P.T., Paramesran, R., Ong, S.H.: Image analysis by krawtchouk moments. Image Processing, IEEE Transactions on **12**(11) (2003) 1367–1377
9. Pippin, C.E.: Dynamic signature verification using local and global features. Georgia Institute of Technology (2004)
10. Afsar, F., Arif, M., Farrukh, U.: Wavelet transform based global features for online signature recognition. In: 9th International Multitopic Conference, IEEE INMIC 2005, IEEE (2005) 1–6
11. Lam, C.F., Kamins, D., Zimmermann, K.: Signature recognition through spectral analysis. In: Acoustics, Speech, and Signal Processing, IEEE International Conference on ICASSP'87. Volume 12., IEEE (1987) 1790–1792
12. El-Henawy, I., Rashad, M., Nomir, O., Ahmed, K.: Online signature verification: State of the art. International Journal of Computers & Technology **4**(2c) (2013)
13. Liao, S.X., Pawlak, M.: On image analysis by moments. Pattern analysis and machine intelligence, IEEE Transactions on **18**(3) (1996) 254–266
14. Khotanzad, A., Hong, Y.H.: Invariant image recognition by zernike moments. Pattern Analysis and Machine Intelligence, IEEE Transactions on **12**(5) (1990) 489–497
15. Chang, C.C., Lin, C.J.: LIBSVM: A library for support vector machines. ACM Transactions on Intelligent Systems and Technology **2** (2011) 27:1–27:27 Software available at http://www.csie.ntu.edu.tw/~cjlin/libsvm

16. Yan Yeung, Hong Chang, Y.X.S.G.R.K.T.M., Rigoll, G.: Online signature database. SVC2004: First International Signature Verification Competition. Proceedings of the International Conference on Biometric Authentication (ICBA), Hong Kong, 15–17 July 2004.) **Svc2004** (2004)

A Bengali Handwritten Vowels Recognition Scheme Based on the Detection of Structural Anatomy of the Characters

Priyanka Das, Tanmoy Dasgupta and Samar Bhattacharya

Abstract The present paper describes the development of a novel scheme for recognising Bengali handwritten vowels using mathematical morphology where the characters are first categorised on the basis of their anatomical features. The scanned images of the characters are passed through a weighted decision tree which is designed to analyse the feature set present in them. Based on the detected features, the algorithms easily recognise the individual characters. The extraction of some features has been carried out by generating the curvature scale spaces for the characters. The treatment is performed on scanned binary images of the handwritten characters to detect anatomical features such as 'bowl', 'lobe' and 'arm'. The scheme developed here is very fast and it does not need training samples to work.

Keywords Handwritten character recognition · Curvature scale space · Mathematical morphology

1 Introduction

Scripts used in many languages often share a common ancestry. For instance, Bengali, which is used by more than a hundred million people in the world [1], uses a script that originated from an ancient derivative of the *Brahmi* script, known as *Siddham*. Brahmi is known to be the origin of almost all Indic scripts. Thus, a closer look into the structures of Bengali letterforms easily reveals the coherent presence of

P. Das (✉)
Department of Electronics and Communication Engineering,
Techno India University, Kolkata, West Bengal, India
e-mail: priyankadas700@gmail.com

T. Dasgupta · S. Bhattacharya
Department of Electrical Engineering, Jadavpur University, Kolkata, West Bengal, India
e-mail: thetdg@live.com

S. Bhattacharya
e-mail: samar_bhattacharya@ee.jdvu.ac.in

© Springer Nature Singapore Pte Ltd. 2018
P.K. Sa et al. (eds.), *Progress in Intelligent Computing Techniques: Theory,
Practice, and Applications*, Advances in Intelligent Systems and Computing 518,
DOI 10.1007/978-981-10-3373-5_24

245

some consistent anatomical features existing in other Indic languages as well. Over the previous few decades, the anatomical structures of different Latin and non-Latin scripts have been extensively studied by many researchers [2–4].

Analysis of the structural features of the letterforms is often successfully used in handwritten character recognition. In 2006, Chowdhury et al. [5] developed a method for recognising Bengali handwritten numerals where the characters are modelled as water reservoirs. In 2011, Mandal [6] developed a Bengali handwritten character recognition scheme based on the analysis of gradient features. Recently, Das et al. developed a method for recognising handwritten Bengali numerals using mathematical morphology [7].

The present work is influenced from the semiotic study on Bengali scripts [1] and also from an approach for curvature estimation based on bicubic spline interpolation [8]. The present paper proposes algorithms that analyse the anatomical features of the handwritten Bengali scripts to recognise them. The terminology used in this work are derived from animal and plant anatomical nomenclatures such as stem, leg, tail and nose. The algorithms also accentuate on an efficient shape descriptor technique called 'morphological curvature scale space' for detecting some anatomical features present in some of the Bengali vowels. The theory of scale space is a fundamental utility for mathematically describing and analysing different shapes [9], and with automatic scale selection, many structural features can be detected [10]. An extensive study was suggested by Sai Anand et al. [11] discussing various types of curvature scale spaces.

The remaining sections of the paper are organised as follows: Starting with a brief overview of a few useful morphological operations, Sect. 2 describes the structural features of the Bengali vowels that are later used for recognising them. Next, the presence of different types of strokes in the scanned images of handwritten characters is determined by generating morphological curvature scale space, which is described in Sect. 3. This is followed by a discussion on the experimentally obtained results in Sect. 4. The conclusive remarks are presented in Sect. 5.

2 Structural Features of Bengali Vowels

2.1 Morphological Operations

Mathematical morphology can be used as a powerful yet simple tool for extracting different features from an image. In character recognition processes, it might sometimes be necessary to connect the broken parts of the handwritten character image using a morphological operation called dilation with the help of a suitable structuring element (SE). The morphological dilation of a set E by another set S (SE) is defined as $E \oplus S = \{z : (\hat{S})_z \cap E^c \neq \phi\}$, where $\hat{S} = \{w : w = -b, \forall b \in S\}$, is the reflection of the set $S \subseteq \mathbb{Z}^2$. Here, the existence of the blobs is detected using the region filling method. Let there be a connected region, and assume that its

8-connected boundary points belong to a set A. To perform the morphological region filling operation, first a point q inside the boundary is assigned to a value of 1. The pixel inside the region is then filled with the same intensity value as that of the point q using $X_k = (X_{k-1} \oplus B) \cap A^c$, for $k \in \mathbb{N}$, where $X_0 = q$, and B a symmetric SE of order 3×3. The condition for the termination of this iterative method at the k-th step is $X_0 = X_{k-1}$. The region-filled set along with its boundary will then be a subset of $X_k \cup A$.

2.2 Anatomy of Bengali Vowels

The list of Bengali vowels and their anatomical features is shown in Figs. 1 and 2, respectively. The present approach uses the typeface *Kalpurush* among the different typefaces available to set up a basis for defining the anatomical features in Bengali vowels. Also, the Bengali vowel set is accompanied with their corresponding vowel sounds to make it more meaningful for non-Bengali readers. This list is accompanied with Table 1 which captures the character features at a glance.

Any scanned image containing handwritten characters cannot always be used reliably as the input to the detection scheme due to the presence of noise in them, and thus, it requires some preprocessing techniques, including the removal of noises using a suitable filter. The image can then be converted to a binary image suitable for feeding into the decision tree.

অ	আ	ই	ঈ	উ	ঊ	ঋ	এ	ঐ	ও	ঔ
a	ā	i	ī	u	ū	ṛ	e	ai	o	ou

Fig. 1 List of the Bengali vowels

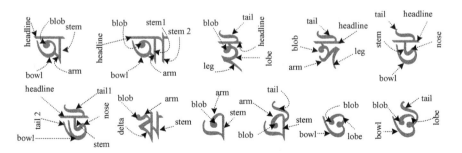

Fig. 2 Anatomical features of the Bengali vowels

Table 1 List of anatomical features of Bengali vowels

Vowel	Matra	Stem	Blobs	Bowl	Lobe	Tail	Arm	Leg	Nose	Delta
a	✓ (1)	✓ (1)	✓ (1)	✓ (1)						
ā	✓ (1)	✓ (2)	✓ (1)	✓ (1)						
i	✓ (1)		✓ (1)		✓ (1)	✓ (1)		✓ (1)		
ī	✓ (1)		✓ (1)			✓ (1)	✓ (1)	✓ (1)		
u	✓ (1)	✓ (1)		✓ (1)		✓ (1)			✓ (1)	
ū	✓ (1)	✓ (1)		✓ (1)		✓ (2)			✓ (1)	
ɽ		✓ (1)	✓ (1)				✓ (1)			✓ (1)
e		✓ (1)	✓ (1)				✓ (1)			
ai		✓ (1)	✓ (1)			✓ (1)	✓ (1)			
o			✓ (1)	✓ (1)	✓ (1)					
ou			✓ (1)	✓ (1)	✓ (1)	✓ (1)				

To construct the feature space of the characters, a 10-digit 'flag $\mathcal{F}$' is designed that can hold the anatomical information. The individual scanned characters undergo a process that separately tests it for the presence of all the pre-described anatomical or structural features. The details of the process are described in Algorithm 1.

3 Curvature Detection

The presence of different anatomical features in the characters are detected using a curvature detection scheme depicted in this section.

Let τ be a smooth parametric two-dimensional curve $\tau(p) = (x(p), y(p))$ with parameter p. It is known that the normalised arc-length independent curvature of $\tau(p)$ is given by $\lambda(p) = (\dot{x}\ddot{y} - \ddot{x}\dot{y}) / (\dot{x}^2 + \dot{y}^2)^{3/2}$. If p is replaced by the normalised arc-length parameter s, then $\lambda(s) = \dot{x}\ddot{y} - \ddot{x}\dot{y}$.

The normalised curvature function can be, thus, computed just from the parametric derivatives of the contours. However, the curvature measure is inversely proportional to the scale. Scale independence can be achieved by normalising the measure by defining the mean absolute curvature as $\lambda'(s) := \frac{\lambda(s)}{\int_0^b |\lambda(s)| ds}$. Thus, the curvature function becomes invariant under any composite homogeneous transformation.

To remove the irregularities in the strokes, a cubic spline interpolation is done, providing C^2 regularity in each point present in the curve.

The curvature function associated with a scanned binary image of a character is considered to be the parametric Euclidean distance function from a fixed point. The location of the fixed point is crucial for the proper detection of the shape of the strokes present in the character. The fixed point is determined to be a point where the centres of curvature for different segments of the character contour tend to converge within

Algorithm 1: Calculation of the flag $\mathcal{F}$ that describes the feature set

Input: Scanned image of input digit
Output: The flag $\mathcal{F}$ containing the feature set of the scanned characters.
begin

  do morphological region filling;

  if *(Headline) a horizontal stem on top of the character is present* **then**
  | change the 1st element of $\mathcal{F}$ to 1;
  end

  if *(Stems) stems on other positions are present* **then**
  | count the number of stems;
  | store the number in the 2nd element of $\mathcal{F}$;
  end

  if *(Blob) blob is present* **then**
  | change the 3rd element of $\mathcal{F}$ to 1;
  end

  if *(Bowl) a curvilinear stroke of around 180 – 360 degrees is present* **then**
  | change the 4th element of $\mathcal{F}$ to 1;
  end

  if *(Lobe) a curvilinear stroke within bound of 90 – 180 degrees is present* **then**
  | change the 5th element of $\mathcal{F}$ to 1;
  end

  if *(Tails) strokes coming out from the main letter part individually are present* **then**
  | count the number of strokes;
  | store the number of strokes in the 6th element of $\mathcal{F}$;
  end

  if *(Arm) a bounded curvilinear stroke of 30–90 degrees is present* **then**
  | change the 7th element of $\mathcal{F}$ to 1;
  end

  if *(Leg) a small balanced stroke coming out from the body part is present* **then**
  | change the 8th element of $\mathcal{F}$ to 1;
  end

  if *(Nose) a junction of two differently oriented strokes is present* **then**
  | change the 9th element of $\mathcal{F}$ to 1;
  end

  if *(Delta) a connected triangular blob is present* **then**
  | change the 10th element of $\mathcal{F}$ to 1;
  end
  else
  | keep $\mathcal{F}$ unchanged;
  end
  Return the flag $\mathcal{F}$;
end

Algorithm 2: Calculation of the curvature distance function

Input: Binary image containing a handwritten Bengali vowel : S
Output: Distance function **k**
begin

    Calculate the curvature along different strokes of the character;

    Calculate the centre of curvature for fixed lengths in the strokes;

    if *The center for a reasonably good portion of the curvature is in a small neighbourhood of a point* **then**

        select the point as the centre of the character C_{xy};

        create a small vicinity of black pixels (intensity = 0) around the centre;

        create a blank image $\mathcal{E}$ of the same size as S having only black pixels (intensity = 0);

        set $k = 0$ (angle of the line to be drawn);

        while $k \leq 360$ **do**

            draw a 2-pixel wide white line (intensity = 1) at k degree angle in $\mathcal{E}$;

            calculate $\mathcal{R} = \mathcal{E} \oplus S$;

            find the locations of pixels (x_i, y_i) with intensity = 1 in $\mathcal{R}$;

            calculate $d[k] = \min \sqrt{(C_{xy}[0] - x_i)^2 + C_{xy}[1] - y_i)^2}$;

            increment the angle of the white line (k) by 1 degree;

        end

    end

    Return the distance function **k**;

end

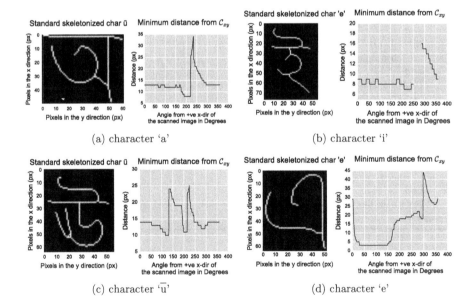

(a) character 'a'

(b) character 'i'

(c) character 'u'

(d) character 'e'

Fig. 3 Curvature distance functions for a few Bengali vowels

a small neighbourhood. Once a suitable fixed point is located, the angular spans of different curve sections of the character contours are determined by creating a blank figure of the same size and spanning a thin line of white pixels over the whole image centring at the fixed point and subsequently ANDing it with the original binary image containing the scanned character. The details of the process are depicted in Algorithm 2.

Once the curvature distance function has been calculated for a scanned character image, the function is tested for detecting the presence of curvilinear strokes spanning a certain amount of angle. This is further used to detect the presence of some features such as lobe, bowl and arm.

Figure 3 depicts the distance function for the Bengali vowels 'a', 'i', 'ū' and 'e'. The distance functions are calculated from the measured centres denoted by C_{xy}.

4 Experimental Results

The above algorithm is implemented using the Python binding of OpenCV. A set of Bengali handwritten vowel samples is collected from a few students and faculty members in the institution the first author is affiliated with. Since Algorithm 1 iterates on the scanned character image at most 10 times and as all the processes are vectorised, the run time of the algorithm is dependent on the features that are being recognised on each iteration as provided in Algorithm 2. Table 2 depicts the runtime and accuracy of recognition of different Bengali vowels while the programmes are executed in a machine with an Intel Celeron processor running at 1.5 GHz.

As the method depicted here does not require any kind of machine learning to work, it is very fast. Also, as the method is based on the analysis of the construction of the letterforms, it is pretty accurate too. In this regard, one can note that the number

Table 2 Experimental detection of the Bengali vowels

Character	Approx. run time (ms)	Accuracy (%)
a	100	96
ā	100	97
i	130	94
ī	150	86
u	150	92
ū	150	93
ṛ	120	100
e	80	97
ai	80	97
o	80	95
ou	120	93

of possible anatomical features present in any Indic script is much higher than that in Latin scripts. So the detection of different curved strokes is much complicated in the present case. A single character contour in Bengali script often contains three or more different curved sections, and thus, the precise identification of the centres of curvature is crucial.

5 Conclusion and Future Work

The method presented here tackles the task of handwritten character recognition in a way that represents how a child learns to read and write. Any character is nothing but an organised collection of simple strokes. This insight can easily be further extended towards developing a complete scheme for Bengali handwritten character recognition that can handle the entire Bengali alphabet.

References

1. Chandra, S., Bokil, P., Udaya Kumar, D.: Anatomy of bengali letterforms: A semiotic study. In Chakrabarti, A., ed.: ICoRD15 Research into Design Across Boundaries Volume 1. Volume 34 of Smart Innovation, Systems and Technologies. Springer India (2015) 237–247
2. Ross, F., Shaw, G.: Non-Latin scripts From metal to digital type. St. Bride Foundation, London (2012)
3. Ghosh, P.K.: An approach to type design and text composition in indian scripts. Technical report, Stanford, CA, USA (1983)
4. Sinha, R.: A journey from indian scripts processing to indian language processing. Annals of the History of Computing, IEEE **31**(1) (2009) 8–31
5. Pal, U., Chaudhuri, B.B., Belaïd, A.: A system for bangla handwritten numeral recognition. IETE Journal of Research, Institution of Electronics and Telecommunication Engineers **52**(1) (2006) 27–34
6. Mandal, S., Sur, S., Dan, A., Bhowmick, P.: Handwritten bangla character recognition in machine-printed forms using gradient information and haar wavelet. In: Image Information Processing (ICIIP), 2011 International Conference on. (2011) 1–6
7. Das, P., Dasgupta, T., Bhattacharya, S.: A novel scheme for bengali handwriting recognition based on morphological operations with adaptive auto-generated structuring elements. In: 2nd International Conference on Control, Instrumentation, Energy and Communication (CIEC16). (2016) 211–215
8. Jalba, A.C., Wilkinson, M.H.F., Roerdink, J.B.T.M.: Shape representation and recognition through morphological curvature spaces. IEEE Transactions On Image Processing **15**(2) (2006) 331–341
9. Lindeberg, T.: Scale-space theory:a basic tool for analyzing structures at different scales. International Journal Of Computer Vision **11**(3) (1993) 283–318
10. Lindeberg: Feature detection with automatic scale selection. International Journal Of Computer Vision **30**(2) (1998)
11. Anand, C. S., Tamilarasan.M, Arjun.P: A study on curvature scale space. International Journal Of Innovative Research in Computer and Communication Engineering **2**(3) (2014)

SVM with Inverse Fringe as Feature for Improving Accuracy of Telugu OCR Systems

Amit Patel, Burra Sukumar and Chakravarthy Bhagvati

Abstract Designing an OCR system with high accuracy is quite a tough task as the system performance gets affected by its component modules. The accuracy and quality of the OCR system depends on impact of each module. The overall system performance changes if there is an improvement in a module. In our work at present, we have developed an OCR system for Telugu (Drishti System). We proposed in our paper SVM algorithm with inverse fringe as feature for Telugu OCR. The idea is to improve the performance of system by increasing recognition accuracy of the developed system. Support vector machines (SVM) was shown by several researchers to deliver high performance on Indic OCRs. SVMs have been applied to Telugu OCR and are tested with different features. In our experiments, we used fringe distance and its complementary version, the inverse fringe as a feature to the SVM. These two features have been used to develop the working model of Telugu OCR with an accuracy approaching 90%. It is shown that the performance is good over more than 300 classes. With inverse fringe as feature, the system with 325 classes is trained with 15543 labeled Telugu characters and tested over 75335 unlabeled Telugu characters; the accuracy of the system is found 99.50%. The SVM-based classifier is tested on our scanned image document corpus of more than 4500 pages and about 5,000,000 symbols. Evaluation of end-to-end system performance is done in our experiments. From the results, it has been depicted that SVM classifier is giving an improvement of approximately 1.24% over the developed Telugu OCR (Drishti System).

Keywords Fringe map · Telugu script · Telugu OCR · System performance · Indian scripts

A. Patel (✉)
Rajiv Gandhi University of Knowledge Technologies, IIIT Nuzvid, Nuzvid,
Krishna 521201, Andhra Pradesh, India
e-mail: amtptl93@gmail.com

B. Sukumar · C. Bhagvati
School of Computer and Information Sciences, University of Hyderabad,
Hyderabad 500046, Telangana, India
e-mail: chakcs@uohyd.ernet.in

© Springer Nature Singapore Pte Ltd. 2018   253
P.K. Sa et al. (eds.), *Progress in Intelligent Computing Techniques: Theory,*
Practice, and Applications, Advances in Intelligent Systems and Computing 518,
DOI 10.1007/978-981-10-3373-5_25

1 Introduction

Designing an OCR system with high accuracy is a challenging task as the performance of the system will be affected by its component modules such as preprocessing (binarization, noise removal, skew detection and correction, etc.), line, word and character segmentation, feature extraction, classification, postprocessing. The overall quality of the system depends on the correctness of each module. If errors occur in any of the modules, they may propagate through the successive modules and may turmoil the overall system performance. In order to get a more robust OCR system [1], there is an urgent need to handle these errors, i.e., we need to handle the errors in any of the modules. In our presented work we considered an OCR system, being developed for Telugu as a part of funded project by Govt. of India. From different kinds of books, a collection of 4500 scanned documents (about 5,000,000 symbols) has been created along with their ground truth. As most of these documents have been taken from old books so they have got degraded. Other attempts to Telugu OCR [2, 3] in the literature have neither have reported an end-to-end performance evaluation nor they have shown results on such large data [1]. We observed from the experiments that the performance of the system is less due to recognition module and even due to improper segmentation. Here, we tries to improve the recognition module by replacing it with SVM instead of K-NN (K-Nearest Neighbor) classifier which is based on the fringe map feature and uses distance measure [4]. We trained SVM with inverse fringe distance map as feature. Finally, as a result we obtained a major improvement in the overall accuracy of the end-to-end system.

Telugu Optical Character Recognizing system (Telugu OCR) [4] is to recognize Telugu character components by labeling them with appropriate labels. Telugu script contains large number of component classes set, which is including consonants, vowels, and combinations of consonant, vowel modifier, combinations of consonants and few other symbols. If we can separate the vowel and consonant modifier part by segmenting, all these components may result in lesser number of classes. As there exists no mechanism to do this perfectly, components to be dealt by the classifier are resulting in large number of classes. We have adopted *connected component strategy* to reduce the possible number of distinct components to be recognized by Telugu OCR, and this resulted in approximately 400 classes.

An OCR system has mainly consists of two phases in it. First is feature extractor and second is classifier. Purpose of feature extractor is to develop features for recognizing classes using available information. Classifier functionality is to classify the input samples and assign the appropriate labels. In the literature, there are many classifiers such as Bayesian networks, decision tree classifiers, support vector machines (SVM) and nearest neighbor classifiers. Among these all, no classifier available which is guaranteed to perform well; we choose SVM for our experiments and tried with different features such as fringe and inverse fringe as feature for the SVM and found that performance is good over even if we increased the number of classes more than 300. The performance of inverse fringe distance map (IFDM) as feature is better than fringe distance map (FDM) as feature.

The paper is structured as follows. Apart from introduction, there are five sections in this paper. Section 2 highlights the overview of the Telugu OCR system. Section 3 is about SVM and SVM library adopted for experiments. In Sect. 4, we discussed about the feature extraction process and recognition accuracy with that feature. In Sect. 5, we outlined about the experimental results and compared with the existing system results. We concluded with Sect. 6.

2 Telugu OCR System

An OCR system is one which consists of several modules such as preprocessing (binarization, noise removal, skew detection, etc.), segmentation of line, word and character, feature extraction, classification, postprocessing. In the survey paper [5], author has discussed about different methods for binarization and found adaptive method shows best performance on our corpus. Based on projection profiles, segmentation of line, word and character is performed [6]. From the experiments performed on our corpus, it is found that the classifiers such as neural networks, SVM [7] do not perform well because corpus contains large number of broken characters/classes with different features [4, 8, 9]. Fringe distance map as a feature [4] for K-NN classifier is used as it is found that K-NN showed the better performance than other classifier on our corpus.

3 Support Vector Machine

In the field of pattern recognition, support vector machine (SVM) [10] is a statistical method which has shown great success in many practical approaches, such as text classification, face recognition, handwritten digit recognition. By projecting data into the feature space and then finding the optimal separate hydroplane [11], SVMs can transform a nonlinear separable problem into linear separable problem with the help of different kernel functions. Initially, this kind of method was used to solve two class classification problems. Few strategies were introduced later to extend this technique to solve multiple class classification problems. In [12], author has discussed a very good and comprehensive theory about SVM.

3.1 SVM Library Adopted in the Experiments

OpenCV SVM (LIBSVM) [13] has been used to build the SVM models in our experiments. OpenCV SVM is quite efficient, simple and easy to use software for the classification using SVM. For the experiments, the SVM classifier was trained to make the prediction using probabilities value assigned to each class. By looking the values of probabilities assigned to each class, we can easily apply postprocessing for decision making whether to reject or accept the character. Kernel function used is Radial Basis Function (RBF). SVM type as NU_SVC function is used for training and prediction. Tenfold cross-validation is done while training.

4 Feature Extraction for the SVM Model

To achieve a good performance from an SVM classifier, we used fringe distance map and inverse fringe distance map as a feature. There are 1024 variables in one feature vector, computed by fringe. These features are described below.

4.1 Fringe Distance Map

We used *fringe distance map* [14] as feature for the SVM, to calculate FDM, we took an image of a character of dimension of 32 × 32, so total number of pixel present in image is 1024 and we assumed every pixel value as a variable, so one feature vector contains 1024 variables. This is also assumed that the glyphs are in black color and background of glyph is of white color. Use L1 metric or city-block distance to compute the distances [4] (Fig. 1). Efficient computation of the fringe distances can be done if at each pixel position of the template, the distances of the nearest black pixel are pre-computed and stored [4].

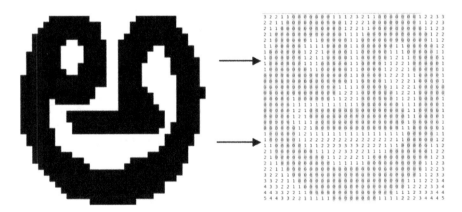

Fig. 1 Example of fringe distance map: A character of Telugu is taken which is represented in over the *white background*; distances of *black pixel* are calculated by L1 metric or city-block distance

We used this calculated value as a feature vector.

"3,2,2,1,1,0,0,0,0,0,0,0,1,1,1,2,3,2,1,1,0,0,0,0,0,0,0,1,2,2,3,3,2,2,1,1,0,0,0,0,0,0,0,0,0,0,0,1,2,2,2,1,0,0,0,0,0,0,0,0,0,0,1,1,2,2
,3,2,1,1,0,0,0,0,0,0,0,0,0,0,0,1,2,2,1,1,0,0,0,0,0,0,0,0,0,1,1,2,3,2,1,0,0,0,0,0,0,0,0,0,0,0,0,1,1,1,1,0,0,0,0,0,0,0,0,0,0,0,0,1
,2,2,2,1,0,0,0,0,0,0,1,1,0,0,0,0,0,1,1,0,0,0,0,0,0,0,0,0,0,0,0,1,1,2,1,1,0,0,0,0,0,1,1,1,1,0,0,0,0,1,1,0,0,0,0,0,0,1,1,0,0,0,0
,0,1,2,1,0,0,0,0,0,0,1,2,2,1,0,0,0,0,0,1,1,0,0,0,0,0,1,1,1,0,0,0,0,0,1,1,1,0,0,0,0,0,0,1,2,1,1,0,0,0,0,1,1,0,0,0,0,0,1,2,1,1,0,0
,0,0,0,1,1,0,0,0,0,0,0,1,1,1,0,0,0,0,0,1,1,0,0,0,0,1,1,2,2,1,0,0,0,0,0,1,1,0,0,0,0,0,0,1,1,0,0,0,0,0,1,1,0,0,0,0,0,1,2,2,2,1,0
,0,0,0,0,1,0,0,0,0,0,0,0,0,0,0,0,0,0,0,1,1,0,0,0,0,1,2,2,2,1,1,0,0,0,0,1,0,0,0,0,0,0,0,0,0,0,0,0,0,0,0,1,1,0,0,0,0,1,1,2,2,2
,1,0,0,0,0,1,0,0,0,0,0,0,0,0,0,0,0,0,0,1,1,1,1,0,0,0,0,0,1,1,2,2,1,0,0,0,0,0,0,0,0,0,0,0,0,0,0,0,1,2,2,1,0,0,0,0,0,0,0,0,1,2
,2,1,0,0,0,0,0,0,0,0,0,1,0,0,0,0,0,0,1,1,2,2,1,0,0,0,0,0,1,1,2,1,0,0,0,0,0,1,0,0,0,0,0,0,1,1,1,1,1,1,1,1,1,1,1,1,1,0,0,0,0,0,0,0
,1,1,1,0,0,0,0,1,0,0,0,0,0,1,2,1,0,0,0,0,0,0,0,0,0,1,1,0,0,0,0,1,0,0,0,0,0,0,1,2,1,0,0,0,0,0,0,0,0,0,0,0,0,0,0,0,0,0,0
,0,0,1,1,0,0,0,0,1,0,0,0,0,0,1,1,1,1,0,0,0,0,0,0,0,0,0,0,0,0,0,0,0,0,0,0,0,1,0,0,0,0,0,1,0,0,0,0,0,0,1,2,1,0,0,0,0,0,0,0,0,0,0,0,0
,0,0,1,1,0,0,0,0,0,1,0,0,0,0,0,1,2,1,1,1,1,1,1,1,1,1,1,1,1,1,1,1,0,0,0,0,1,1,1,0,0,0,0,1,1,2,2,2,2,2,2,2,2,2,2,2,2,2
,2,2,1,1,0,0,0,0,0,1,2,1,1,0,0,0,0,0,1,1,1,2,2,2,3,3,3,3,3,2,2,2,2,1,1,1,0,0,0,0,0,1,1,2,2,1,0,0,0,0,0,0,1,1,1,2,2,3,3,2,2,1,1
,1,1,0,0,0,0,0,0,1,2,2,2,1,0,0,0,0,0,0,1,1,2,2,2,2,1,1,0,0,0,0,0,0,0,1,2,3,2,1,1,0,0,0,0,0,0,0,1,1,1,1,1,1,1,0
,0,0,0,0,0,0,0,0,1,1,2,3,2,2,1,1,0,0,0,0,0,0,0,0,0,0,0,0,0,0,0,0,0,0,0,0,1,2,2,3,3,2,2,1,1,0,0,0,0,0,0,0,0,0,0,0,0,0
,0,0,0,0,0,0,0,0,1,1,2,3,3,3,3,2,2,1,1,0,0,0,0,0,0,0,0,0,0,0,0,0,0,0,1,1,1,2,2,3,4,4,3,3,2,2,1,1,0,0,0,0,0,0,0,0,0
,0,0,0,0,0,0,0,0,1,2,2,2,3,3,4,4,4,4,4,3,3,2,2,1,1,0,0,0,0,0,0,0,0,0,0,0,0,1,1,1,2,3,3,3,3,4,4,5,4,4,3,3,2,2,1,1,1,1,1,1,0,0,0,0
,0,0,0,0,1,1,1,1,2,2,2,3,4,4,4,5"

4.2 Inverse Fringe Distance Map

We used *inverse fringe distance map* [14] as feature for the SVM, to calculate IFDM, we took an image of a character of dimension of 32 × 32, so total number of pixel present in image is 1024 and we assumed every pixel value as a variable, so one feature vector contains 1024 variables. This is also assumed that the glyphs are in white color and background of glyph is of black color. Use L1 metric or city-block distance to compute the distances [4] (Fig. 2). Efficient computation of the fringe distances can be done if at each pixel position of the template, the distances of the nearest black pixel are pre-computed and stored [4].

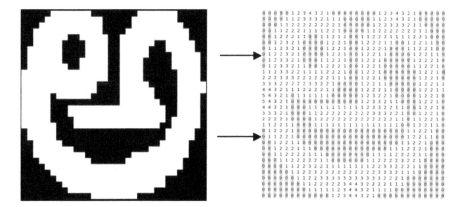

Fig. 2 Example of inverse fringe distance map: A character of Telugu is taken which is represented in over the *black background*; distances of *black pixel* are calculated by L1 metric or city-block distance

We used this calculated value as a feature vector.

```
"0,0,0,0,0,1,2,3,4,3,2,1,0,0,0,0,0,0,0,0,0,1,2,3,4,3,2,1,0,0,0,0,0,0,0,0,0,0,1,1,2,3,3,3,2,1,1,1,0,0,0,0,0,1,1,2,3,4,3,2,1,0,0,0,0
,0,0,0,0,1,1,2,2,2,2,2,2,2,2,1,0,0,0,0,0,1,2,2,3,3,3,2,1,1,0,0,0,0,0,0,1,1,2,2,2,1,1,1,1,2,2,1,0,0,0,0,1,1,2,2,2,2,2,2,2,1,1,0
,0,0,0,0,1,2,2,2,1,1,0,0,1,1,2,1,1,0,0,0,1,1,2,2,2,1,1,1,1,1,2,2,1,0,0,0,0,0,0,1,2,3,2,1,0,0,0,0,1,2,2,1,0,0,0,1,2,2,2,1,1,0,0,1,2,2,1
,1,0,0,0,1,1,2,3,2,1,0,0,0,0,0,1,2,2,1,0,0,0,1,2,2,2,1,0,0,0,0,1,1,2,2,1,0,0,0,1,2,2,3,2,1,0,0,0,0,0,1,2,2,1,0,0,1,2,2,1,1,0,0,0,0,1,2
,2,1,1,0,0,0,1,2,3,3,2,1,0,0,0,0,1,1,2,2,1,0,0,1,2,2,1,0,0,0,0,0,0,1,2,2,2,1,0,0,1,2,3,3,2,1,1,0,0,1,2,2,2,1,0,0,0,1,2,2,1,0,0,0,0,0,1
,1,2,2,1,0,1,1,1,2,3,3,2,2,1,1,1,1,2,2,2,1,0,0,1,2,2,1,0,0,0,0,0,0,1,2,2,1,0,2,2,2,3,3,3,2,2,2,2,2,2,1,1,1,0,0,1,2,2,1,0,0,0,0,0
,0,1,2,2,1,0,3,3,3,2,2,2,2,2,3,3,2,1,0,0,0,0,1,2,2,1,1,0,0,0,0,0,1,2,2,1,1,4,4,3,2,1,1,1,2,2,2,1,1,0,0,0,0,1,2,2,1,1,0,0
,0,0,1,2,2,1,1,5,4,3,2,1,0,1,1,1,1,1,1,1,0,0,0,0,0,1,2,3,2,1,0,0,0,0,1,2,2,1,0,5,4,3,2,1,0,0,0,0,0,0,0,0,0,0,0,0,0,1,2,3,3,2,1,1
,0,0,0,1,2,2,1,0,4,4,3,2,1,0,0,0,1,1,1,1,1,1,1,1,1,1,1,2,3,3,2,1,1,0,0,1,2,2,1,0,3,3,3,2,1,0,0,0,1,1,2,2,2,2,2,2,2,2,2,2,3,3,3,2
,2,1,0,0,1,2,2,1,0,2,2,2,2,1,0,0,0,0,1,2,2,2,2,2,2,2,2,2,2,2,2,2,1,1,0,1,1,2,2,1,0,1,1,2,2,1,0,0,0,0,1,1,1,1,1,1,1,1,1,1,1,1,1
,1,1,0,0,1,2,2,1,1,0,0,1,2,2,2,1,0,0,0,0,0,0,0,0,0,0,0,0,0,0,0,0,1,1,2,2,1,0,0,0,1,1,2,2,1,0,0,0,0,0,0,0,0,0,0,0,0,0,0,0,0,0,0,0
,0,0,0,0,1,2,2,1,1,0,0,0,0,1,2,2,1,1,0,0,0,0,0,0,0,0,0,0,0,0,0,0,0,0,1,1,2,2,1,0,0,0,0,1,2,2,2,1,1,1,0,0,0,0,0,0,0,0,0,0,0,0
,0,0,1,1,1,2,2,2,1,0,0,0,0,0,0,1,1,2,2,2,2,1,1,1,0,0,0,0,0,0,0,0,0,1,1,1,1,2,2,2,2,1,1,0,0,0,0,0,0,1,1,2,2,2,2,2,1,1,0,0,0,0,0,0,1
,1,2,2,2,2,3,2,2,1,0,0,0,0,0,0,0,0,0,0,1,1,2,2,3,2,2,1,1,1,1,1,1,1,1,1,1,2,2,3,3,2,2,2,2,1,1,0,0,0,0,0,0,0,0,0,0,0,0,0,1,1,2,2,3,2,2,2,2,2,2,2,2
,2,2,3,3,3,2,1,1,1,1,0,0,0,0,0,0,0,0,0,0,0,0,1,1,2,2,3,3,3,3,3,3,3,3,3,3,3,3,2,2,2,1,0,0,0,0,0,0,0,0,0,0,0,0,0,0,0,0,1,1,2,2,2,2,2,3,4,4
,3,2,2,2,2,1,1,1,0,0,0,0,0,0,0,0,0,0,0,0,0,0,0,1,1,1,1,1,1,2,3,4,4,3,2,1,1,1,1,0,0,0,0,0,0,0,0,0,0,0,0,0,0,0,0,0,0,0,0,0,0,0,1,2,3,4
,4,3,2,1,0,0,0,0,0,0,0,0,0,0,0,0,0"
```

4.3 Experiments with Fringe as Feature

We started our experiment with 20 classes. In this process, first we took an image of character and calculated the FDM and stored all the values of pixel as a variable in a feature vector, so one feature vector consists of 1024 variables. For training, we took 445 sample characters from different 20 classes and 1132 characters for test data and found the average accuracy of the system is 98.93%. The next experiment is conducted with 45 classes; we trained SVM with 2436 characters and tested 19582 characters and found the recognition accuracy of the SVM is 99.45%. The results are shown in Table 1.

4.4 Experiments with Inverse Fringe as Feature

We started the experiment with the 20 classes, trained the SVM with small dataset and tested the recognition accuracy of the SVM. In this process, we did the same as we did in experiment 4.3. For training, we took 445 character samples from

Table 1 Comparison of the results of SVMs with FDM and IFDM as features

No. of classes	Classifier's feature	Train data set size	Test data set size	Correct recognition	Incorrect recognition	Accuracy
20	FDM	445	1132	1120	12	98.93
20	IFDM	445	1132	1132	0	100.00
45	FDM	2436	19582	19475	107	99.45
45	IFDM	2436	19582	19562	20	99.89

Table 2 Results of SVMs using IFDM as feature

No. of classes	Training data Set size	Test data set size	Correct recognition	Incorrect recognition	Accuracy
20	445	1132	1132	0	100.00
45	2436	19582	19562	20	99.89
75	3862	26246	26208	38	99.85
100	5239	37534	37438	96	99.74
125	7796	44480	44367	113	99.74
175	10900	62306	62059	247	99.60
208	12287	67323	66990	333	99.50
325	15543	75335	74960	375	99.50

different 20 classes and 1132 characters for test data, and found the average accuracy of the system is 100.00% which is better than the SVM in which we used fringe distance map as a feature. The next experiment is conducted with the 45 different classes. We trained the SVM with a dataset of 2436 character samples and tested on large dataset consisting of 19582 characters and found the recognition accuracy of the SVM is 99.89%. The results are shown in Table 1.

4.5 Experimental Results

From the above two experiments, we found that using inverse fringe distance map as a feature gives good result as compared to fringe distance map as a feature. Table 1 shows the comparison between the two SVMs and their recognition accuracy.

Table 2 shows the results of all the experiments done by using inverse fringe distance as feature vector. First, we started the experiment with less number of classes, and gradually, we increased the number of classes; as we increased the number of classes, the recognition accuracy is decreased by small value, but the SVM shows good performance in handling the large number of classes with IFDM as feature.

5 Experimental Results and Analysis

After integrating the SVM into our developed OCR system [1], we have compared both the previous OCR system and improved OCR system performances over 27 novels on around 4525 pages in the corpus. Computation of error rate for a given page can be done by using a traditional string matching algorithm, Levenshtein edit distance (LED) [15]. We need to convert one string to other for a given two strings by using LED as it gives minimum number of insertion, deletion and substitution.

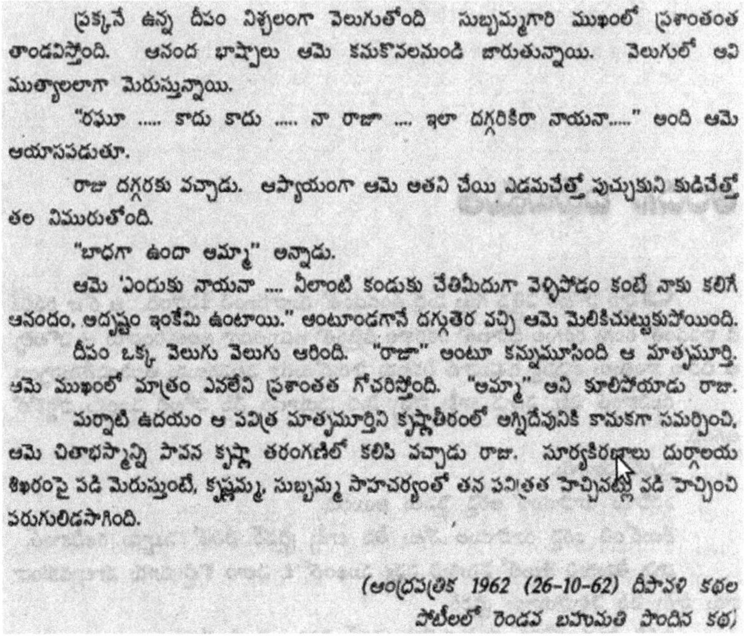

ప్రక్కనే ఉన్న దీపం విశ్వలంగా వెలుగుతోంది సుబ్బమ్మగారి ముఖంలో ప్రశాంతంత తాండవిస్తోంది. ఆనంద భాష్పాలు ఆమె కనుకొనలమండి జారుతున్నాయి. వెలుగులో ఆవి ముత్యాలలాగా మెరుస్తున్నాయి.

"రఘూ ..... కాదు కాదు ..... నా రాజా ... ఇలా దగ్గరికిరా నాయనా....." అంది ఆమె ఆయాసపడుతూ.

రాజ దగ్గరకు వచ్చాడు. ఆప్యాయంగా ఆమె ఆతని చేయి ఎడమచేత్తో పుచ్చుకుని కుడిచేత్తో తల నిమురుతోంది.

"బాధగా ఉందా ఆమ్మా" అన్నాడు.

ఆమె 'ఎందుకు నాయనా .... నీలాంటి కండుకు చేతిమీదుగా వెళ్ళిపోడం కంటే నాకు కలిగే ఆనందం, ఆదృష్టం ఇంకేమి ఉంటాయి." ఆంటూండగానే దగ్గతెర వచ్చి ఆమె మెలికెచుట్టుకుపోయింది. దీపం ఒక్క వెలుగు వెలుగు ఆరింది. "రాజా" ఆంటూ కన్నుమూసింది ఆ మాతృమూర్తి. ఆమె ముఖంలో మాత్రం ఎనలేని ప్రశాంతత గోచరిస్తోంది. "ఆమ్మా" అని కూలిపోయాడు రాజ. మర్నాటి ఉదయం ఆ పవిత్ర మాతృమూర్తిని కృష్ణాతీరంలో ఆగ్నిదేవునికి కామకగా సమర్పించి, ఆమె చితాభస్మాన్ని పావన కృష్ణా తరంగణీలో కలిపి వచ్చాడు రాజ. సూర్యకిరణాలు దుర్గాలయ శిఖరంపై పడి మెరుస్తుంటే, కృష్ణమ్మ సుబ్బమ్మ సాహచర్యంలో తన పవిత్ర సాచ్చినట్లు పడి పొచ్చింది పరుగులిడసాగింది.

(ఆంధ్రప్రభ 1962 (26-10-62) దీపావళి కథల పోటీలలో రెండవ బహుమతి పొందిన కథ)

Fig. 3 Input image for previous and improved OCR system

For a given page, we are taking all the UNICODES of ground truth text into one string and all UNICODE of OCR output text as another string. The computation of LED can be done between these strings for a given page. The overall error rate for a given page can be obtained by the ratio of LED to the number of UNICODES in the ground truth text. The errors in the text output are ultimately reflected as the errors of any module of OCR system.

The example shown in Figs. 3 and 4 is input and output of the previous OCR system and improved OCR system. In the previous OCR system, there are 428 glyphs can be recognized by K-NN as they were present in template, but in improved OCR system only 325 glyphs can be recognized by using SVM as SVM is trained with only 325 different glyphs. For the given input page, recognition accuracy is increased by 6.88%. Figure 4 shows the results which incorrectly classified (red box) by previous system and correctly classified (green box) by improved system.

Table 3 shows the performance of the previous OCR and improved OCR system. We took different pages from different books and evaluated the performance over those pages; some of the good improvements are shown in Table 3.

Table 4 contains the previous OCR error rate and improved OCR error rate and their differences (performance improvement) on 4525 pages of 27 Novels. Table 4 clearly depicts that the overall performance of the improved OCR system is increased and improved by 1.22%. For some books, the performance is good and

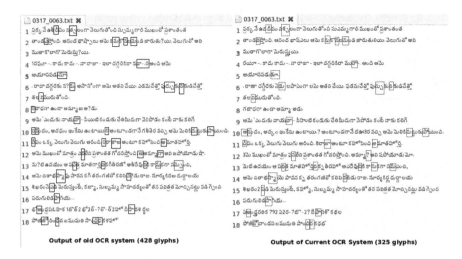

Fig. 4 Output image for input image in Fig. 3, given to previous and improved OCR system

Table 3 Performance of previous and improved OCR system on some pages

Page number	Novel name	Err-rate previous OCR	Err-rate improved OCR	Performance improvement
0320_0293.tifUNIocr	OoragaayaNavvindi	46.32	37.87	8.44
0320_0261.tifUNIocr	OoragaayaNavvindi	20.59	12.42	8.16
0320_0191.tifUNIocr	OoragaayaNavvindi	16.29	9.06	7.22
0320_0079.tifUNIocr	OoragaayaNavvindi	14.88	7.96	6.92
0317_0063.tifUNIocr	KRKMohankathalu	21.42	14.53	6.88
0317_0118.tifUNIocr	KRKMohankathalu	20.69	13.84	6.85
0317_0167.tifUNIocr	KRKMohankathalu	16.54	9.82	6.71
0317_0010.tifUNIocr	KRKMohankathalu	18.89	12.25	6.63
0011_0005.tifUNIocr	RambabuDiarypart1	13.38	5.71	7.67
0011_0122.tifUNIocr	RambabuDiarypart1	17.86	12.09	5.76
0011_0075.tifUNIocr	RambabuDiarypart1	14.36	9.65	4.70
0011_0042.tifUNIocr	RambabuDiarypart1	16.34	11.70	4.63
0120_0103.tifUNIocr	GurajadaRachanalu	71.42	50.00	21.42
0120_0079.tifUNIocr	GurajadaRachanalu	50.00	38.46	11.53
0120_0167.tifUNIocr	GurajadaRachanalu	17.88	10.12	7.75
0120_0032.tifUNIocr	GurajadaRachanalu	16.45	10.04	6.41

for some it is bad, the reason for this is that the numbers of glyphs present in improved OCR system are less as improved OCR system contains only 325 glyphs, but the previous system contains 428 glyphs.

Table 4 Results and comparison of previous and improved OCR system on different books

Book labels	Novel name	Err-rate previous OCR	Err-rate improved OCR	Performance improvement
0317	KRKMohankathalu	16.01	12.39	3.61
0320	OoragaayaNavvindi	18.06	14.72	3.33
0318	Vaikunthapalli	20.02	16.97	3.05
0001	Aashyapadham	13.04	10.83	2.20
0044	BankimchandraChatterjee	16.19	14.09	2.10
0319	Pillalakathalu	23.24	21.38	1.86
0010	DivamVaipu	7.39	5.61	1.78
0120	GurajadaRachanaluKavithalu	19.39	17.70	1.69
0011	RambabuDiarypart1	12.93	11.36	1.57
0321	GVSNavalalukathalu	21.95	20.42	1.53
0105	Gangajaatara	7.69	6.34	1.35
0045	Chadapurugu	8.36	7.13	1.23
0324	PalnatiViracharithra	8.75	7.55	1.20
0323	Daankwikostsaahasayaatralu	10.44	9.27	1.17
0042	RaniLakshmibai	19.73	18.58	1.15
0322	BalaGayyaluGeyakathalu	12.60	11.56	1.04
0063	Rajasekharacharitra	7.77	6.80	0.97
0103	GurajadaRachanaluKathanikalu	8.01	7.80	0.93
0357	AnnamayyaSankeerthanalu	14.15	13.41	0.74
0013	Jeevansmruthulu	9.79	9.45	0.34
0325	VijayaVilasamu	11.48	11.33	0.15
0015	RambabuDairypart2	14.36	14.23	0.13
0125	Kalaapoornodayamu	9.94	10.07	−0.13
0122	PanduranagaMahatyam	10.20	10.68	−0.48
0127	SrungaraNyshadam	8.19	8.80	−0.61
0136	Vasucharitramu	27.41	28.09	−0.68
0137	SassankaVijayamu	30.24	33.27	−3.03

6   Conclusion

In this paper, we proposed the SVM with fringe as feature for Telugu OCR. The idea is to improve the recognition accuracy of the existing system. The existing OCR system with nearest neighbor algorithm was not giving good performance over some book as they contain broken characters. Our proposed algorithm with inverse fringe as a feature gives good results for maximum books available in dataset. Experiments in Sect. 4 show that the performance of SVM is better if we use inverse fringe distance map as a feature instead of fringe distance map. From the experiments, we found that SVM is also good for handling large number of classes.

Then, we integrated our proposed SVM with inverse fringe distance map as a feature to the previous OCR system and tested over large dataset and found that the system performance is improved and increased.

References

1. P. Pavan Kumar, Chakravarthy Bhagvati, Atul Negi, Arun Agarwal, Bulusu Lakshmana Deekshatulu: Towards Improving the Accuracy of Telugu OCR Systems. ICDAR pp. 910–914, 2011.
2. C. V. Lakshmi and C. Patvardhan: An optical character recognition system for printed Telugu text. Pattern Analysis and Applications, vol. 7, no. 2, pp. 190–204, 2004.
3. C. V. Lakshmi, R. Jain, and C. Patvardhan: OCR of printed Telugu text with high recognition accuracies. Computer Vision, Graphics and Image Processing, pp. 786–795, 2006.
4. Atul Negi, Chakravarthy Bhagvati, B. Krishna: An OCR System for Telugu. IEEE Document Analysis and Recognition, Sixth International Conference, 2001.
5. O. Trier and A. K. Jain: Goal-directed evaluation of binarization methods. IEEE Transactions on Pattern Analysis and Machine Intelligence, vol. 17, no. 12, pp. 1191–1201, 1995.
6. K. Y. Wong, R. G. Casey, and F. M. Wahl: Document analysis system. IBM Journal of Res. Develop., vol. 26, no. 6, pp. 647–656, 1982.
7. V. Govindaraju and S. Srirangaraj: Guide to OCR for Indic Scripts. Advances in Pattern Recognition, Springer 2010.
8. S. Rajasekaran and B. Deekshatulu: Recognition of printed Telugu characters. Computer Graphics and Image Processing, vol. 6, no. 4, pp. 335–360, 1977.
9. A. K. Pujari, C. D. Naidu, M. S. Rao, and B. C. Jinaga: An intelligent character recognizer for Telugu scripts using multiresolution analysis and associative memory. Image and Vision Computing, vol. 22, no. 14, pp. 1221–1227, 2004.
10. C. J. Burges: A tutorial on support vector machines for pattern recognition. Data Mining and Knowledge Discovery, 2(2): 121–167, 1998.
11. Xiao-Xiao Niu, Ching Y. Suen: A novel hybrid CNN-SVM classifier for recognizing handwritten digits. Centre for Pattern Recognition and Machine Intelligence, Concordia University, Suite EV003.403, 1455 de Maisonneuve Blvd. West, Montreal, Quebec, Canada H3G 1M8.
12. S. Abe: Support vector machines for pattern classification. Springer-Verlag, London 2005.
13. C. C. Chang and C. J. Lin: LIBSVM: a library for support vector machines. 2001, Software available at http://www.csie.ntu.edu.tw/~cjlin/libsvm.
14. R. L. Brown: The Fringe Distance Measure: An Easily Calculated Image Distance Measure with Recognition Results Comparable to Gaussian Blurring. IEEE TRANSACTIONS ON SYSTEMS, MAN, AND CYBERNETICS, VOL. 24, NO. I, JANUARY 1994.
15. P. Pavan Kumar, C. Bhagvati, A. Negi, A. Agarwal, and B. L. Deekshatulu: Towards improving the accuracy of Telugu OCR systems. 2011 International Conference on Document Analysis and Recognition, pages 910–914, 2011.

Continuous Emotion Recognition: Sparsity Analysis

Neeru Rathee

Abstract Continuous emotion recognition is the key concern of researchers working in the field of facial behavior analysis and human–computer interaction. An attempt for continuous emotion recognition is made along with the sparsity analysis. In the presented work, Gabor filters are used to extract features from facial images. Before applying Gabor filters, preprocessing is done on facial images so as to reduce the variance due to illumination, scaling, and rotation. The Gabor magnitude as well as Gabor phase is used to represent facial features. The features were applied to relevance vector regression for continuous emotion recognition. The sparsity analysis is done by analyzing the support vectors to ensure its application in real time. The proposed approach was evaluated on extended Cohn-Kanade database. The results represent the efficacy of the presented approach.

Keywords Relevance vector regression · Emotion recognition · Facial expression analysis · Gabor filters · Sparsity

1 Introduction

Emotion results in changes in facial expressions corresponding to the different moods. Ekman and Friesen [5] defined six basic emotions: anger, fear, disgust, sad, surprise, and happy. Earlier facial expression recognition was a key concern of researchers, but the pioneering work of [2] introduced facial expressions as a challenge to the researcher working in the field of computer vision.

Continuous emotion recognition can be achieved either directly (recognizing the emotion directly using facial features) or indirectly (first recognizing the facial action unit and then recognizing the emotion). The accuracy of both the methods is controversial due to the dependence of emotion recognition methods on several

N. Rathee (✉)
Maharaja Surajmal Institute of Technology, (affiliated to Guru Gobind Singh Indraprastha University), Janakpuri C-4, New Delhi, India
e-mail: neeru1rathee@gmail.com

© Springer Nature Singapore Pte Ltd. 2018   265
P.K. Sa et al. (eds.), *Progress in Intelligent Computing Techniques: Theory,*
Practice, and Applications, Advances in Intelligent Systems and Computing 518,
DOI 10.1007/978-981-10-3373-5_26

Fig. 1 Block diagram representation of continuous emotion recognition

factors such as feature extraction method, preprocessing applied, and the classification algorithm.

The emotion recognition approaches proposed so far cater for the emotion recognition accuracy by presenting the simulation results and have not commented on the sparsity of the data. The sparsity can be roughly estimated by computing the relevance or support vectors used by the regression method (used for continuous emotion estimation). The larger the vectors, the lesser is the sparsity, and ultimately, it will lead to large computation load. In the presented approach, sparsity analysis is presented for continuous emotion recognition.

In the presented approach, face is extracted from the given image frames using the approach mentioned in [13]. The extracted face images are preprocessed so that the effect of illumination variation and scaling may be reduced. The preprocessed images are applied to Gabor filters to extract features. The features so extracted are applied to relevance vector regression (RVR) for continuous emotion recognition. The motivation behind using RVR is the use of relevance vector machine for facial expression recognition [3].

The organization of the paper is as follows: The preprocessing step is briefly explained in Sect. 2. Section 3 gives details of Gabor feature extraction. The basic RVR is described in brief in Sect. 4. The database used to evaluate the proposed approach is mentioned in Sect. 5. The experimental results are discussed in briefly in Sect. 6. Section 7 lists conclusion and future scope. The block diagram of the proposed approach is shown in Fig. 1.

2 Preprocessing

Emotion recognition methods are highly dependent on facial feature extraction methods. The face images are extracted from the given image by computing the integral image from the given image and then adopting the method introduced by [12]. The two important tasks performed in preprocessing include histogram equalization and scaling. Histogram equalization helps in removing the effects of non-uniform illumination by normalizing the contrast to the complete range of available pixels. Due to varying distance between subject and camera, and varying image resolution, the actual information is suppressed. To remove these effects, scaling is done and the images are resized to a fix size.

3 Gabor Feature Extraction

The pixel-based features or texture features are extracted by many techniques. The most popular techniques include LBP, HOG, and Gabor wavelets. The Gabor features are extracted by applying a bank of filters to an image. To extract the detailed information, features were extracted at 8 different orientations and 5 spatial frequencies being frequently used for extraction of texture features. To extract the texture information, filter bank with different characteristic frequencies and orientations is implemented for feature extraction. This resulted in decomposition of an image [9], which is popularly used for facial feature extraction [7].

4 Relevance Vector Regression

Relevance vector regression, proposed by Tipping [11], is applied for continuous emotion recognition. The linear model in feature space is given by

$$f(x, w) = \sum_{n=1}^{N} w_n . K(x, x_n) + w_0 \tag{1}$$

where w_0 represents the bias term and $K(x, x_n)$, $n = 1, 2.....N$ represents a set of nonlinear transformations. RVR uses a loss function $L(t, f(x, w))$ that is used to measure the quality of the estimation. RVR uses a loss function called ε-insensitive loss function.

$$L(t, f(x, w)) = \begin{cases} 0 & p = |t - f(x, w)| - \varepsilon \leq 0 \\ |t - f(x, w)| & otherwise \end{cases}$$

RVR uses ε-insensitive loss to perform linear regression in the high-dimensional feature space while reducing model complexity by minimizing $w^2|$. This is done by introducing slack variables ξ_n, ξ_n^*, $n = 1, 2, ...N$ to measure the deviation of training samples outside the ε-sensitive zone. So, RVR is formulated as minimization of the following function:

$$min \frac{1}{2}||w||^2 + C \sum_{n=1}^{N} (\xi_n + \xi_n^*) \tag{2}$$

subjected to the condition

$$\begin{cases} t_n - f(x_n, w) \leq \varepsilon + \xi_n^* \\ f(x_n, w) - t_n \leq \varepsilon + \xi_n \\ \xi_n, \xi_n^* \geq 0, n = 1, 2, ...N \end{cases}$$

The constant $C > 0$ determines the trade-off between the flatness of f and the values up to which deviations greater than ε are tolerated.

5 Database

We experimented with the Extended Cohn-Kanade (CK+) [6] dataset as this dataset is the most frequently used in the literature. The CK+ dataset contains 327 image sequences of seven expressions: happy, sad, fear, surprise, disgust, surprise, and contempt, performed by 118 subjects. Recordings in this dataset end at the apex of the expression. We have used only six basic expressions defined by [4] and had not included images of 'contempt' expression in our experiments. The first frame of every sequence represents a neutral expression and last frame represents emotion expression. The regression analysis is done on the complete sequence of an emotion.

6 Experimental Results

The facial images were preprocessed and Gabor filters were applied to extract facial features. To remove the redundant data, independent component analysis (ICA) was applied on the extracted features. The ICA mapped data was applied to RVR and SVR for training. The three most popular kernels have been adopted in our experiments: radial basis function kernel, linear kernel, and polynomial kernel.

To train the RVR, cross-validation (CV) is adopted due to availability of smaller data. The two main types of cross-validation are leave one subject out CV and tenfold CV. In leave one subject out CV, features of one subjects are used for testing and rest are used for training. In tenfold CV, the whole data is distributed in 10 parts equally. Out of 10 parts, nine parts are used for training and one for testing. A set of ten experiments is repeated to compute tenfold CV [1]. The performance of the proposed approach is also affected by image size. In preprocessing, images were rescaled to different sizes, which resulted in variation of performance of the proposed approach.

Performance criterion: The measures that have been used for evaluating the proposed approach include mean squared error (MSE), number of relevance vectors, and correlation coefficient (Corr). MSE represents the difference between actual and expected values. To avoid this, squared correlation coefficient was used as a measure of performance. Correlation coefficient refers to the degree of similarity between two datasets. Sparsity refers to the nonzero values that represent the sparsity of the data. The number of relevance vectors represents the sparsity of the data. For an efficient system, MSE should be low, Corr should be high, and number of support vectors should be low.

The Gabor features were extracted after resizing the image to three different sizes and performance was computed for three different feature sets. The results are represented in Table 1.

Table 1 Performance evaluation of RVR on different image sizes

Sr.no.	Image size	MSE	Correlation	Number of relevance vectors
1	70 × 70	1.132	0.754	385
2	80 × 80	1.041	0.773	296
3	90 × 90	1.184	0.753	319

Table 2 Performance evaluation of the SVR on different image sizes

Sr.no.	Image size	MSE	Correlation	Number of support vectors
1	70 × 70	1.112	0.779	435
2	80 × 80	1.001	0.792	308
3	90 × 90	1.165	0.760	382

From the above table, it may be concluded that image size 80 × 80 is the optimum size. The number of relevance vectors is directly related to sparsity of the data so the sparsity analysis along with other measures for support vector regression was also computed and presented in Table 2.

The experimental results represent that the number of support is minimally required for 80 × 80 resolution. Along with this, MSE is minimum and correlation is maximum at this particular resolution.

7 Conclusion

An approach for continuous emotion recognition is presented by extracting Gabor features from facial images. To make the proposed system applicable in real time, sparsity analysis is also presented by computing the number of support vectors during regression. The two regression techniques, SVR and RVR, have been explored for sparsity analysis. Along with sparsity, accuracy is also compared for both the techniques at various resolutions of images. The experimental results shows that accuracy of both the techniques is almost the same. The optimum resolution is 80 × 80 in terms of MSE, correlation coefficient, and number of support vectors for RVR. In the presented approach, only Gabor features have been explored. The combination of features may result in better accuracy. Moreover, the proposed approach may be evaluated on recent popular database to prove its robustness.

References

1. Ahonen, T., Hadid, A., Pietikainen, M.: Face description with local binary patterns: Application to face recognition. Pattern Analysis and Machine Intelligence, IEEE Transactions on 28(12), 2037–2041 (2006)
2. Darwin, C.: The expression of the emotions in man and animals. Oxford University Press (1998)
3. Datcu, D., Rothkrantz, L.J.: Facial expression recognition with relevance vector machines. In: Multimedia and Expo, 2005. ICME 2005. IEEE International Conference on. pp. 193–196 (2005)
4. Ekman, P., Friesen, W.: Facial Action Coding System: A Technique for the Measurement of Facial Movement. Consulting Psychologists Press, Palo Alto (1978)
5. Ekman, P., Friesen, W.V.: Measuring facial movement. Environmental psychology and nonverbal behavior 1(1), 56–75 (1976)
6. Kanade, T., Cohn, J.F., Tian, Y.: Comprehensive database for facial expression analysis. In: Automatic Face and Gesture Recognition, 2000. Proceedings. Fourth IEEE International Conference on. pp. 46–53 (2000)
7. Li, Y., Mavadati, S.M., Mahoor, M.H., Zhao, Y., Ji, Q.: Measuring the intensity of spontaneous facial action units with dynamic bayesian network. Pattern Recognition 48(11), 3417–3427 (2015)
8. Littlewort, G., Bartlett, M.S., Fasel, I.R., Susskind, J., Movellan, J.R.: Dynamics of facial expression extracted automatically from video. Image and Vision Computing 24, 615–625 (2006)
9. Tian, Y.l., Kanade, T., Cohn, J.F.: Evaluation of Gabor-wavelet-based facial action unit recognition in image sequences of increasing complexity. In: Automatic Face and Gesture Recognition, 2002. Proceedings. Fifth IEEE International Conference on. pp. 229–234. IEEE (2002)
10. Tipping, M., Bishop, C.: Variational relevance vector machine (Apr 12 2005), US Patent 6,879,944
11. Tipping, M.E.: Sparse bayesian learning and the relevance vector machine. The journal of machine learning research 1, 211–244 (2001)
12. Viola, P., Jones, M.: Robust Real-time Object Detection. International Journal of Computer Vision (2001)
13. Viola, P.A., Jones, M.J.: Rapid Object Detection using a Boosted Cascade of Simple Features. In: Computer Vision and Pattern Recognition. vol. 1, pp. 511–518 (2001)

A Novel Approach to Gesture Recognition in Sign Language Applications Using AVL Tree and SVM

Sriparna Saha, Saurav Bhattacharya and Amit Konar

Abstract Body gesture is the most important way of non-verbal communication for deaf and dumb people. Thus, a novel sign language recognition procedure is presented here where the movements of hands play a pivotal role for such kind of communications. Microsoft's Kinect sensor is used to act as a medium to interpret such communication by tracking the movement of human body using 20 joints. A procedural approach has been developed to deal with unknown gesture recognition by generating in-order expression for AVL tree as a feature. Here, 12 gestures are taken into consideration, and for the classification purpose, kernel function-based support vector machine is employed with results to gesture recognition into an accuracy of 88.3%. The foremost goal is to develop an algorithm that act as a medium to human–computer interaction for deaf and dumb people. Here, the novelty lies in the fact that for gesture recognition in sign language interpretation, the whole body of the subject is represented using a hierarchical balanced tree (here AVL).

Keywords Gesture recognition · Sign language · AVL tree · Support vector machine · Kinect sensor

S. Saha (✉) · S. Bhattacharya · A. Konar
Electronics & Tele-Communication Engineering Department,
Jadavpur University, Kolkata, West Bengal, India
e-mail: sahasriparna@gmail.com

S. Bhattacharya
e-mail: saurav.mtechiar@gmail.com

A. Konar
e-mail: akonar@etce.jdvu.ac

© Springer Nature Singapore Pte Ltd. 2018   271
P.K. Sa et al. (eds.), *Progress in Intelligent Computing Techniques: Theory,
Practice, and Applications*, Advances in Intelligent Systems and Computing 518,
DOI 10.1007/978-981-10-3373-5_27

1 Introduction

Body gesture plays the most important and guiding aspects in non-verbal communication between two persons. Hence, the main principle lies in recognition of body gesture that imparts certain information as a sign language. For recognition of body gestures, Microsoft's Kinect [1] plays an important role as it aims to detect human body using 20 joints in 3D coordinate space.

Ren et al. [2] developed Finger's earth mover distance-based technique for hand gesture recognition. Li [3] exposed an approach to human gesture recognition where a specific depth information-based threshold value is assigned. The threshold value helps in detection of hand from body, and clustering is carried out by k-means clustering algorithm. Oszust and Wysocki [4] have proposed a method for Kinect sensor-based sign language recognition. Skeletal images of the body as well as shape and position of the hands are the variants of this work. In this paper, Polish Sign Language (PSL) words are studied using k-nearest neighbor (kNN) classifier. This work helps impaired people to hear and interact globally. Le et al. [5] approached to recognize human posture using Kinect for health monitoring framework. The experiment is being performed with four main postures of standing, sitting, bending and lying using support vector machine (SVM). Biswas and Basu [6] also proposed an approach for gesture recognition using Microsoft's Kinect where multiclass SVM is used to categorize eight gestures. Patsadu et al. [7] proposed a comparison study for human gesture recognition using Kinect. Here, backpropagation neural network (BPNN) is used as a classifier. There exists certain shortcoming as the author considered only few specified gesture, but all other possible gestures are not considered.

Proposed methodology is to process the skeleton obtained from Kinect sensor to produce feature in terms of in-order expression from an AVL tree [8]. For this purpose, weight adaptation is executed on the 3D coordinate value acquired for each joint. Moreover, the weight is empirically calculated in such a way such that the z coordinate gets the maximum importance over all other coordinate weightage values (i.e., x and y). Reason is that z coordinate basically deals with the depth value, which is ultimately the area of interest to our proposed work. The values obtained through this operation are basically 20 weighted values for 20 3D joints with respect to a frame. With this, a balanced AVL tree is framed. Since human body structure is balanced with head as a parent node and all other parts as its children nodes, thus in-order traversal to the tree structure is performed. With these features as inputs, SVM [9, 10] classifier based on kernel function is used for recognition of unknown gestures related to sign language with 88.3% accuracy.

2 Proposed Work

The work under taken can be segregated into four stages:

2.1 Detection of Human Body Gesture Using Kinect Sensor

For the implementation of our proposed work for gesture recognition in sign language applications, we have used Microsoft's Kinect sensor [1]. This device has the ability to represent the human body in skeletal form with 20 body joint coordinates in 3D space using software development kit (SDK) v. 1.6. From these 20 joints, we have studied that while recognizing the gestures, all the body joints do not have equal importance due to their motion while displaying the sign languages. Thus, based on this factor, we have differentiated these joints in two different groups namely static (s) and dynamic (d) joints. The s joints are those in which very little variation (nearly equal to zero) is observed in successive frames. Here, d joints are those where a large amount of variations are noticed in the consecutive frames. If there are total S and D number of joints, then

$$S + D = 20 \text{ where } 1 \leq s \leq S \text{ and } 1 \leq d \leq D \qquad (1)$$

2.2 Weight Adaptation for Static and Dynamic Body Joints

Our aim is to draw a balanced binary tree from 20 joints where we have empirically adjusted weightage value to the 3D coordinates. The weightage value assigned to each coordinate is based on the contribution of that coordinate to impart fruitful information so as to recognize the translation of the joints from previous frames. Thus, appropriate weights need to be given to the x, y and z coordinate values, and weighted sum of these coordinates is taken as an input to form a node of the AVL tree. The adaptation of these weights for s and d joints is done differently, but a same constraint is followed such that the summation of the weights (respectively, w_x, w_y and w_z) for these three directions is

$$\left(w_x + w_y + w_z\right) \approx 1 \qquad (2)$$

Suppose for s and d joints the corresponding 3D coordinate values are (x_s, y_s, z_s) and (x_d, y_d, z_d) correspondingly. Thus, the respective values given to form the nodes $(n_s \text{ and } n_d)$ of the AVL tree are

$$n_s \leftarrow \left(x_s \times w_x^s + y_s \times w_y^s + z_s \times w_z^s \right) \tag{3}$$

$$n_d \leftarrow \left(x_d \times w_x^d + y_d \times w_y^d + z_d \times w_z^d \right) \tag{4}$$

where superscript for weight w is given to indicate the type of the joints s and d.

2.3 Feature Extraction Using In-Order Expression from AVL Tree

For human body, the joints obtained using Kinect sensor are balanced from the laws of nature. So as to implement the structure of human body to a computer terminology, binary tree approach is implemented. Binary tree is an unbalanced tree since whenever a new data is being added to it, the height of left and right sub-trees of all the nodes differs. This disadvantage is overcome by introducing a balance factor to any node, which ultimately results into an AVL tree [8]. Here lies our novelty as we have taken the in-order expression for the AVL tree for each skeleton. The in-order expression is given priority when compared with its counter parts, e.g., post-order and pre-order due to the following reason. For instance, Kinect sensor defines shoulder left as a parent node followed by elbow left, wrist left, hand left as its components. In case of in-order evaluation, the same kind of definition is observed in which the parent node represents the vital components and its children are sub-part of the parent node. Thus, to represent the body joints as a balanced tree structure, we have implemented in-order traversal to form our feature space.

Suppose there are total t numbers of frames obtained to express a particular gesture g. Thus, for a particular frame $i(1 \leq i \leq t)$, the in-order expression is IO_i, which is comprised of the arrangement of 20 joints to form the balanced tree. The total feature space becomes,

$$IO = [IO_1 \, IO_2 \, \ldots \, IO_i \, \ldots \, IO_{t-1} \, IO_t]^T \tag{5}$$

2.4 Classification Using SVM

Support vector machine (SVM) computes as a non-probabilistic binary linear classifier that separates input vectors into two classes. In case of a linear SVM [11], it is performed by building a hyper-plane depending on the support vectors. This phenomenon is only applicable when the data are linearly separable. However, this condition is accomplished accounting a kernel function [9, 10]. The overall work flow is presented in Fig. 1.

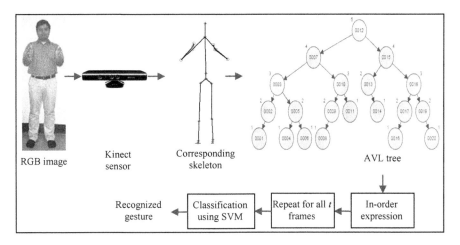

Fig. 1 Block diagram for gesture recognition in sign language applications using AVL tree and SVM

3 Experimental Results

The detailed results obtained while performing the experiment are given in this section. The 12 gestures related to sign language applications are: Move down to up, Move up to down, Move left to right, Move right to left, Pushing, Grabbing, Chirognomy, Waving, Quenelle, Self-clasping, Applause and Fist.

Each gesture is taken for 3-s duration, i.e., t is 90 (=3 s × 30 frames/s) frames. For the implementation of the proposed work, we have created three different datasets after accumulating data from three distinct age groups 25 ± 5 yrs, 30 ± 5 yrs and 35 ± 5 yrs.

The dimension of feature space IO for each gesture is 20 × 90. For the proposed work, the empirically calculated values are $s = 12$, $d = 8$, $w_x^s = w_y^s = w_z^s = 0.333$ and $w_x^d = 0.275$, $w_y^d = 0.275$, $w_z^d = 0.450$. For the unknown RGB and skeleton, images for frame number 30 are given in Fig. 2 and the in-order expression generated for the AVL tree (provided in Fig. 2 using OpenCV software) is $IO_{54} = $ [0.803(ankle left) − 0.843(knee left) − 0.974(ankle right) − 1.043(knee right) − 1.149(hip left) − 1.031(hip right) − 1.049(hand right) − 1.050(foot left) − 1.237 (foot right) − 1.149(hip center) − 1.075(spine) − 1.041(shoulder left) − 0.748(wrist right) − 0.570(shoulder center) − 0.437(elbow left) − 0.393(wrist left) − 0.797 (elbow right) − 0.651(head) − 0.532(shoulder right) − 0.506(hand left)]. Here, the exact coordinate values obtained after weight adaptation with corresponding joint names are given. The unknown gesture is correctly recognized as 'Applause.'

The comparative framework includes k-nearest neighbor (kNN) [4], ensemble decision tree (EDT) and backpropagation neural network (BPNN) [7] with respect to precision, recall, accuracy, error rate and timing complexity. The analysis results

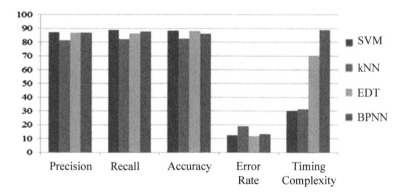

Fig. 2 Performance analysis results with respect to precision, recall, accuracy, error rate and timing complexity (in 10^{-1}s)

Table 1 Friedman test

Algorithm	Dataset 1	Dataset 2	Dataset 3	Average ranking	χ_F^2
SVM	1	2	1	1.3	8.2
kNN	4	4	4	4.0	
EDT	2	1	2	1.7	
LMA-NN	3	3	3	3.0	

are provided in Fig. 2, from where it is evident that SVM is the best choice for our proposed work.

The results obtained after Friedman test are provided in Table 1. The null hypothesis has been overruled, as $\chi_F^2 = 8.2 > 7.8$, the critical value of Chi-square distribution at probability of 0.05.

4 Conclusion

Kinect sensor-based sign language recognition dealing with both hand movement is developed in this work using AVL tree as a feature extraction procedure and SVM as a classifier. This framework yields a recognition rate of 88.3%. The data acquisition is performed by Kinect sensor for 3-s duration. The novelty of the framework lies in representing the complete human body joints in a form of balanced binary tree. Upon which in-order traversal is executed that ultimately yields as an input to SVM as a classifier. This paper has a pitfall that movement is only restricted to upper part of human body, as movement of whole body yields joints value that affects the balanced factor of AVL tree. Thus, our future advancement lies in dealing with movement of both upper and lower part of human body structure.

Acknowledgements The research work is supported by the University Grants Commission, India, University with Potential for Excellence Program (Phase II) in Cognitive Science, Jadavpur University and University Grants Commission (UGC) for providing fellowship to the first author.

References

1. K. Khoshelham and S. O. Elberink, "Accuracy and resolution of kinect depth data for indoor mapping applications," *Sensors*, vol. 12, no. 2, pp. 1437–1454, 2012.
2. Z. Ren, J. Yuan, J. Meng, and Z. Zhang, "Robust part-based hand gesture recognition using kinect sensor," *Multimedia, IEEE Trans.*, vol. 15, no. 5, pp. 1110–1120, 2013.
3. Y. Li, "Hand gesture recognition using Kinect," in *Software Engineering and Service Science (ICSESS), 2012 IEEE 3rd International Conference on*, 2012, pp. 196–199.
4. M. Oszust and M. Wysocki, "Recognition of signed expressions observed by Kinect Sensor," in *Advanced Video and Signal Based Surveillance (AVSS), 2013 10th IEEE International Conference on*, 2013, pp. 220–225.
5. T.-L. Le, M.-Q. Nguyen, and T.-T.-M. Nguyen, "Human posture recognition using human skeleton provided by Kinect," in *Computing, Management and Telecommunications (ComManTel), 2013 International Conference on*, 2013, pp. 340–345.
6. K. K. Biswas and S. K. Basu, "Gesture recognition using microsoft kinect®," in *Automation, Robotics and Applications (ICARA), 2011 5th International Conference on*, 2011, pp. 100–103.
7. O. Patsadu, C. Nukoolkit, and B. Watanapa, "Human gesture recognition using Kinect camera," in *Computer Science and Software Engineering (JCSSE), 2012 International Joint Conference on*, 2012, pp. 28–32.
8. R. W. Irving and L. Love, "The suffix binary search tree and suffix AVL tree," *J. Discret. Algorithms*, vol. 1, no. 5, pp. 387–408, 2003.
9. C. Cortes and V. Vapnik, "Support vector machine," *Mach. Learn.*, vol. 20, no. 3, pp. 273–297, 1995.
10. J. A. K. Suykens and J. Vandewalle, "Least squares support vector machine classifiers," *Neural Process. Lett.*, vol. 9, no. 3, pp. 293–300, 1999.
11. T. M. Mitchell, "Machine learning and data mining," *Commun. ACM*, vol. 42, no. 11, pp. 30–36, 1999.

Probability-Induced Distance-Based Gesture Matching for Health care Using Microsoft's Kinect Sensor

Monalisa Pal, Sriparna Saha and Amit Konar

Abstract Detection of 14 healthcare-related gestures due to pain at different body parts is the target area of this work using Kinect sensor. The novelty of our work lies in suppressing the problem of compensation by the use of probability while using similarity matching technique for gesture recognition. The adopted method enhances the matching accuracy for all the similarity measures. A shared probability and similarity measure-based metric has been defined as the matching index. This unique technique contributes to field of health care under static gesture recognition as an application of machine learning with a high accuracy of 99.1071% in 0.0126 s using probability-induced city-block distance.

Keywords Distance matching · Health care · Kinect sensor · Probability

1 Introduction

Gesture represents a static or dynamic orientation of the structural components of a human being, aimed at communicating certain message to others. The gestures due to certain diseases, such as lumbosacral strain, rotator cuff tear of shoulder, knee sprain, are the inputs of the proposed system. The reasons behind these diseases are injury, fatigue and aging. The lifestyle of the affected people which involves constantly being in the same posture due to the nature of deskbound jobs exaggerates the problems.

M. Pal (✉) · S. Saha · A. Konar
Electronics & Tele-Communication Engineering Department,
Jadavpur University, Kolkata, West Bengal, India
e-mail: monalisap90@gmail.com

S. Saha
e-mail: sahasriparna@gmail.com

A. Konar
e-mail: akonar@etce.jdvu.ac.in

© Springer Nature Singapore Pte Ltd. 2018
P.K. Sa et al. (eds.), *Progress in Intelligent Computing Techniques: Theory, Practice, and Applications*, Advances in Intelligent Systems and Computing 518, DOI 10.1007/978-981-10-3373-5_28

There are a few works on gesture recognition using Kinect sensor present in the literature. The procedure for gesture recognition for health care is given in [1] using a neural network and in [2] using fuzzy c-means algorithm. Parajuli et al. have used Kinect sensor for senior health monitoring by predicting fall using support vector machine (SVM) [3]. Le et al. have also demonstrated a technique for health care based on SVM [4]. Here, angle-based feature space has been constructed. Kinect has tremendous applicability in several other fields, e.g., children tantrum analysis [5], recognition of Parkinson's disease [6].

Here, the authors aim to build a system for detection of muscle and joint pains in the concerned subject by tracking the subject using Kinect sensor. For this purpose, 14 related body gestures are considered and 14 angle features are learned for each gesture. Similarity matching is used to detect the category for query gestures which is same as the procedure used in [7] by the authors. However, the compensation relationship of the distance metric has been identified as a penalizing factor in this context. Probability is combined to overcome this problem while defining a new metric called the matching index. This proposed method produces an accuracy of 99.1071% for city-block metric responding in 0.0126 s to a query.

2 Principals and Methodology

This work addresses diseases related to 14 different gestures, affecting lower back, knee, calf muscle, shoulder and neck. These gestures have been selected after consulting with several doctors of reputed medical college such that these gestures significantly represent the diseases. For the data acquisition and the feature extraction (14 angles) procedure, the work in [7] has been consulted.

2.1 Similarity Matching and Compensation Relationship

Similarity matching is performed following the algorithm mentioned in [7]. For this purpose, here as well, the seven distance metrics viz. Euclidean distance (D_E), Standardized Euclidean distance (D_{SE}), City-block metric (D_{CB}), Chebyshev distance (D_{CC}), Mahalanobis distance (D_M), Cosine distance (D_{CS}) and Correlation distance (D_{CR}) are considered.

Suppose Euclidean distance is considered as an example. If the difference between $X(k)$ and $Y(k)$, the k-th attributes of two l-dimensional feature vectors, is very large compared to other $(k - 1)$ attributes, then the k-th attribute dictates the value of the distance between X and Y using any of the seven metrics.

The existing similarity measures are greatly dominated by this compensation relationship [8]. However, the compensation is not useful in every context. For our work, suppose a person having neck pain is massaging the neck with one hand while the other hand is engaged in some other work, i.e., not in the relaxed state.

Table 1 Illustration of compensation relationship

Gesture (G)	Training gestures	Query gesture (Q)	Similarity measure, S (approximate values)						
			D_E	D_{SE}	D_{CB}	D_{CC}	D_M (10^7)	D_{CS}	D_{CR}
1	[106, 95, 102, 183]	[49, 91, 45, 72]	137	3	229	111	4.27	0.0671	1
2	[47, 15, 42, 78]		76	2	87	76	4.27	0.1761	1
3	[62, 103, 57, 56]		27	1	53	16	4.27	0.0156	0
Decided G (minimum S)			3	3	3	3	1/2/3	3	3

This particular gesture has not been recorded in our database. Although the rest of the gesture apart from the work-engaged hand is quite similar to the recorded ones for the concerned disease, this may lead to wrong detection. Also, in a high-dimensional space, similarity measures face the curse of dimensionality as considering the full-dimensional distance makes the distance of the farthest and the nearest neighbor almost similar.

Table 1 explains how the compensation relationship can lead to false alarm. For the shown example with four features, intuition leads to the conclusion that the query vector has maximum match with the training vector from second gesture (G) with respect to all the features except the second feature, but every similarity measure does not agree with this conclusion. Thus, there is a need to overcome this compensation relationship.

2.2 Matching Index

In our work, given a query gesture $Q(j)$, for each of the 14 features we obtain a particular disease closest (minimum difference in feature values) to the query gesture based on that feature. This yields a projected gesture vector, PG, as shown in (1), where $j = 1$ to q, $k = 1$ to 14 and $G = 1$ to 28 (There are 14 unique disease-related gestures. However, while constructing the dataset, we consider pain at either side of the body. This provides $2 \times 14 = 28$ gestures to be addressed.). For a given query gesture $Q(j)$, PG is of the dimension similar to a feature vector, i.e., 1×14, and consists of only integer values in the range [1, 28].

$$PG_{Q(j)}(k) = G \text{ if } |T(x, k) - Q(j, k)|$$
$$= \min_{i=1}^{t} |T(i, k) - Q(j, k)| \quad \text{and} \quad x \in G \tag{1}$$

Table 2 Probabilistic matching

Gesture (G)	Training gestures				Query gesture (Q)				Feature-wise difference (fd)			
	f_1	f_2	f_3	f_4	f_1	f_2	f_3	f_4	fd_1	fd_2	fd_3	fd_4
1	106	95	102	183	49	91	45	72	57	4	57	111
2	47	15	42	78					2	76	3	6
3	62	103	57	56					13	12	12	16
Projected gesture, PG (minimum fd)									2	1	2	2
Gesture (G)	1				2				3		Decided G = 2	
P(Q/G)	1/4 = 0.25				3/4 = 0.75				0/4 = 0.00			

From this vector, we obtain the probability of the query gesture vector matching a given diseased gesture, $P(Q(j)|G)$, using (2). This value is stored for all the instances of training gestures related to that particular diseased gesture G. Thus, we have a $t \times q$ probability matrix, P, where $P(i, j)$ denotes the probability of query gesture $Q(j)$ belonging to a disease G whose one of the training instances is $T(i)$. The formation of P using the described procedure is illustrated in Table 2 for one query vector (same example as used for Table 1) where features are labeled from $f1$ to $f4$, and absolute values of the feature-wise difference between the training and query gesture are labeled $fd1$ to $fd4$.

$$P\left(\frac{Q(j)}{G}\right) = \frac{\text{No. of indices where } PG_{Q(j)} = G}{\text{No. of features}} \tag{2}$$

When instead of the similarity matrix, S, the probability matrix, P, is used, the accuracy is observed to deteriorate in our experiment. This is due to the fact that the joint occurrence of all the features is not considered rather the features are treated in an independent manner. However, when the information of the similarity measure is used along with the probability, the accuracy improves as described in Table 3 (same example as used in Tables 1 and 2). The probability reduces the uncertainty

Table 3 Probabilistic induced matching

Gesture (G)	Training gestures	Query gesture (Q)	Matching index, $M = 100 \times P/S$ (approximate values)						
			[a]M_E	M_{SE}	M_{CB}	M_{CC}	M_M (10^{-7})	M_{CS}	M_{CR}
1	[106, 95, 102, 183]	[49, 91, 45, 72]	0.18	8.33	0.11	0.23	5.85	372	25
2	[47, 15, 42, 78]		0.99	37.5	0.86	0.99	17.5	426	75
3	[62, 103, 57, 56]		0	0	0	0	0	0	NaN
Decided G (maximum M)			2	2	2	2	2	2	2 (ignoring NaN values)

[a]M_X denotes matching index where S is calculated using D_X and $x = \{E, SE, CB, CC, M, CS, CR\}$

Table 4 Matching of right knee sprain ($G = 9$) with training gestures in standing position

Query	Train									Decided gesture
Gestures	6	7	8	9	10	11	12	13	14	
Similarity measure										
D_E	124.26	**68.80**	106.37	116.46	113.79	96.73	150.38	155.04	178.26	7
D_{SE}	6.44	**3.27**	6.30	6.47	6.07	5.05	7.69	5.93	7.55	7
D_{CB}	471.78	**222.92**	337.90	355.05	363.19	287.50	508.25	497.05	529.99	7
D_{CC}	72.98	**33.39**	54.95	59.14	55.51	41.12	75.74	105.45	95.51	7
D_M	8.94	12.56	10.77	9.73	10.93	**8.18**	10.94	9.05	10.98	11
D_{CS}	0.03	0.04	**0.01**	0.02	0.02	0.01	0.04	0.05	0.05	8
D_{CR}	0.30	0.26	**0.07**	0.25	0.27	0.11	0.40	0.48	0.65	8
Matching index										
M_E	0	0.15	0.06	**0.17**	0.06	0	0	0.08	0.02	9
M_{SE}	0	**3.14**	1.01	3.05	1.18	0	0	2.12	0.51	7
M_{CB}	0	0.05	0.02	**0.06**	0.02	0	0	0.02	0.01	9
M_{CC}	0	0.31	0.12	**0.33**	0.13	0	0	0.12	0.04	9
M_M	0	0.82	0.59	**2.03**	0.66	0	0	1.39	0.35	9
M_{CS}	0	256.85	638.41	**986.28**	360.41	0	0	251.09	76.50	9
M_{CR}	0	39.51	**91.20**	78.90	26.70	0	0	26.15	5.88	8

that gets incorporated from the compensation relationship. For probability, higher the dimension, more is the uncertainty that is captured. Thus, the use of probability trades-off the curse of dimensionality as well as provides better performance.

The probability being directly proportional to the matching and the similarity measures being inversely proportional, to quantify the matching for the probability-induced method, matching index, M, is defined as shown in (3) where 100 is a scaling factor and the division operation is done element-wise. Thus, M is also of dimension $t \times q$, and higher the value of the matching index, more close is the query to the training vector.

$$M = 100 \times \frac{P}{S} \tag{3}$$

3 Experimental Results

The matching of an unknown gesture by noting the corresponding similarity scores and the matching indices with gestures of different diseases used for training is shown in Table 4. We note from the results that while all the similarity measures fail in identifying the unknown gesture as a gesture showing knee sprain, the most of the matching indices succeeds in doing so, thereby, validating the fact that probability-fused gesture matching is a more robust technique than the existing similarity measures. The performance of the proposed work is analyzed using six performance metrics of a multi-class classification algorithm with $C = 28$ classes and results are given in Table 5. This application shows highest accuracy with the city-block metric in both the matching techniques. Hence, a threshold of 0.14 is chosen, keeping a safe margin, for gesture segmentation from an online stream of gestures.

Table 5 Performance for probabilistic matching for 50 independent runs

Parameters	Recall (%)	Precision (%)	Accuracy (%)	F1 score (%)	Error rate (%)	Time (ms)
M_E	91.35	93.96	98.57	92.64	1.43	12.77
M_{SE}	99.06	97.18	98.93	98.11	1.07	13.67
M_{CB}	99.67	**98.71**	**99.11**	**99.19**	**0.89**	12.65
M_{CC}	92.63	94.20	97.86	93.41	2.14	12.75
M_M	94.46	93.10	96.43	93.77	3.57	13.99
M_{CS}	99.03	93.16	98.57	96.00	1.43	13.06
M_{CR}	**99.68**	96.02	98.93	97.82	1.07	13.29

4 Conclusion and Future Directions

This work identifies and demonstrates the intrinsic problem of the existing methods related to compensation relationship which is reimbursed by integrating probability into the similarity measure. Among the several distance metrics, probability-induced city-block metric yields superior accuracy of 99.11% in 0.0126 s in this application.

Acknowledgements The research work is supported by the University Grants Commission, India, University with Potential for Excellence Program (Phase II) in Cognitive Science, Jadavpur University and University Grants Commission (UGC) for providing fellowship to the second author.

References

1. S. Saha, M. Pal, A. Konar, and R. Janarthanan, "Neural Network Based Gesture Recognition for Elderly Health Care Using Kinect Sensor," in *Swarm, Evolutionary, and Memetic Computing*, Springer, 2013, pp. 376–386.
2. M. Pal, S. Saha, and A. Konar, "A Fuzzy C Means Clustering Approach for Gesture Recognition in Healthcare," *Knee*, vol. 1, p. C7.
3. M. Parajuli, D. Tran, W. Ma, and D. Sharma, "Senior health monitoring using Kinect," in *Communications and Electronics (ICCE), 2012 Fourth International Conference on*, 2012, pp. 309–312.
4. T.-L. Le, M.-Q. Nguyen, and T.-T.-M. Nguyen, "Human posture recognition using human skeleton provided by Kinect," in *Computing, Management and Telecommunications (ComManTel), 2013 International Conference on*, 2013, pp. 340–345.
5. X. Yu, L. Wu, Q. Liu, and H. Zhou, "Children tantrum behaviour analysis based on Kinect sensor," in *Intelligent Visual Surveillance (IVS), 2011 Third Chinese Conference on*, 2011, pp. 49–52.
6. B. Galna, G. Barry, D. Jackson, D. Mhiripiri, P. Olivier, and L. Rochester, "Accuracy of the Microsoft Kinect sensor for measuring movement in people with Parkinson's disease," *Gait Posture*, 2014.
7. M. Pal, S. Saha and A. Konar, "Distance Matching Based Gesture Recognition For Healthcare Using Microsoft's Kinect Sensor," in *International Conference on Microelectronics, Computing and Communication*, IEEE, 2016, (to be published).
8. G. Yu, S. Shao, and B. Luo, "Mining Crime Data by Using New Similarity Measure," in *Genetic and Evolutionary Computing, 2008. WGEC'08. Second International Conference on*, 2008, pp. 389–392.

Gesture Recognition from Two-Person Interactions Using Ensemble Decision Tree

Sriparna Saha, Biswarup Ganguly and Amit Konar

Abstract The evolution of depth sensors has furnished a new horizon for human–computer interaction. An efficient two-person interaction detection system is proposed for an improved human–computer interaction using Kinect sensor. This device is able to identify twenty body joint coordinates in 3D space among which sixteen joints are selected and those have been adapted with certain weights to form four average points. The direction cosines of these four average points are evaluated followed by the angles made by x, y and z axes, respectively, i.e., twelve angles have been constructed for each frame. For recognition purpose, ensemble of tree classifiers with bagging mechanism is used. This novel work is widely acceptable for various gesture-based computer appliances and yields a recognition rate of 87.15%.

Keywords Human–computer interaction · Kinect sensor · Direction cosines · Ensemble decision tree

1 Introduction

Human body tracking is a well-studied topic in today's era of human–computer interaction (HCI) [1], and it can be formed by the virtue of human skeleton structures. These skeleton structures have been detected successfully due to the smart progress of some devices, used to measure depth (e.g., Sony PlayStation, Kinect sensor). Human body movements have been viewed using these depth

S. Saha (✉) · B. Ganguly · A. Konar
Electronics & Tele-Communication Engineering Department,
Jadavpur University, Kolkata, West Bengal, India
e-mail: sahasriparna@gmail.com

B. Ganguly
e-mail: biswarupgangulyee24@gmail.com

A. Konar
e-mail: akonar@etce.jdvu.ac

© Springer Nature Singapore Pte Ltd. 2018
P.K. Sa et al. (eds.), *Progress in Intelligent Computing Techniques: Theory,
Practice, and Applications*, Advances in Intelligent Systems and Computing 518,
DOI 10.1007/978-981-10-3373-5_29

sensors which can provide sufficient accuracy while tracking full body in real-time mode with low cost.

In reality action and reaction activities are hardly periodic in a multi-person perspective situation. Also, recognizing their complex aperiodic gestures are highly challenging for detection in surveillance system. Interaction detections like pushing, kicking, punching, exchanging objects are the essence of this work. Here, two-person interactions have been recognized by an RGB-D sensor, named as Kinect [2, 3].

Park and Aggarwal [4] have presented two-person interactions via natural language descriptors at a semantic level. They have adapted linguistics for representing human actions by forming triplets. Static poses and dynamic gestures are recognized through hierarchical Bayesian network. Yao et al. [5] have indicated velocity features to posses the best accuracy while recognizing single-person activity. A comparative study between SVM and multiple-instance learning boost classifier is drawn. Yun et al. [6] have recognized several interactions acted by two persons using an RGB-D sensor. The color and depth image information is extracted from the skeleton model captured from the sensor. Six different body pose features are considered as the feature set where the joint features work better than others. Saha et al. [7] have proposed a superior approach of two-person interaction model where eight interactions have been modeled and recognition is achieved through multi-class support vector machine (SVM) with rotation invariance case. In the last two literature survey works, the right person is in action while the left one is initially static and gradually reacting to the situation. But in our approach, we built such a model that any person can act or react according to the interaction delivered by any person. Thus, the possible number of gesture interaction reduces in such a case. If two persons are asked to perform eight interactions, the total interactions would be twenty ($=^{8}C_2 - 8$) and here lies the novelty of our work.

Here, we have implemented Kinect sensor to identify eight interactions using skeleton structures with some predefined features. As Kinect sensor captures twenty body joint coordinates in 3D, out of which sixteen have been adapted with some specified weights to form four mean points. The direction cosines of these four average points, i.e., twelve angles per frame, are the feature set of our proposed work. For a specific interaction, 3s video stream is captured to detect the skeleton. Ensemble decision tree (EDT) [8, 9] with bagging technology is employed for recognition purpose with a accuracy of 87.15%.

2 Fundamental Ideas

Kinect [2, 3] is a combination of a camera, an infrared (IR) emitter–receiver, a microphone block and a tilt motor. The RGB camera captures three-dimensional data at 30 frames per second in a 640 × 480 resolution. The IR camera estimates the reflected beam and measures the depth of the subject from the Kinect sensor in 1.2–3.5 m range.

Ensemble decision tree is a well-known pattern recognition classifier [8, 9]. Here, 'tree' classifier is employed as base classifier. The predictions of each base classifier are associated to determine the class of the test samples. Bagging mechanism is carried out for classification. In bagging, classifiers are trained by various datasets, obtained from the original dataset via bootstrapping. The divergence among the weak learners is examined by this re-sampling process, repeated T times. Then majority voting is taken to predict the class of the unknown. Here, T has been chosen to be 100, and bootstrap size (n) is taken 30% of the total dataset.

3 Proposed Algorithm

The block diagram of the proposed algorithm is given in Fig. 1. Suppose for our proposed algorithm, number of subjects to be chosen are N and number of actions to be executed for a single person are G. Thus, the total interactions possible between the two persons are GG. Now when one specific subject n ($1 \leq n \leq N$) is asked to perform a particular action g ($1 \leq g \leq G$), we have captured a total number of T frames. Now for each tth ($1 \leq t \leq T$) frame, we have twenty 3D body joints information out of which sixteen joints are selected for this suggested work. These selected joints are shoulder left (SL), shoulder right (SR), elbow left (EL), elbow right (ER), wrist left (WL), wrist right (WR), hand left (HaL), hand right

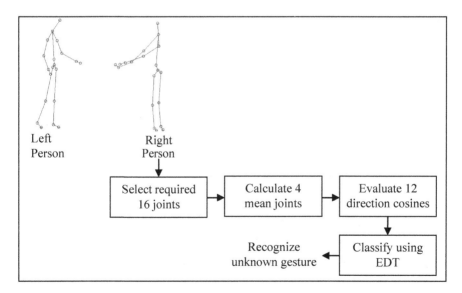

Fig. 1 Block diagram for gesture recognition for two-person interactions (as same steps need to be followed for both the persons; thus, steps required to be followed for *right* person are given only)

(*HaR*), hip left (*HL*), hip right (*HR*), knee left (*KL*), knee right (*KR*), ankle left (*AL*), ankle right (*AR*), foot left (*FL*) and foot right (*FR*).

Now four mean joints are formulated taking four body joints from arm or leg at a time. But while calculating these mean joints, we have given weights to the corresponding body joints based on the distances between them [7]. Considering left arm, the distance between *SL* and *EL* is the highest, while the distance between *WL* and *HaL* is the lowest. The weightage given to the sixteen body joints should be according to these ratios. Finally, the four mean joints become,

$$J_1^t = \frac{w^{SL} \times SL^t + w^{EL} \times EL^t + w^{WL} \times WL^t + w^{HaL} \times HaL^t}{4} \tag{1}$$

$$J_2^t = \frac{w^{SR} \times SR^t + w^{ER} \times ER^t + w^{WR} \times WR^t + w^{HaR} \times HaR^t}{4} \tag{2}$$

$$J_3^t = \frac{w^{HL} \times HL^t + w^{KL} \times KL^t + w^{AL} \times AL^t + w^{FL} \times FL^t}{4} \tag{3}$$

$$J_4^t = \frac{w^{HR} \times HR^t + w^{KR} \times KR^t + w^{AR} \times AR^t + w^{FR} \times FR^t}{4} \tag{4}$$

where w is the respective weights given to the body joints according to the superscript values. These weight values are same irrespective of time or frame number. The weights have been adopted in such a fashion that $w^{SL} + w^{EL} + w^{WL} + w^{HaL} \approx 1$, $w^{SR} + w^{ER} + w^{WR} + w^{HaR} \approx 1$, $w^{HL} + w^{KL} + w^{AL} + w^{FL} \approx 1$, $w^{HR} + w^{KR} + w^{AR} + w^{FR} \approx 1$ [10]. From each J_i ($1 \leq i \leq 4$) bearing 3D coordinate information, direction cosines ($\cos\alpha_{Ji}^t$, $\cos\beta_{Ji}^t$, $\cos\gamma_{Ji}^t$) are evaluated followed by the angles α, β, γ; that the mean joints make with positive x, y and z axes, respectively, using Eqs. (5)–(7).

$$\alpha_{Ji}^t = \cos^{-1} \frac{x_{Ji}^t}{\sqrt{(x_{Ji}^t)^2 + (y_{Ji}^t)^2 + (z_{Ji}^t)^2}} \tag{5}$$

$$\beta_{Ji}^t = \cos^{-1} \frac{y_{Ji}^t}{\sqrt{(x_{Ji}^t)^2 + (y_{Ji}^t)^2 + (z_{Ji}^t)^2}} \tag{6}$$

$$\gamma_{Ji}^t = \cos^{-1} \frac{z_{Ji}^t}{\sqrt{(x_{Ji}^t)^2 + (y_{Ji}^t)^2 + (z_{Ji}^t)^2}} \tag{7}$$

where x, y, z represents the 3D axes. Thus, for a specific subject n to interact with a particular action g, we have a total of twelve angles (three angles from each J_i) for a particular frame, which forms the feature space of our modeled work. For each action performed by each person, we have T number of frames and twelve features have been extracted per frame. Therefore, the dimension of the feature space becomes $T \times 12$. Since the total training dataset is composed of N subjects and eight interactions, the total dimension is $N \times 8 \times T \times 12$. Whenever an unknown interaction is delivered by two persons, we segregate the two actions performed by two subjects and each subject's body gestures are recognized using EDT already specified in Sect. 2.

4 Experimental Results

For the proposed work, we have granted $GG = {}^8C_2 - 8 = 20$ two-person interactions while each person is showing $G = 8$ actions, namely approaching, departing, exchanging, hugging, shaking hands, punching, pushing and kicking. The number of subjects in the dataset is taken as $N = 70$ in the age group 25–35 yrs, and each interaction is taken for 3s, i.e., $T = 30$ fps $\times 3$ s $= 90$ frames. The calculation procedure of cosine angles is explained in Table 1. The weights, by which the joints are adjusted, are $w^{SL} = w^{SR} = 0.271$, $w^{EL} = w^{ER} = 0.449$, $w^{WL} = w^{WR} = 0.149$, $w^{HaL} = w^{HaR} = 0.131$, $w^{HL} = w^{HR} = 0.348$, $w^{KL} = w^{KR} = 0.437$, $w^{AL} = w^{AR} = 0.119$, $w^{FL} = w^{FR} = 0.096$ [10]. The recognized interaction GG is approaching–hugging, i.e., the left person gesture is approaching, while the same for right person is hugging. The comparison of proposed method is done with support vector machine (SVM), k-nearest neighbor (k-NN) and back-propagation neural network (BPNN) as given in Fig. 2.

Table 1 Procedure for calculation of feature space for the subject in the right side from Fig. 2

3D coordinates obtained							Direction cosines		
SL^{48}	−0.423	0.440	3.048	J_1^{48}	−0.109	0.049	0.736		
EL^{48}	−0.468	0.199	2.933					$\cos\alpha_{J1}^{48}$	−0.146
WL^{48}	−0.411	−0.009	2.878					$\cos\beta_{J1}^{48}$	0.066
HaL^{48}	−0.388	−0.086	2.858					$\cos\gamma_{J1}^{48}$	0.986
SR^{48}	−0.258	0.407	3.289	J_2^{48}	−0.059	0.029	0.812		
ER^{48}	−0.233	0.110	3.161					$\cos\alpha_{J2}^{48}$	−0.072
WR^{48}	−0.212	−0.136	3.321					$\cos\beta_{J2}^{48}$	0.036
HaR^{48}	−0.218	−0.179	3.386					$\cos\gamma_{J2}^{48}$	0.996
HL^{48}	−0.336	0.050	3.039	J_3^{48}	−0.091	−0.090	0.743		
KL^{48}	−0.379	−0.451	2.980					$\cos\alpha_{J3}^{48}$	−0.120
AL^{48}	−0.407	−0.808	2.865					$\cos\beta_{J3}^{48}$	-0.120
FL^{48}	−0.355	−0.876	2.840					$\cos\gamma_{J3}^{48}$	0.985
HR^{48}	−0.260	0.037	3.179	J_4^{48}	−0.072	−0.094	0.783		
KR^{48}	−0.272	−0.486	3.202					$\cos\alpha_{J4}^{48}$	−0.091
AR^{48}	−0.374	−0.807	2.865					$\cos\beta_{J4}^{48}$	−0.119
FR^{48}	−0.340	−0.844	2.975					$\cos\gamma_{J4}^{48}$	0.989

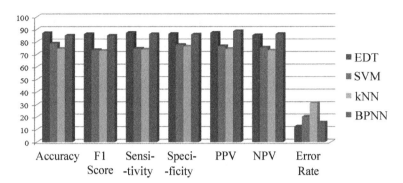

Fig. 2 Performance analysis with standard classifiers

5 Conclusion

Our proposed scheme recognizes twenty two-person interactions to explore an improved human–computer interaction in an efficient technique using Kinect sensor. Previous research [7] has been gone through calculating only average points, but in our proposed work, the body joints have been adapted with some weights based upon the distance between different body joints of human anatomy. We have obtained a better recognition rate of 87.15%. Therefore, this proposed method discovers its application from vision-based gesture recognition to public place surveillance. In the upcoming days, we will figure out to enhance our dataset comprising of some complicated interactions between two persons and recognized them with some statistical models like hidden Markov model, hidden conditional random field.

Acknowledgements The research work is supported by the University Grants Commission, India, University with Potential for Excellence Program (Phase II) in Cognitive Science, Jadavpur University and University Grants Commission (UGC) for providing fellowship to the first author.

References

1. S. S. Rautaray and A. Agrawal, "Vision based hand gesture recognition for human computer interaction: a survey," *Artif. Intell. Rev.*, vol. 43, no. 1, pp. 1–54, 2015.
2. M. R. Andersen, T. Jensen, P. Lisouski, A. K. Mortensen, M. K. Hansen, T. Gregersen, and P. Ahrendt, "Kinect depth sensor evaluation for computer vision applications," *Tech. Rep. Electron. Comput. Eng.*, vol. 1, no. 6, 2015.
3. T. T. Dao, H. Tannous, P. Pouletaut, D. Gamet, D. Istrate, and M. C. H. B. Tho, "Interactive and Connected Rehabilitation Systems for E-Health," *IRBM*, 2016.
4. S. Park and J. K. Aggarwal, "Event semantics in two-person interactions," in *Pattern Recognition, 2004. ICPR 2004. Proceedings of the 17th International Conference on*, 2004, vol. 4, pp. 227–230.

5. A. Yao, J. Gall, G. Fanelli, and L. J. Van Gool, "Does Human Action Recognition Benefit from Pose Estimation?.," in *BMVC*, 2011, vol. 3, p. 6.
6. K. Yun, J. Honorio, D. Chattopadhyay, T. L. Berg, and D. Samaras, "Two-person interaction detection using body-pose features and multiple instance learning," in *Computer Vision and Pattern Recognition Workshops (CVPRW), 2012 IEEE Computer Society Conference on*, 2012, pp. 28–35.
7. S. Saha, A. Konar, and R. Janarthanan, "Two Person Interaction Detection Using Kinect Sensor," in *Facets of Uncertainties and Applications*, Springer, 2015, pp. 167–176.
8. T. G. Dietterich, "An experimental comparison of three methods for constructing ensembles of decision trees: Bagging, boosting, and randomization," *Mach. Learn.*, vol. 40, no. 2, pp. 139–157, 2000.
9. T. G. Dietterich, "Ensemble methods in machine learning," in *Multiple classifier systems*, Springer, 2000, pp. 1–15.
10. R. Drillis, R. Contini, and M. Maurice Bluestein, "Body segment parameters 1," *Artif. Limbs*, p. 44, 1966.

Integrating Liveness Detection Technique into Fingerprint Recognition System: A Review of Various Methodologies Based on Texture Features

Jayshree Kundargi and R.G. Karandikar

Abstract Automatic fingerprint recognition systems can be deceived by spoof attack wherein an artificial fingerprint from synthetic finger fabricated using material like silicone, latex, is presented for verification. This is a serious issue as an adversary can impersonate a legitimate user of the system, especially when fingerprint technology is used as a security measure instead of conventional passwords. The solution is to integrate a liveness detection technique into the fingerprint recognition system to ensure the presence of a legitimate user. One approach is to use a dedicated hardware module with liveness detection capability, but it is intrusive, non-flexible, costly and needs user co-operation. Another approach is to use a dedicated software module with liveness detection capability, which is non-intrusive and user-friendly. Among the software-based methods, single image-based methods are simple, faster, cheaper, user-friendly and adaptable. Use of texture features for image analysis and classification is an important field in machine vision applications. Due to the fact that live and spoof fingerprints exhibit different textural properties, the proposed methods with several texture features from the literature have offered significant improvement in liveness detection accuracy. This paper presents a review on the existing texture features-based fingerprint liveness detection methods.

Keywords Biometrics · Spoof attack · Fingerprint liveness detection · Texture features

J. Kundargi (✉) · R.G. Karandikar
K. J. Somaiya College of Engineering, Mumbai, Maharashtra, India
e-mail: jmkundargi@somaiya.edu

R.G. Karandikar
e-mail: rameshkarandikar@somaiya.edu

© Springer Nature Singapore Pte Ltd. 2018   295
P.K. Sa et al. (eds.), *Progress in Intelligent Computing Techniques: Theory,
Practice, and Applications*, Advances in Intelligent Systems and Computing 518,
DOI 10.1007/978-981-10-3373-5_30

1 Introduction

Biometrics refers to an automated individual identification using a certain physiological or behavioral trait [1]. Fingerprint is the oldest, mature and most commonly used biometric because of uniqueness, permanence and ease of use [2]. In the last two decades, fingerprint systems have been widely and incrementally used for security and commercial applications. It was demonstrated that these systems are susceptible to spoof attacks where an artificial replica of a live finger allowed illegitimate entry into the system [3]. The attacks created the awareness to secure the fingerprint systems by incorporating liveness detection, to detect whether the captured fingerprint image is from a live finger or an artificial finger.

A fingerprint consists of a specific arrangement of ridges and valleys of fingertip area. A live finger is characterized by the presence of pores (diameter 80–200 μm) along the ridges through which sweat is released due to perspiration process. The pores are somewhat circular in shape and can be open or closed. Open pores appear bright, and closed pores appear dark in the captured live fingerprint images. Locations, distribution and appearance of pores in a live fingerprint are different from those in a spoof fingerprint. The sweat from the pores travels along and across the ridges changing the wetness of these regions. The wet regions appear dark, whereas dry regions appear bright, in the captured fingerprints. The perspiration phenomenon does not occur in artificial fingers due to which gray-level distribution of a spoof fingerprint is different from that of a live fingerprint.

Spoof fingers can be fabricated with the user consent (consensual method), from latent fingerprint of the user (non-consensual method), or using synthetic fingerprint generator. A spoof finger is fabricated by pressing a finger into a soft material to create a mold, which is then filled with a material like gelatin, silicone or latex.

A fingerprint recognition system, categorized as an image processing system, consisting of sensor, feature extractor and matcher modules, identifies specific patterns in the input images. A liveness detection capability is integrated into fingerprint recognition systems to prevent an adversary from entering the system fraudulently. Recently, an exhaustive survey on fingerprint liveness detection techniques appeared in [4]. Hardware approaches include an additional hardware at the sensor level to detect liveness from biological or physiological signals such as temperature, odor, pulse oximetry [5–7]. S. Shuckers [8] pointed out that hardware methods based on live finger properties can be fooled by using spoof fingerprints created with properties similar to those of live fingerprints. Software-based methods are widely researched in the last few years. These methods use additional signal processing techniques for liveness detection. They are user-friendly, noninvasive and flexible. In dynamic features-based software methods, multiple images, separated by 2 or 5 s, of the same finger, are acquired to detect the changes in fingerprint image properties due to perspiration [9, 10], or skin deformation [11, 12], which can occur only in a live finger. The change is quantified using appropriately designed metrics to distinguish live fingers from spoof. Performance of these methods depends on the pressure of the presented finger, environmental conditions, user skills and material used in spoof finger fabrication. Static features-based software methods process the information from a single fingerprint image. These are classified into perspiration-based [13], pore-based [14], quality-based [15] and texture-based [16] methods. Perspiration-based methods use signal processing techniques to quantify perspiration phenomenon for liveness detection. However, the dynamic process of perspiration cannot be accurately captured by a single image. Pore-based methods count the number of pores for liveness detection. Requirement of high-resolution images (1000 dpi) to detect pores makes the method unsuitable for general-purpose applications. Quality-based methods define measures like ridge strength, clarity and continuity to detect quality from structural and statistical properties of images to differentiate between live and spoof fingerprints. Nowadays, a large number of liveness detection methods use texture features.

A fingerprint is characterized by abundant and strong textural information. The textural properties of a live fingertip surface are dependent upon skin elasticity, pore distribution and perspiration phenomenon. As a result, the pixels along and around the ridges of a live fingerprint exhibit wide and random variations in gray-level values. The material and physical characteristics of spoof fingers are constant. The live and spoof fingerprints differ in ridge width, inter-ridge distances, ridge frequency and gray-level distribution [16]. Therefore, texture features-based methods which can capture these variations from fingerprint image properties are expected to perform better.

The purpose of this work is to focus on the existing methods that use texture features to detect liveness in the presented fingerprints.

Section 2 contains brief discussion on the existing texture features-based fingerprint liveness detection methods followed by conclusion and reference sections, respectively.

2 Texture Features-Based Fingerprint Liveness Detection Methods

Spatial domain pixel intensity value variations create texture patterns in an image. One of the applications of image texture is image analysis using texture properties. The existing texture features-based fingerprint liveness detection methods are grouped into following categories:

2.1. Global texture features
2.2. Local texture features
2.3. Hybrid (global and local) texture features

2.1 Global Texture Features-Based Methods

Global texture features are computed over the entire image to represent it in a compact form. They essentially capture macrostructures in an image. Any standard classifier can be used due to compact feature size. Existing global texture features-based fingerprint liveness detection methods are described below-

Fingertip Surface Texture Coarseness

Compared to live fingers, the surface texture of spoof fingers is coarse due to synthetic material properties. Moon et al. [17] used high-resolution (1000 dpi) images of fingerprints to capture the coarseness. They used a wavelet-based image denoising method to eliminate the coarseness. The difference between the original image and the denoised image is high for spoof fingerprints due to large pixel value fluctuations in a coarse texture compared to live images of smooth texture. They used standard deviation as a metric to quantify this difference.

Power Spectrum Features

Though ridge-valley periodicity is not altered in a spoof finger fabrication process, some microcharacteristics, responsible for high-frequency features, are less defined due to roughness of the material and the increase in the ridgelines thickness. This results in loss of high-frequency details in a spoof fingerprint, whereas the power spectrum of a live fingerprint image exhibits significant high-frequency characteristics. Coli et al. [18] differentiated live and spoof fingerprints based on the power spectrum magnitude computed over the defined region of high frequencies.

A two-dimensional discrete Fourier transform of a fingerprint image produces two concentric rings around the origin. Jin et al. [19] used magnitude of energy in the inside ring, outside ring and the entire image energy, to differentiate between live and spoof fingerprints. A live fingerprint is characterized by high energy in the inside and outside ring due to clear ridge-valley structure. The method offered better classification accuracy compared to the existing dynamic features-based methods.

Spatial Domain Texture Features

Marasco and Sansone [20] used multiple texture features, pore spacing along the ridges, surface coarseness, first-order statistical texture features and gray-level intensity ratios, computed over the entire image, to test the classification accuracy on LivDet 2009 database [21]. The authors reported that the classification performance was different for each sensor for each texture feature.

Multiscale Transform Subband Energy Feature

Nikam and Agarwal performed classification on MNIT database [4] using feature vector consisting of energies of multiscale high-frequency subbands of wavelet transform [22], curvelet transform [23] and contourlet transform [24]. Multiple classifiers were used individually and in cascade. Contourlet and curvelet transform capture curve features well and offered better results than wavelet. However, for LivDet2011 database [25], the performance was not satisfactory [26].

Summary

Surface coarseness detection using wavelet denoising may help to reduce sensor-induced noise required for sensor independent evaluation at the cost of high-resolution systems. Fourier domain presents rich information about the signal. Particularly, phase information can be captured using complex transforms to distinguish between live and spoof finger textures. Fusion of global texture features reported better but variable performance for different sensors. It indicates that some features specific to an individual sensor are not captured by global features. Multiscale transform subband coefficients convey information about saliency in the images which corresponds to texture information. Appropriate features extracted from these coefficients may improve the classification accuracy.

2.2   Local Texture Features-Based Methods

Recently, fingerprint liveness detection methods using local texture features that are investigated in a variety of machine vision applications, are increasingly explored. Local texture features are derived by processing small patches of images and are represented by histogram of features. Some form of high-pass filtering is performed over image patches to capture local microtextures to improve discrimination capability compared to global features. Local features suffer from high dimensionality. Many of them have the advantages of rotation and illumination invariance. Existing local texture features-based fingerprint liveness detection methods are described below-

Local Binary Patterns

Use of local texture feature, Local Binary Pattern (LBP), was proposed by Nikam and Agarwal [27]. The LBP operator encodes the gray-level differences between the central pixel and the surrounding pixels over the defined region to assign a label to each image pixel. The histogram of all the labels constitutes the feature vector. LBP features capture coarseness and orientation of microtextures present in a fingerprint image. LBP performance was superior compared to the contemporary methods for a limited database. Experimentation on LivDet2011 database [25] reported different classification accuracy for each sensor and each type of spoof material [26].

   X. Jia et al. [28] proposed Multiscale Block local Ternary Pattern [29] based on the difference between a central pixel and the average of a block to reduce sensitivity to noise. Unlike LBP, the ternary patterns reflect the difference between selected pixels and threshold. Experiments on LivDet2011 database [25] reported the lowest classification error among the methods considered. The same authors in [30] proposed two types of Multiscale Local Binary Patterns to take into account large-scale dependencies in spatial domain. In the first type, the radius of the LBP operator was increased to collect intensity information from a large area by using a Gaussian low-pass filter. In the second type, they applied a set of mean filters to the image. The comparison between pixels in the original LBP was replaced by the comparison between average values of pixels in each subregion. Experimental results on LivDet2011 database [25] reported lowest classification error rate for Biometrika, Digital Persona and Sagem sensors, while higher error rate for Italdata sensor.

Local Phase Quantization (LPQ)

Ghiani et al. [31] proposed LPQ, based on the finding that live and spoof finger images exhibit different frequency domain characteristics [18]. In frequency domain, phase represents many important features of a signal compared to amplitude. LPQ feature vector is constructed from de-correlated and quantized phase of four low-frequency discrete Fourier transform coefficients computed at each pixel of image using windowing technique. Histogram of LPQ feature vectors computed for the entire image compactly represents the image. LPQ, proposed for texture classification, is insensitive to image blurring and illumination variations as only phase information is used. Performance of LPQ was similar to that of LBP, but LPQ and LBP concatenated together offered significantly low error rate for all sensors, indicating they complement each other. LPQ is insensitive to small blur, but its performance drops for large amount of blur [32]. This suggests performance of LPQ depends on spoof material.

Binarized Statistical Image Feature (BSIF)

LBP and LPQ use heuristically designed filters to compute the features. Ghiani et al. [33] proposed BSIF features generated using filters that are designed from a large number of natural images to extract meaningful information from data. Learning-based methods allow flexibility in filter design with respect to descriptor length and image characteristics. The thresholded and binarized response of each filter computes a bit in BSIF code of the image. Different filter is used for each bit. The histogram of pixel BSIF code values represents the image. Experiments on LivDet2011 database [25] reported significantly low error rate for all the four sensors compared to LBP and LPQ but at the cost of large feature dimension.

Weber Local Descriptor (WLD)

Gragnaniello et al. [34] proposed WLD texture feature which consists of two components to represent contrast and orientation, respectively. Contrast is represented by the ratio of differences between central pixel and the surrounding pixel values to the central pixel value. The orientation is represented by the gradient computed at the current pixel location. Both the components are encoded to reduce feature dimension and capture high-frequency details. By concatenating the two components, a 2-D histogram is constructed to represent the image. Experimental results on LivDet2011 database [25] reported higher error rate compared to LBP and LPQ. Concatenation of WLD with LBP and LPQ offered superior results.

Improvement in the results can be achieved with the increase in code length at the cost of high feature dimension.

Wavelet Markov Local Descriptor (WMLD)

Gragnaniello et al. proposed WMLD [35] texture feature, based on Markov features, generated from transition probability matrices, to capture joint dependencies among wavelet coefficients across position, scale and orientation. Original image is decomposed to generate 13 subbands including the original image. Residue of each subband coefficient is obtained by subtracting predicted value from original value to capture local deviation due to image characteristics. Multiple Markov features constructed from transition probability matrices of residue coefficients capture dependency among wavelet coefficients to characterize microtextures. Experiments conducted on LivDet 2009 database [21] reported the lowest classification error for Crossmatch and Identix sensors. Orientation features were reported to have highest discrimination capability.

Histogram of Invariant Gradients (HIG)

Gottschlich et al. [36] proposed a gradient-based texture feature, HIG. Image was divided into circular regions of 16 pixels radius around the detected minutiae. Alternatively, the image can be divided into subblocks. For each identified local region, a histogram is constructed which counts occurrences of gradient orientation and magnitude. All computed local histograms constitute image feature. Experiments on LivDet2013 database [37] reported average classification accuracy similar to the winner of LivDet 2013 competition winner for Biometrika, Italdata and Swipe sensors and improved results for Crossmatch sensor.

Local Contrast Phase Descriptor (LCPD)

Gragnaniello et al. [38] proposed LCPD, which describes both spatial and frequency domain features to increase discrimination capability. In [31], performance of LPQ feature alone was worse and improved jointly with LBP and WLD. Local phase information captured by LPQ is complemented by concatenating with a modified differential excitation component of WLD. Instead of using a high-pass filter, Laplacian of Gaussian was used to construct differential excitation component. The infinite range of differential excitation ratio was encoded into finite values using nonlinear quantization. Experimental results on LivDet 2011 database [25] reported lowest overall average classification error.

Local Processing Using Ridgelet Transform

Nikam and Agarwal [39] proposed ridgelet transform, better at detecting line singularities than wavelet transform, for liveness detection. Image is partitioned into blocks; ridgelet transform is applied to each block. Energy and texture features derived from subband coefficients were tested individually on MNIT [4] database. Results were poor than wavelet transform-based results. As transform is applied locally, we classified the method as local feature based.

Summary

Local texture features have shown promising results, but their performance varies with sensor and spoof finger material. Fusion of multiple features has improved error rate and can be explored further. Increasing the support of local features has shown improvement in results as information is captured over large area.

2.3 Hybrid-Global and Local-Texture Features-Based Methods

Global features and local features capture different image characteristics because the support over which features are computed varies. Global features capture macro-textures, whereas local features capture local microtextures. Combined use of both is likely to improve the classification performance. However, the features need to be used judiciously to ensure improvement in the performance. The feature dimensionality needs to be taken care of. Use of both features is likely to improve fingerprint liveness detection accuracy and efficiency. The existing methods in this group are described below-

Multiscale Transform Subband-Based Local Texture Features

Nikam and Agarwal [27] proposed wavelet subband energy feature concatenated with Local Binary Pattern histogram features for fingerprint liveness detection. Wavelets capture image characteristics at multiple scales, and LBP captures fine microtextures. Experimental results on MNIT database [4] reported higher classification accuracy rate than the contemporary methods. However, the results in [26] on LivDet2011 database [25] reported relatively poor performance. Curvelet transform captures line and curve singularities better than wavelet transform. The same authors in [23] proposed curvelet transform and the gray-level co-occurrence matrix of subbands coefficients to generate texture features. Experimental results reported better classification accuracy than wavelet transform but poor performance compared to other methods [26]. In [24] texture features derived from gray-level co-occurrence matrix of contourlet transform subbands coefficients reported classification rate better than wavelet transform but almost same as curvelet transform.

Spatial Surface Coarseness Analysis (SSCA)

Periera et al. [40] complemented fingerprint coarseness computed at global level with local spatial information for images of 500 dpi resolution, followed in common commercial scanners, circumventing the need for high-resolution images in [17]. The residual noise image is divided into number of partitions. As the coarseness of each partition varies, the standard deviation of each partition is computed to produce deviation map. Deviation map is divided into sections; the histograms of all sections concatenated together represent the image feature. Experimental results on LivDet2011 database [25] conducted only for Sagem sensor show significant improvement over [17] but almost the same as offered by LivDet2011 competition winner.

Summary

A fingerprint image contains information from coarse to fine scales. An appropriate combination of global and local features using techniques that match fingerprint signal characteristics are likely to offer the desired results.

3 Conclusion

Fingerprint images of live and spoof fingers possess different textural characteristics. Single image-based software methods using local texture features have reported promising capabilities and results. Some of them are computationally intensive and have high feature dimensionality. They cannot capture gradual variations in the signal as computations are local. Global texture features-based methods are computationally economical, have compact feature size but fail to capture fine variations in the signal. It is desirable to have a liveness detection method whose performance is independent of sensor and spoof finger material with low spoof finger acceptance rate as well as low live finger rejection rate. Appropriate fusion of global and local features can provide more information to achieve the desired results.

References

1. A. K. Jain, A. Ross, S. Pankanti: Biometrics: A Tool for Information Security. In: IEEE Transactions on Information Forensics and Security, vol. 1, pp. 125–140, (2006).
2. A. K. Jain: Biometric Recognition. In: NATURE, vol. 449, pp. 38–40, (2007).
3. T. Matsumoto, H. Matsumoto, K. Yamada, and S. Hoshino: Impact of Artificial Gummy Fingers on Fingerprint Systems. In: Proc. of SPIE, vol. 4677, pp. 275–289, (2002).
4. Emanuela Marasco, Arun Ross: A Survey on Anti-Spoofing Schemes for Fingerprint Recognition Systems. In: ACM Comput. Surv., 47, 2, Article A (2014).
5. D. Osten, H. M. Carin, M. R. Arneson, and B. L. Blan.: Biometric Personal Authentication System, U. S. Patent # 571, 9950, (1998).
6. D. Baldissera, A. Franco, D. Maio, and D. Maltoni: Fake Fingerprint Detection by Odor Analysis. In: Proc. of International Conference on Biometric Authentication, (2006).
7. P.V. Reddy, A. Kumar, S.M.K. Rahman, and T.S. Mundra.: A New Antispoofing Approach for Biometric Devices. In: IEEE Transactions on Biomedical Circuits and Systems, vol. 2, no. 4, pp. 328–337, (2008).
8. S. Schuckers.: Spoofing and Anti-spoofing Measures. Information Security Technical Report, pp. 56–62, (2002).
9. S.T.V. Parthasaradhi, R. Derakhshani, L.A. Hornak and S.A.C. Schuckers.: Time-Series Detection of Perspiration as a Liveness Test in Fingerprint Devices. In: IEEE Transactions on Systems, Man, and Cybernetics, Part C: Applications and Reviews, vol. 35, pp. 335–343, (2005).

10. A. Abhyankar and S. Schuckers.: Fingerprint Liveness Detection Using Local Ridge Frequencies and Multiresolution Texture Analysis Techniques. In: IEEE International Conference on Image Processing, pp. 321–324, (2006).
11. A. Antonelli, R. Cappelli, D. Maio and D. Maltoni.: Fake Finger Detection by Skin Distortion Analysis. In: IEEE Transactions on Information Forensics and Security, vol. 1, pp. 360–373, (2006).
12. M. Drahansky, R. Notzel and W. Funk.: Liveness Detection Based on Fine Movements of the Fingertip Surface. In: IEEE Information Assurance Workshop, pp. 42–47, (2006).
13. S. Schuckers and B. Tan.: Liveness Detection for Fingerprint Scanners Based on the Statistics of Wavelet Signal Processing. In: IEEE Computer Vision and Pattern Recognition Workshop (CVPR), 26, (2006).
14. S Memon, N Manivannan, and W Balachandran.: Active Pore Detection for Liveness in Fingerprint Identification Systems. In: IEEE Telecommunications Forum (TELFOR), 619–622, (2011).
15. J. Galbally, F. Alonso-Fernandez, J. Firrez, and J. Ortega-Garcia.: A High Performance Fingerprint Liveness Detection Method Based on Quality Related Features. In: Future Generation Comp. Syst., pp. 311–321, (2012).
16. S. Nikam and S. Agarwal.: Texture and Wavelet-Based Spoof Fingerprint Detection for Fingerprint Biometric Systems. In: First International Conference on Emerging Trends in Engineering and Technology, pp. 675–680, (2008).
17. Y. Moon, J. Chen, K. Chan, K. So., and K. So. Woo.: Wavelet based Fingerprint Liveness Detection. In: IEE Electronic Letters, vol. 41, pp. 1112–1113, (2005).
18. P. Coli, G. Marcialis, and F. Roli.: Power Spectrum-based Fingerprint Vitality Detection. In: IEEE Int. Work. on Automatic Identification Advanced Technologies (AutoID) (2007).
19. C. Jin, H. Kim, and S. Elliott.: Liveness Detection of Fingerprint based on Band-Selective Fourier Spectrum. In: Information Security and Cryptology, vol. 4817, pp. 168–179, (2007).
20. E. Marasco and C. Sansone.: An Anti-spoofing Technique using Multiple Textural Features in Fingerprint Scanners. In: IEEE Workshop on Biometric Measurements and Systems for Security and Medical Applications (BioMs), PP. 8–14, (2010).
21. G. L. Marcialis *et al.*: First International Fingerprint Liveness Detection Competition— LivDet 2009. In: Image Analysis and Processing, Berlin, Germany: Springer-Verlag, pp. 12–23, (2009).
22. S. Nikam and S. Aggarwal.: Wavelet Energy Signature and GLCM Features-Based Fingerprint Anti-Spoofing. In: IEEE Int. Conf. On Wavelet Analysis and Pattern Recognition, (2008).
23. S. Nikam and S. Agarwal.: Fingerprint Liveness Detection using Curvelet Energy and Co-occurrence Signatures. In: IEEE Fifth International Conference on Computer Graphics, Imaging and Visualisation (CGIV), pp. 217–222, (2008).
24. S. Nikam and S. Agarwal.: Contourlet-Based Fingerprint Antispoofing. In: Citeseer, CS & IT-CSCP, pp. 153–160, (2013).
25. D. Yambay, L. Ghiani, P. Denti, G. L. Marcialis, F. Roli, and S. Schuckers.: LivDet 2011— Fingerprint Liveness Detection Competition 2011. In: Proc. 5th IAPR/IEEE Int. Conf. Biometrics, pp. 208–215, (2012).
26. L. Ghiani, P. Denti, and G. Marcialis.: Experimental Results on Fingerprint Liveness Detection. In: AMDO'12 - 7th international conference on Articulated Motion and Deformable Objects, Mallorca, Spain, pp. 210–218, (2012).
27. S. Nikam and S. Agarwal.: Local Binary Pattern and Wavelet-based Spoof Fingerprint Detection. In: International Journal of Biometrics, vol. 1, pp. 141–159, (2008).
28. X. Jia, X. Yang, Y. Zang, N. Zhang, R. Dai, J. Tian, and J. Zhao. Multi-scale Block Local Ternary Patterns for Fingerprints Vitality Detection. In: International Conference on Biometrics (ICB), 2013, pages 1–6, (2013).
29. B. T. Xiaoyang Tan.: Enhanced Local Texture Feature Sets for Face Recognition Under Difficult Lighting Conditions. In: Analysis and Modeling of Faces and Gestures, 4778:168–182, (2007).

30. X. Jia, X. Yang, K. Cao, Y. Zang, N. Zhang, R. Dai, X. Zhu and J. Tian.: Multi-scale Local Binary Pattern with Filters for Spoof Fingerprint Detection. In: Information Sciences, vol. 268, pp. 91–102, (2014).
31. L. Ghiani, G. Marcialis, and F. Roli.: Fingerprint Liveness Detection by Local Phase Quantization. In: 21st International Conference on Pattern Recognition, pp. 1–4, (2012).
32. Priyanka Vageeswaran.: Blur and Illumination Robust Face Recognition via Set Theoretic Characterization. In: Master of Science thesis, Department of Electrical and Computer Engineering, University of Maryland, (2013).
33. L. Ghiani, A. Hadid, G. Marcialis, and F. Roli.: Fingerprint Liveness Detection using Binarized Statistical Image Features. In: IEEE Biometrics: Theory, Applications, and Systems (BTAS), pp. 1–6, (2013).
34. D. Gragnaniello, G. Poggi, C. Sansone, and L. Verdoliva.: Fingerprint Liveness Detection based on Weber Local Image Descriptor. In: IEEE Workshop on Biometric Measurements and Systems for Security and Medical Applications (BioMs), pp. 1–5, (2013).
35. D. Gragnaniello, G. Poggi, C. Sansone, and L. Verdoliva.: Wavelet-Markov local descriptor for detecting fake fingerprints. In: Electron. Lett., vol. 50, no. 6, pp. 439–441, (2014).
36. C. Gottschlich, E. Marasco, A. Y. Yang, and B. Cukic.: Fingerprint Liveness Detection Based on Histograms of Invariant Gradients. In: Proc. IEEE Int. Joint Conf. Biometrics, pp. 1–7, (2014).
37. Ghiani, Luca, David Yambay, Valerio Mura, Simona Tocco, Gian Luca Marcialis, Fabio Roli, and Stephanie Schuckers.: Livdet 2013- Fingerprint Liveness Detection Competition 2013. In: IEEE International Conference on Biometrics, pp. 1–6, (2013).
38. D. Gragnaniello, G. Poggi, C. Sansone, and L. Verdoliva.: Local Contrast Phase Descriptor for Fingerprint Liveness Detection. In: Pattern Recognit., vol. 48, pp. 1050–1058, (2015).
39. S. Nikam and S. Agarwal.: Ridgelet-Based Fake Fingerprint Detection. In: Neurocomputing, 72, 2491–2506, (2009).
40. Pereira, L.F.A., Pinheiro, H.N.B., Cavalvanti G.D.C. and Ren T.I.: Spatial Surface Coarseness Analysis: technique for fingerprint spoof detection. In: Electron. Lett., vol. 49, pp. 260–261, (2013).

Part IV
Computational Intelligence Algorithms, Applications, and Future Directions

Process and Voltage Variation-Aware Design and Analysis of Active Grounded Inductor-Based Bandpass Filter

Vikash Kumar, Rishab Mehra, Debosmit Majumder, Shrey Khanna, Santashraya Prasad and Aminul Islam

Abstract An active inductor based on voltage differencing voltage transconductance amplifier (VDVTA) as an active element is presented. Using the active inductor, a bandpass filter is designed and the effect of process and current variations on the characteristics of the active grounded inductor-based bandpass filter is demonstrated. The bandpass filter shows its robustness against process and current variations. The simulation of the presented circuit is done using Virtuoso Analog Design Environment of Cadence @ 45-nm CMOS model parameters.

Keywords Active filter · Voltage differencing voltage transconductance amplifier · Bandpass filter

1 Introduction

CMOS-based spiral inductors suffer from a lot of problems. Drawbacks include low quality factor, large silicon area, low self-resonant frequency and lack of electronic adjustability [1]. Active inductors with electronically controllable (tunable) resonant frequency and quality factor are desirable for fully integrated circuit operations such as filters, oscillators.

In recent past, a number of active building blocks like op-amps [2–4], current feedback operational amplifiers (CFOAs) [5, 6], voltage differencing buffered amplifiers (VDBA) [7], current conveyors [8] and current differencing transconductance amplifiers (CDTA) [9] have been used for analog signal processing applications like filters, oscillators. However, these building blocks have several drawbacks. They (i) contain more number of active/passive elements (ii) make use of floating passive components (iii) lack independent tunability. Voltage differencing voltage transconductance amplifier (VDVTA) has been a new advancement

V. Kumar (✉) · R. Mehra · D. Majumder · S. Khanna · S. Prasad · A. Islam
Department of Electronics and Communication Engineering, Birla Institute
of Technology, Mesra 835 215, Jharkhand, India
e-mail: vikashkr@bitmesra.ac.in

© Springer Nature Singapore Pte Ltd. 2018 309
P.K. Sa et al. (eds.), *Progress in Intelligent Computing Techniques: Theory,
Practice, and Applications*, Advances in Intelligent Systems and Computing 518,
DOI 10.1007/978-981-10-3373-5_31

in active building block families [10–12]. VDVTA consists of an auxiliary node, regulated by the voltage difference of two input nodes, and it also consists of two different transconductance gains. Hence, VDVTA-based design provides electronically tunable active synthesis. In this paper, an active grounded inductor-based bandpass filter is realized using VDVTA as an active element.

As CMOS technology nodes are continuously scaling down, unavoidable variations in process parameters and supply voltage are posing a major design challenges on the circuit performance and thus reducing the manufacturing yield. The variation in process parameters such as oxide thickness, impurity concentration densities, channel length and diffusion depths leads to different threshold voltages of the transistors across the chip [13]. In this paper, the impact of process variations on the design metrics of grounded active inductor-based bandpass filter circuit is analyzed at different design corners, i.e., FF (fast-fast) MOSFETs, NN (nominal-nominal) MOSFETs and SS (slow-slow) MOSFETs. Due to aggressive scaling of the CMOS technology, the power supply voltage also scales down. At such low voltage, the variations in the supply voltage are much more significant. In order to observe the variations of power supply voltage on the bandpass filter, the bias currents are varied and its effect on the AC response characteristics is presented in the paper.

2 Concept of Voltage Differencing Voltage Transconductance Amplifier (VDVTA)

VDVTA is symbolized in Fig. 1. Figure 2 represents a CMOS realization of VDVTA with the node 'v' grounded [14]. AGT-I and AGT-II denote positive and negative transconductances, respectively. VDVTA contains as high impedance input nodes p, n, v and z, x^+, x^- as high impedance output nodes. VDVTA is characterized by the following matrix:

Fig. 1 Electronic symbol of VDVTA

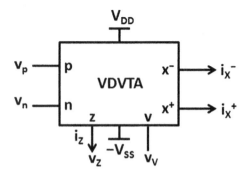

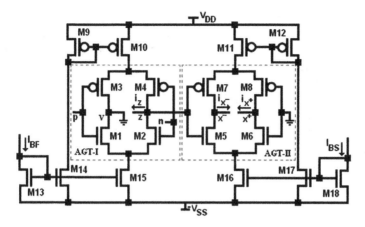

Fig. 2 CMOS realization of a VDVTA

$$\begin{bmatrix} i_z \\ i_{x^+} \\ i_{x^-} \end{bmatrix} = \begin{bmatrix} g_{m_1} & -g_{m_1} & 0 \\ 0 & 0 & g_{m_2} \\ 0 & 0 & -g_{m_2} \end{bmatrix} \begin{bmatrix} v_p \\ v_n \\ v_z - v_v \end{bmatrix} \tag{1}$$

The port equations obtained from the aforementioned matrix are

$$i_z = g_{m_1}(v_p - v_n), \tag{2}$$

$$i_{x^+} = g_{m_2}(v_z - v_v) \tag{3}$$

$$i_{x^-} = -g_{m_2}(v_z - v_v) \tag{4}$$

where g_{m1} and g_{m2} are the Arbel–Goldminz transconductances (AGTs) given in [14].

3 Grounded Inductor Configuration of VDVTA and Its Application as Bandpass Filter

Figure 3a gives VDVTA-based grounded inductor circuit. It has one VDVTA and a capacitor (C_1). Based on equations obtained from (2–4) and approximating we get the input impedance (Z_{in}) of the circuit as:

$$Z_{in} = \frac{sC_1}{g_{m_1}g_{m_2}} \tag{5}$$

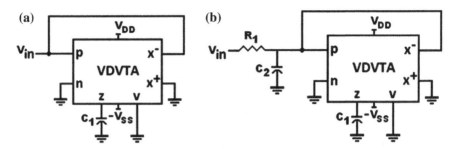

Fig. 3 **a** VDVTA-based grounded inductor. **b** Bandpass filter circuit realized with VDVTA

The equivalent input impedance of the grounded inductor is

$$Z_{in} = sL_{eq} \tag{6}$$

Comparing (5) with (6), the equivalent inductance is found to be

$$L_{eq} = \frac{C_1}{g_{m_1} g_{m_2}} \tag{7}$$

It can be seen that the inductance (L_{eq}) is directly proportional to capacitance and inversely proportional to g_{m1} and g_{m2} which are transconductance of the transistors. The inductor obtained from Fig. 3 is electronically tunable with both g_{m1} and g_{m2}.

As an application example of active grounded inductor, an active RLC bandpass filter is implemented using VDVTA and is shown in Fig. 3b. The resonant frequency (ω_0) of the bandpass filter is given by

$$\omega_0 = \frac{1}{\sqrt{L_{eq} C_2}} \tag{8}$$

where L_{eq} is the equivalent inductance obtained from the input impedance of the VDVTA network and C_2 is the capacitance of the RLC network. Substituting the value of L_{eq} from (7), the resonant frequency (ω_0) is obtained as

$$\omega_0 = \sqrt{\frac{g_{m_1} g_{m_2}}{C_1 C_2}} \tag{9}$$

4 Simulation Results and Discussion

The simulated active inductor bandpass filter circuit was analyzed at dual supply of ±0.9 V. All the simulations were performed using Cadence Virtuoso ADE at 45-nm model parameters. By using different component values of the active inductor and active bandpass filter characteristics, the simulated inductance (L_{eq}) was found to be 8.5 µH and the resonant frequency (f_0) was 17.53 MHz.

4.1 Impact of Input Bias Current (I_{BIAS}) Variation on the Resonant Frequency of the Voltage Differencing Voltage Transconductance Amplifier-Based Bandpass Filter

Voltage variations are attributed due to random or repetitive variations in the magnitude of the power supply voltage. Usually these variations do not exceed about 10% of the nominal supply voltage. These variations lead to change in the bias currents which modulate the operating point of the transistors and result in change in circuit performance different than the theoretical value. In this paper, ±0.9 V supply voltage is taken, and at such low voltage the effect of voltage variations (or current variations) is even more significant. In this paper, the supply voltage is kept constant whereas the bias currents are varied about 10% of the nominal value. Figure 4 plots the AC gain response of the active bandpass filter at multiple input bias currents, and the resonant frequency (ω_0) is almost insensitive to variations in currents.

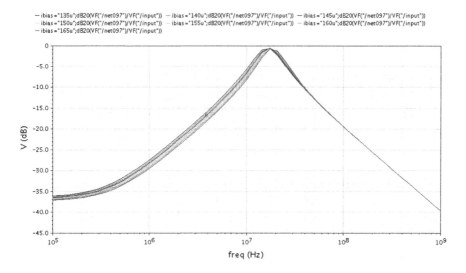

Fig. 4 Variation on the resonant frequency (f_0) of the voltage differencing voltage transconductance amplifier-based bandpass filter with respect to input bias current (I_{BIAS})

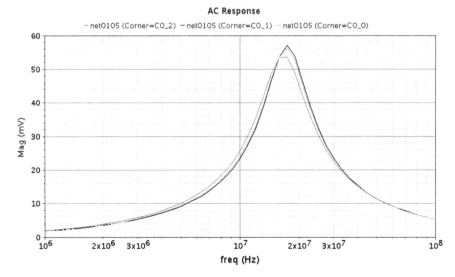

Fig. 5 Variation on the resonant frequency (f_0) of the voltage differencing voltage transconductance amplifier-based bandpass filter at different process corners C0_2–SS, C0_1–TT and C0_0–FF

Table 1 Resonant frequency values at three different model corners

Model corners	FF	NN	SS
Resonant frequency (in MHz)	17.38	17.378	17.336

4.2 Impact of Process Variations on the Resonant Frequency of the Voltage Differencing Voltage Transconductance Amplifier-Based Bandpass Filter

Process variations accounts for deviations in the process parameters such as oxide thickness, impurity concentration, channel length and diffusion depths during the semiconductor fabrication process. Such deviations can cause variations in threshold voltage across different dies and within a die itself, and the resulting chip performance varies significantly from the theoretical value after fabrication. In this paper, the process variations are estimated through Corner analysis. In Corner analysis, the simulated active bandpass filter is analyzed at different process corners, S–S, N–N and F–F. The corners have been defined as: C0_2–SS, C0_1–NN and C0_0–FF as observed in Fig. 5, and their values are reported in Table 1.

5 Conclusion

This paper presents an active inductor bandpass filter-based voltage differencing voltage transconductance amplifier (VDVTA) configuration which can be electronically tuned by varying the transconductances g_{m1} and g_{m2}. Two types of analysis voltage and process variability analysis have been performed to check for the robustness of the circuit. In voltage variability analysis, the simulated circuit is subjected to different values bias currents, and it is observed that there is hardly any deviation in the resonant frequency of the active bandpass filter circuit. The process variability analysis has also been performed at different process corners, and the circuit performance is stable at these process corners. The active inductor circuit finds its application in different analog signal processing applications and data communication networks.

References

1. Fei Yuan, "CMOS Active Inductors and Transformers," New York: Springer, 2008, ch. 2.
2. R.L. Ford and F.E.J Girling, "Active filters and oscillators using simulated inductance," Electronics Letters, vol. 2, no. 2, 1996, pp. 481–482.
3. A.J. Prescott, "Loss compensated active gyrator using differential input operational amplifier," Electronics Letters, vol. 2, no. 7, 1966, pp. 283–284.
4. S.C. Dutta Roy, "On operational amplifier simulation of grounded inductance," Archiv fuer Elektronik and Uebertragungstechnik, vol. 29, 1975, pp. 107–115.
5. E. Yuce and S. Minaei, "A modified CFOA and its applications to simulated inductors, capacitance multipliers, and analog filters," IEEE Trans. Circuits and Systems, vol. 55, no. 1, 2008, pp. 254–263.
6. F. Kacar and H. Kuntman, "CFOA-based lossless and lossy inductance simulators," Radioengineering, vol. 20, no. 3, 2011, pp. 627–631.
7. A. Yesil, F. Kacar and K. Gurkan, "Lossless grounded inductance simulator employing single VDBA and its experimental band-pass filter application," International Journal of Electronics and Communication (AEU), vol. 68, no. 2, 2014, pp. 143–150.
8. R. Senani, "Active simulation of inductors using current conveyors," Electronics Letters, vol. 14, 1978, pp. 483–484.
9. D. Prasad, D.R. Bhaskar and A.K. Singh, "New grounded and floating simulated inductance circuits using current differencing transconductance amplifiers," Radioengineering, vol. 19, no. 1, 2010, pp. 194–198.
10. Prasad D., Bhaskar D. R. 2012 "Grounded and floating inductance simulation circuits using VDTAs," Scientific Research, Circuits and Systems 3 342–347.
11. Gupta, G., Singh, S. and Bhooshan, S., "VDTA Based Electronically Tunable Voltage-Mode and Trans-Admittance Biquad Filter," Circuits and Systems, 2015, pp. 93–102.
12. Arbel, A. F., Goldminz, L. Output stage for current-mode feedback amplifiers, theory and applications. Analog Integrated Circuits and Signal Processing, 1992, vol. 2, no. 3, p. 243–255.
13. A. Islam, Mohd. Hasan, "A technique to mitigate impact of process, voltage and temperature variations on design metrics of SRAM cell," *Microelecctronics Reliability,* vol. 5, no. 2, pp. 405–411, Feb 2012.
14. A. Yesil, F. Kacar and H. Kuntman, "New Simple CMOS Realization of Voltage Differencing Transconductance Amplifier and Its RF Filter Application," Radioengineering, Vol. 20, No. 3, pp. 632–637, 2011.

Mining Closed Colossal Frequent Patterns from High-Dimensional Dataset: Serial Versus Parallel Framework

Sudeep Sureshan, Anusha Penumacha, Siddharth Jain,
Manjunath Vanahalli and Nagamma Patil

Abstract Mining colossal patterns is one of the budding fields with a lot of applications, especially in the field of bioinformatics and genetics. Gene sequences contain inherent information. Mining colossal patterns in such sequences can further help in their study and improve prediction accuracy. The increase in average transaction length reduces the efficiency and effectiveness of existing closed frequent pattern mining algorithm. The traditional algorithms expend most of the running time in mining huge amount of minute and midsize patterns which do not enclose valuable information. The recent research focused on mining large cardinality patterns called as colossal patterns which possess valuable information. A novel parallel algorithm has been proposed to extract the closed colossal frequent patterns from high-dimensional datasets. The algorithm has been implemented on Hadoop framework to exploit its inherent distributed parallelism using MapReduce programming model. The experiment results highlight that the proposed parallel algorithm on Hadoop framework gives an efficient performance in terms of execution time compared to the existing algorithms.

Keywords Frequent patterns · Closed patterns · Minimum support · Closed colossal frequent patterns · High-dimensional datasets · Hadoop · MapReduce

S. Sureshan (✉) · A. Penumacha · S. Jain · M. Vanahalli · N. Patil
National Institute of Technology Karnataka, Surathkal, Mangalore 575025,
Karnataka, India
e-mail: sudeeps.nitk@gmail.com
URL: http://www.nitk.ac.in/

A. Penumacha
e-mail: anusha.penumacha@gmail.com

S. Jain
e-mail: sid.j1501@gmail.com

M. Vanahalli
e-mail: manjunath.k.vanahalli@gmail.com

N. Patil
e-mail: nagammapatil@nitk.ac.in

© Springer Nature Singapore Pte Ltd. 2018   317
P.K. Sa et al. (eds.), *Progress in Intelligent Computing Techniques: Theory,
Practice, and Applications*, Advances in Intelligent Systems and Computing 518,
DOI 10.1007/978-981-10-3373-5_32

1   Introduction

Exhaustive research has been carried out in the field of frequent pattern mining. A lot of algorithms were proposed to mine frequent patterns from datasets. A frequent pattern is a pattern which appears in user-specified number of rows r. The user-specified threshold is called as minimum support, *minsup*. Frequent pattern mining has applications in multiple fields including association rule mining and bioinformatics. These algorithms face problems when a number of frequent patterns explode. New information cannot be retrieved from any subpattern of a frequent pattern which is also frequent. The algorithms spend time mining these patterns without discovering any new information. To tackle this problem, closed frequent patterns were introduced. A frequent pattern is called as a closed frequent pattern if and only if there exists no superpattern with identical support. Closed frequent pattern mining decreases the number of patterns to be mined without any information loss.

The development in bioinformatics has contributed to a different form of dataset called as a high-dimensional dataset. High-dimensional datasets are characterized by a large number of features and relatively less number of rows. Traditional algorithms are inefficient when the transaction length increases. The problem with high-dimensional datasets is that because of high dimensions, a lot of frequent patterns of small length can be found which are harder to analyze because of the sheer number. A cardinality threshold is introduced here which is basically the length of the pattern. To tackle this, colossal patterns were introduced and are mined instead of small patterns. Colossal patterns, as they are large, are more informative also. The existing algorithms for mining closed colossal frequent patterns are serial algorithms. Mining closed colossal frequent patterns from the existing serial algorithms is slow and not effective in terms execution time. Parallel algorithms can provide the required efficiency that is lost in a serial version for mining. This research also introduces the concept of user-defined cardinality threshold, where the user defines the length of patterns that they want to mine.

In Hadoop, data is stored on HDFS (Hadoop Distributed File System). Initially, while storing into HDFS, the huge data file is split into many smaller chunks called blocks, which are then stored on the datanodes, also replicating them on multiple nodes based on the replication factor, to ensure robustness and high availability.

The paper is organized as follows: Related work has been described in Sect. 2. The data preprocessing is been highlighted in Sect. 3. Section 4 explains the proposed methodology using the parallel framework. Section 5 highlights the result and analysis part. Section 6 elaborates conclusion followed by references.

2   Related Work

Apriori algorithm was the primary algorithm for mining frequent patterns. It uses prior knowledge of patterns that are frequent to mine complete set of frequent patterns from the database. Apriori algorithm can be considered similar to brute force way of mining frequent patterns. At each iteration, it simply adds another itemset to already found a frequent pattern in the previous iteration and then checks whether this new pattern is frequent or not. It requires multiple database scans to do this and exploits the fact that "subsets of a frequent itemset should also be frequent" so if a pattern is not frequent, any other pattern containing it will not be frequent as well. It finds frequent patterns using candidate generation. Rui Chang et al. [1] proposed an improved version of Apriori algorithm. They used hash structure and efficient horizontal data representation to save space. A flag is introduced in the new mixed type structure to indicate whether the store set is complement set or the original pattern. By using a suitable hash function, the new algorithm directly generates second-level scan from one scan of the database.

Jiawei Han et al. [2] proposed a technique of mining frequent patterns without candidate generation. Two database scans are required to build a novel tree data structure called as frequent pattern (FP-tree). The database information is condensed into this structure without any loss, and hence, multiple database scans can be avoided. To avoid costly candidate generation process, pattern fragment growth method was adopted. The overall problem is tackled using a divide and conquer algorithm wherein we divide the mining task to a smaller set of tasks to mine the confined patterns in conditional databases.

Nicolas Pasquier et al. [3] introduced idea of frequent closed patterns for association rules. One of the big problems faced by pattern mining algorithms was a huge number of patterns. If itemset is big, a number of patterns to be mined can go exponential. To alleviate this problem, closed frequent itemsets were introduced. A pattern which is frequent and does not appear in an additional superpattern consisting of identical support is called as a closed frequent pattern. So if a pattern or an itemset is considered to be closed, there is no need to check whether its subsets are frequent or not as just checking for that pattern itself will inherently check for the subsets as well.

Jian Pei et al. [4] proposed an efficient closed frequent pattern mining algorithm. FP-tree structure was followed to represent a database as a set of closed frequent patterns. With this, conditional pattern base was developed as well to reduce the search space. This was done on top of closed patterns, and hence, space efficiency was notably improved. Also, the search branches were pruned to improve time efficiency as well. To scale up the algorithm for larger databases, conditional bases were directly constructed without FP-tree.

Mohammed J. Zaki et al. [5] proposed efficient column enumeration-based closed frequent pattern mining algorithm using a top-down approach. A new data structure called itemset-tidset tree (IT-tree) using which they were able to simultaneously explore itemset space and transaction space. To make search more efficient, a hybrid search method was introduced which was able to skip levels of IT-tree and still find the frequent closed itemsets without going over all the possible subsets. The hash-based approach of overall algorithm eliminates non-closed itemsets, hence reducing the search space further.

Feng Pan et al. [6] tackled the difficulty of mining closed frequent patterns from high-dimensional biological datasets. The methods present till then did not concentrate on high-dimensional data. Bioinformatics datasets possess a huge number of features and relatively lesser number of rows. The algorithms till then used to perform column enumeration which will give a poor performance with bioinformatics dataset because of the reason mentioned above. Algorithm called Carpenter is designed to perform row enumeration rather than column enumeration to reduce computation as high-dimensional datasets have a smaller number of rows compared to a huge number of features. To make Carpenter even more robust, a different search pruning technique is used which complements the row-based enumeration.

The length of a pattern is directly proportional the information. Feida Zhu et al. [7] proposed a novel colossal frequent pattern mining algorithm called as pattern fusion. Colossal patterns are very large patterns found in databases. The algorithm fuses small core patterns to get as close as it can to colossal patterns. By grouping multiple patterns, this algorithm is able to jump in the tree structure to discover colossal patterns. Some patterns can be lost in this algorithm because of the jumps.

To mine colossal patterns from the high-dimensional dataset, Mohammad Karim Sohrabi et al. [8] described a new method. A novel vertical bottom-up technique was proposed, which makes mining colossal patterns efficient, and bit vector representation is chosen to reduce the space required.

Nurul F. Zulkurnain [9] again tackled the problem of large cardinality itemset by proposing a novel data structure called as compact row tree (CR Tree) for storing the itemsets. Itemsets with large cardinality can be more informative, and hence, it is important to mine them rather than the itemsets with small cardinality. A minimum cardinality threshold is used to avoid unnecessary processing of patterns of length smaller than required. Memory optimizations improved the overall performance of the algorithm. The bottom-up row enumeration search starts by searching for large items in a row with small support threshold and then proceeds by increasing the support threshold and hence building smaller itemsets.

3 Proposed Methodology

The intended work flow has been depicted in the block diagram of Fig. 1, showing the entire mining process for closed colossal frequent patterns on both serial and parallel platforms.

3.1 Preprocessing

The work has been done considering two datasets. A two-phase approach has been taken to mine number of colossal patterns. Initial phase is preprocessing of datasets (discretization). The second phase is execution of serial and parallel algorithms to find number of colossal patterns.

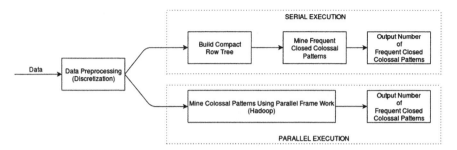

Fig. 1 Work flow of mining closed colossal frequent patterns

Table 1 Datasets

Dataset no.	Dataset	No. of rows	No. of columns
1	Leukemia	72	7129
2	Lung cancer	181	12,533

3.2 Datasets

The work focuses on mainly on the bioinformatics field. So we have used two standard DNA microarray datasets for experimentation. Table 1 summarizes information about these datasets where number of rows signify number of tissue samples taken and number of columns correspond to genes considered. Leukemia dataset was used by T.R. Golub et al. in [15], and Lung Cancer dataset was used by Gavin J. Gordon et al. in [16] for research purposes.

3.3 Data Preprocessing

The datasets considered for the research are continuous valued datasets, but the information is required in a discretized format. Using various discretization methods, the continuous values were mapped to either zero or one. A zero value means that the gene in that particular sample is dormant or inactive while one means that the gene contributes actively for that particular sample. Following techniques have been used for discretization:

1. Z-Score Normalization: First, the values are normalized using Z-score normalization as depicted in Eq. (1) and then processed to get a discretized output.

$$v' = \frac{v - \bar{X}}{\sigma_X} \tag{1}$$

Normalization is done using Eq. (1). v' represents normalized value, while v is the original value. [11] mentions that the method uses mean ($\bar{X}$) and standard deviation (σ_X) for the attribute to normalize.

2. Min-Max Normalization: [12] describes Min-Max normalization. Using this, we can map the original data range to a new user required range. It gives us a continuous valued output. To convert these values to discrete values, mid-value of the new range is used as pivot. Any value less than mid-value is mapped to zero or otherwise one. Min-Max normalization is performed based on the formula illustrated by Eq. (2), where v' represents the new value, v represents the original value, and min_X and max_X represent minimum and maximum values, respectively, of the original range, while $newMin_X$ and $newMax_X$ depict the new minimum and maximum values, respectively, of the new range.

$$v' = \frac{v - min_X}{max_X - min_X} * (newMax_X - newMin_X) + newMin_X \qquad (2)$$

This research introduces two new methods to discretize:

1. Mean Discretization: Here, the mean of all values in a dataset is calculated. If a particular value of an attribute is less than mean, it is mapped to zero, otherwise one.
2. Attribute (Gene)-wise Mean Discretization: The mean for each gene attribute (column) is calculated separately and then used as a comparison parameter for that particular attribute. If particular instance value of a gene is more than or equal to the average value of that attribute, it is mapped to one else zero.

The research also aims to find which discretization technique gives better results based on number of colossal patterns obtained. The techniques have been implemented in R programming language.

3.4 Serial Framework: Finding Number of Closed Colossal Frequent Patterns Serially

The algorithm by [9] has been implemented for serial platform. It is based on CR Tree data structure. With certain optimizations as mentioned in [9], discovering closed colossal patterns becomes memory efficient. But as the dataset has high dimensions, the algorithm takes a large amount of time to produce results.

4 Parallel Framework: Mining Closed Colossal Frequent Patterns Using Hadoop

The focus of this paper is to improve the performance as compared with the serial mining process. As the dataset is huge, one possible solution was to mine closed frequent colossal patterns in a parallel fashion with the help of Hadoop framework using MapReduce paradigm [13].

The whole process of mining in parallel has the following steps involved. First, we start by discretization of the dataset and convert it into discretized format. Then, we implement the algorithm in lines of the technique as mentioned in [14]. Finally, we apply the closeness checking after all the required patterns are mined. This is the additional step we implement in order to mine the required closed frequent colossal patterns.

The implementation is done using Java APIs as MapReduce programs are written in Java. The algorithm uses MapReduce concept two times in total. Once it is used to collect the support count of all items and once again while building conditional patterns. There is performance benefit in terms of time required for mining closed colossal frequent patterns. The output of all the tasks are key–value pairs, and finally, the output is stored on HDFS. The mapper task works on only a portion of the entire data, i.e., a block of data located locally on the datanode. Followed by a shuffle and sort operation, finally the reducer task comes into action where the required patterns are merged from their respective conditional patterns and closeness checking is done as the final step. The mined closed frequent colossal patterns are stored on HDFS. There exists a runner class which acts a single wrapper and invokes the other classes such as mapper, combiner, and reducer classes sequentially one after other. The entire Java application is packaged and exported a single JAR file which is submitted as a job request by the client. As a result, a job execution is evoked by namenode service on the Hadoop cluster.

Also, there can be one mapper and one reducer case where nothing is distributed, which is as good as sequential.

5 Result and Analysis

The experiments for serial version were implemented in Python 2.7.10 and executed on a 64-bit machine with 8 GB-installed RAM, 2.5 GHz speed, and i7-4710HQ processor. For the distributed parallel framework, a small Hadoop cluster comprising of 5 nodes, in which each is a commodity hardware of 64-bit machine with i3 processor at 1.3 GHz speed and having 4 GB RAM, was used.

Figure 2 shows the time taken for mining process alone, and it excludes the construction time of CR Tree and loading the dataset into it. The time required for mining closed frequent colossal patterns from existing serial algorithms is more compared to what is obtained as a result of running it on the parallel platform. As minimum

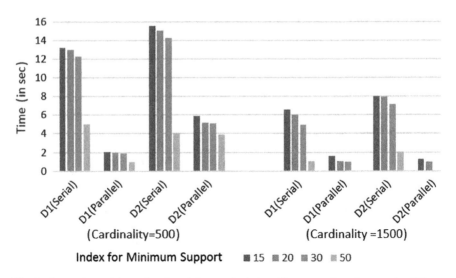

Fig. 2 Comparison of time taken for mining required closed frequent colossal pattern for different minimum support and cardinality threshold values (both on serial and on parallel platforms)

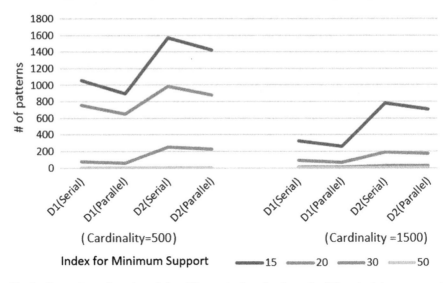

Fig. 3 Comparison of number of closed frequent colossal patterns for different minimum support and cardinality threshold values (both on serial and on parallel platforms)

support value increases, the time taken to mine the required patterns reduces as it is highly unlikely to have a colossal pattern with huge value of minimum support terminating the search quickly.

Further analysis is depicted in Fig. 3, though the parallel execution takes comparatively less time but turns out that the number of patterns mined parallel is not

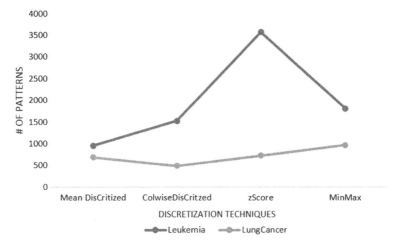

Fig. 4 Number of closed colossal patterns mined for different discretization techniques

exhaustive in nature which is the trade-off involved. But, still the accuracy on an average is around 84% of the serial results in case of first dataset and around 90% in case of the second dataset.

In an effort to identify the best discretization technique, Fig. 4 shows us a comparison of how the number of mined closed colossal patterns varies across different discretization techniques for a given value of minimum support and cardinality threshold. It is clear from the graph that we cannot generalize on any one single technique as the best and effective, since it is highly dependent on nature of dataset involved.

6 Conclusion

The proposed work presents a methodology and comparative study to investigate the performance of mining closed colossal frequent patterns from high-dimensional datasets. The biggest challenge faced was keeping the accuracy of mining number of colossal patterns and at the same time minimizing the execution time required for mining. The results indicate a gain in performance in terms of time required to execute when implemented on the parallel platform, but the patterns mined are not exhaustive as the serial implementation. This eventually turns out to be an inevitable trade-off between performance and accuracy, which we are willing to pay as obtained results vouch for the fact that the obtained results are having on an average around 85%–90% confidence, with a 1%–2% (depending on the benchmark) confidence interval, in terms of number of colossal patterns mined in parallel as compared with the serial counterpart.

Additionally, on further analysis of the results, we also notice that the choice of discretization technique to be employed for best performance is dependent on the nature of dataset involved.

References

1. Chang, Rui, and Zhiyi Liu. "An improved apriori algorithm." Electronics and Optoelectronics (ICEOE), 2011 International Conference on. Vol. 1. IEEE, 2011.
2. Han, Jiawei, Jian Pei, and Yiwen Yin. "Mining frequent patterns without candidate generation." ACM Sigmod Record. Vol. 29. No. 2. ACM, 2000.
3. Pasquier, Nicolas, et al. "Discovering frequent closed itemsets for association rules." Database Theory ICDT99. Springer Berlin Heidelberg, 1999. 398–416.
4. Pei, Jian, Jiawei Han, and Runying Mao. "CLOSET: An Efficient Algorithm for Mining Frequent Closed Itemsets." ACM SIGMOD workshop on research issues in data mining and knowledge discovery. Vol. 4. No. 2. 2000.
5. Zaki, Mohammed J., and Ching-Jui Hsiao. "Efficient algorithms for mining closed itemsets and their lattice structure." Knowledge and Data Engineering, IEEE Transactions on 17.4 (2005): 462–478.
6. Pan, Feng, et al. "Carpenter: Finding closed patterns in long biological datasets." Proceedings of the ninth ACM SIGKDD international conference on Knowledge discovery and data mining. ACM, 2003.
7. Zhu, Feida, et al. "Mining colossal frequent patterns by core pattern fusion." Data Engineering, 2007. ICDE 2007. IEEE 23rd International Conference on. IEEE, 2007.
8. Sohrabi, Mohammad Karim, and Ahmad Abdollahzadeh Barforoush. "Efficient colossal pattern mining in high dimensional datasets." Knowledge-Based Systems 33 (2012): 41–52.
9. Zulkurnain, Nurul F., David J. Haglin, and John A. Keane. "DisClose: discovering colossal closed itemsets via a memory efficient compact row-tree." Emerging Trends in Knowledge Discovery and Data Mining. Springer Berlin Heidelberg, 2012. 141–156.
10. The Data Mining & Research Blog,. "An Introduction To Frequent Pattern Mining - The Data Mining & Research Blog". N.p., 2013. Web. 6 Feb. 2016. http://data-mining.philippe-fournier-viger.com/introduction-frequent-pattern-mining/.
11. Howto.commetrics.com,. "How Raw Data Are Normalized Howto.Commetrics". N.p., 2016. Web. 7 Feb. 2016. http://howto.commetrics.com/methodology/statistics/normalization/.
12. Normalization, Data. "Data Mining Blog: Data Preprocessing Normalization". Intelligencemining.blogspot.in. N.p., 2009. Web. 7 Feb. 2016. http://intelligencemining.blogspot.in/2009/07/data-preprocessing-normalization.html.
13. Prekopcsk, Zoltn, et al. "Radoop: Analyzing big data with rapidminer and hadoop." Proceedings of the 2nd RapidMiner community meeting and conference (RCOMM 2011). 2011.
14. Itkar, Suhasini A., and Uday V. Kulkarni. "Distributed Algorithm for Frequent Pattern Mining using HadoopMap Reduce Framework." (2013).
15. Golub, Todd R., et al. "Molecular classification of cancer: class discovery and class prediction by gene expression monitoring." Science 286.5439 (1999): 531–537.
16. Gordon, Gavin J., et al. "Translation of microarray data into clinically relevant cancer diagnostic tests using gene expression ratios in lung cancer and mesothelioma." Cancer research 62.17 (2002): 4963–4967.

Parallelization of String Matching Algorithm with Compaction of DFA

Apurva Joshi and Tanvi Shah

Abstract String matching algorithms are widely acknowledged due to its use in many areas such as digital forensics, intrusion detection system, plagiarism checking, bioinformatics. For improving the efficiency of the string matching, speed of matching the strings must be elevated. Hence, an approach has been proposed which would significantly reduce the time for matching the strings. Ternary content addressable memory (TCAM) has been used by many for reducing the time requirement. But TCAM has many disadvantages such as high cost, very high power dissipation, problem due to pipelining. Small-scale applications may not be able to bear all the disadvantages associated with TCAM. Hence, an approach has been proposed which would overcome all the disadvantages associated with TCAM. Modern CPUs have multicore facility. These multiple cores have been exploited to provide parallelism. Parallelism greatly helps to increase the speed of matching the string. Apart from this, reducing the memory requirement for string matching algorithm is also necessary. When reduction in memory requirement and parallelization are applied simultaneously, it provides improved results. High response time would be obtained by using this approach.

Keywords Aho–Corasick algorithm (AC algorithm) · Finite automaton (FA) · Parallelization · String matching

1 Introduction

String matching algorithms are the ones which find whether the given string is present within the larger string. These algorithms are at the heart of many important applications such as intrusion detection system (IDS), bioinformatics, digital

A. Joshi (✉) · T. Shah
Department of Computer Engineering, VJTI, Mumbai, Maharashtra, India
e-mail: apurva.joshi91@gmail.com

T. Shah
e-mail: tanvishahvcet06@gmail.com

© Springer Nature Singapore Pte Ltd. 2018   327
P.K. Sa et al. (eds.), *Progress in Intelligent Computing Techniques: Theory,
Practice, and Applications*, Advances in Intelligent Systems and Computing 518,
DOI 10.1007/978-981-10-3373-5_33

forensics, plagiarism checking. When using string matching in IDS, the security factor and time factor along with accuracy must be considered, and when using in plagiarism checking and bioinformatics, correctness is important. Due to high criticality of applications using string matching, we can conclude that string matching applications are very critical and it is necessary to improve its efficiency.

The Aho–Corasick algorithm [1] is a widely used algorithm, used for the purpose of string matching. There are many factors which have led to wide spread use of this algorithm. AC has deterministic performance, and this is because each symbol scan results in a state transition. This does not depend on some sort of specific input, due to which it is not vulnerable to various attacks. Apart from this, this algorithm has a property that more than one string can be matched in a single pass. As an example, in the word share, there are three keywords, viz. share, hare and are. By using AC algorithm, all the three keywords are obtained in a single pass. Due to these characteristics of AC algorithm, it is widely adopted among all the string matching algorithms. AC algorithm is basically divided in major two stages, viz. (I) construction of finite machine (FA) which would act as a string matching machine and (II) traversal of string using the FA. For the purpose of construction of FA, goto, failure transitions and output function are calculated. goto function helps us create a simple FA. Failure function is the longest suffix of the string which is also the prefix of some node. The goal of the failure function is to restrict the algorithm from transiting to any state more than once. Failure functions would redirect us to correct transition in case the word is present, and if the word is not present, it would give a failure. Output function gives the output of a particular state, i.e. the keyword obtained on reaching a particular state. After the creation of FA with the failure states, the next stage is to inspect the string letter by traversing the FA.

Traditional approach for storage of FA requires storing one rule per transition. This leads to much wastage of memory. For the compaction of FA, [2] proposes such an algorithm which makes it possible for us to store one rule per state which in turn reduces the memory requirement to great extent. This algorithm suggests that the states should be encoded in a way such that all the transitions to a single state should be represented by a single prefix. Due to this, the problem boils down to longest prefix match first. The algorithm mainly consists of following stages, viz. (i) state grouping which finds longest common suffix for each state, (ii) common suffix tree which involves creation of tree by encoding each state with smaller number of bits and (iii) state and node encoding which creates a table containing current state, symbol and next state. Using this approach, compaction of FA has been made.

Parallelism is the process in which we divide the given task into chunks and perform those tasks simultaneously. This helps us reduce response time considerably. Parallelism has reached in various nooks and corners of technical fields. The technique of parallelism has been applied even in the field of string matching, but a hardware approach has been proposed which has its own disadvantages. To overcome those disadvantages, a software approach has been proposed which would exploit the multicore architecture of the CPU and give much faster results than its serial counterpart. In this approach, we assign one thread to each letter and make the traversal. Each thread would traverse the FA individually. The threads assigned to the letters which do not start from starting state are terminated immediately. Due to this, there are not many threads running at the same time and there is not much overhead on the system.

2 Related Work

Many intensive efforts have been made for improving the efficiency of AC algorithm. Methods for compaction of DFA have been proposed. Also, the use of TCAM and IP lookup chips has been proposed. These algorithms have greatly contributed in making the string matching more efficient and time saving. For the real-time applications like IDS, use of ternary content addressable memory is very useful which has very good speed. But this comes with its own disadvantages.

Many efforts have been made to improve the efficiency of string matching keeping the AC algorithm as base. [3] suggests the elimination of failure transitions from AC algorithm. [4] suggests a technique of compression which in turn reduces the memory bandwidth required for processing the string. When processing a string which is of length N, this approach would require minimum 2N traversals. [5–8] propose a mechanism for reducing the space required to store the DFA. Some use Chinese remainder theorem, while some make use of more compact data structure for storage of the larger strings. [8] proposes use of NFA, i.e. nondeterministic finite automaton instead of using a DFA. [9, 10] have suggested methods for reducing the time required to match the string with the larger string by proposing the use of TCAM, i.e. ternary content addressable memory. The most considerable contribution is done by [2] which suggest compact DFA. The DFA used by AC algorithm uses one rule per transition as a result; there are many rules per state. This causes high requirement of memory for storage of DFA. Compact DFA suggests use of only one rule per state. As a result, the storage requirement would reduce, which would ultimately lead to improved throughput. These rules are stored as a triplet of [11]:

$$\text{Current State Field} \rightarrow \text{Symbol Field} \rightarrow \text{Next State Field}$$

3 Implementation

As referred above, the basic intention here is to increase the efficiency of the system, i.e. bring improvement in response time. For that purpose, following steps are followed: An Aho–Corasick machine is created. The technology used for implementation of the AC machine is CPP. goto function, failure function and

Fig. 1 Flow chart for the
proposed system

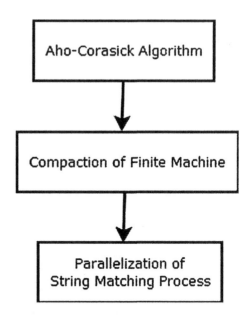

construction of output function for FA are done using CPP. This leads to the
completion of the phase of construction of FA. After that compaction of the FA,
which is created is done. Each state is considered separately, and longest prefix for
individual state is calculated which follows construction of common suffix tree
based on the results of first step and then follows node encoding in which the actual
encoding of each and every node is done. This process gives us a rule set which is
much compact than its AC algorithm counterpart (Fig. 1).

This compact DFA structure has a table which contains current state, symbol and
next state. Current state gives us the state where the transition has reached currently,
symbol is the letter which causes transition from one state to another, and next state
gives us the state where the transition would reach if the symbol found is correct.
This compact data structure is much lesser in size and thus can be stored in faster
memory. Current state for start state is always encoded by ***** and the states
following the start state. For the purpose of parallelization, OpenMP with MPI has
been used. OpenMP is a memory multiprocessing API which can be used with C,
CPP, Fortran, etc. This API takes advantage of many cores available on the CPU.
MPI helps to make distributed computing possible. It helps us to make use of
several cores of different computers for the purpose of computations. Nowadays,
every CPU is multicore. Using this API, one thread can be assigned to each core
and program could be made to run in parallel. Each thread is assigned a single letter
of the larger string. In the first pass, it checks whether the letter matches with the
symbol led created in the above table, having current state as *****. If it does, the
thread proceeds with its transitions, else it terminates then and there. This approach
ensures the property of AC algorithm that all substrings could be identified in a
single pass is maintained. As an example, in the word share, there is a substring

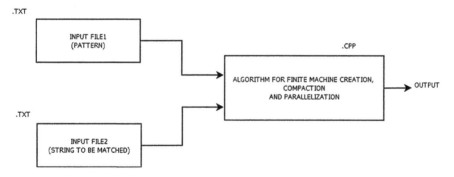

Fig. 2 Architecture for the proposed system

share, hare and are. All these three words could be identified in a single pass with AC algorithm. Along with preserving the property, it also helps us reduce the response time. A small improvement would be to use shared memory for storing the variables instead of using global memory. Due to this, the overhead involved in accessing the global variables will decrease greatly.

Figure 2 shows the architecture. All the patterns are already stored in a text file. These patterns are passed as an input to the program which implements AC algorithm. The program takes these patterns one line at a time and creates a FA and applies compaction on it. After that, the larger string, which should be searched into, is passed as an input to the same program. The program extracts words in parallel (as many threads are assigned through the program) and performs string matching. This explains the total working and architecture of the proposed work.

4 Results

Evaluation of the technique proposed in this paper has been done by checking it on a pattern set which contains about 50,000 different patterns. In the experiment, about 50,000 plus strings to be matched are taken and the response time is obtained. Here, the response time is the addition of time required for construction of FA and time for traversal of FA for finding the matches. The response time required is very less and satisfies the real-time requirements. In comparison with its serial counterpart, the response time reduces effectively. Suppose there are n threads which have been assigned to the system to perform matching, then the decrease in response time of the system is multiple of n.

It is proved fact that the smaller the memory size, higher is the speed of the system. When we perform compaction of FA, the FA could be stored in more compact memory than the amount of memory needed when it was not compacted. Hence, we could conclude that the speed of the system increases significantly. Apart from this, as we are performing task of matching in parallel, the response

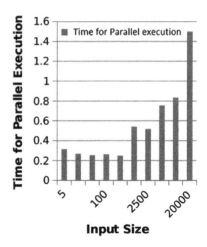

time would decrease further. There is no overhead involved in matching the string in parallel; as a result, the response time decreases without involvement of inner costs. There is no pipelining involved nor there is any stalling of the process of parallelization.

For parallel processing, Fig. 3 shows us the results on string matching. For serial processing, each 100th increase in the input size matters and the response time increases additively for such increase in input size. Figure 4 shows us the results of serial processing on string matching.

Figure 5 Shows Graph representing comparison in response time of serial and parallel execution. Below figure shows the comparison between the parallel

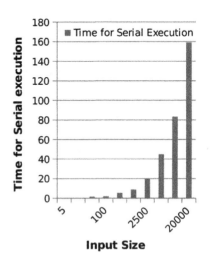

Fig. 5 Graph representing comparison in response time of serial and parallel execution

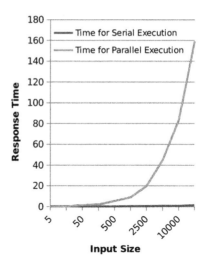

algorithm and its serial counterpart. For small inputs, serial processing takes less time as compared to parallel. This is because of the overhead involved in breaking up the input data set in subsets so that one thread is assigned to each subset. Also, overhead involved in assigning thread to each subset is countable. As the input size increases, the disparity between the response times required for serial and parallel processing increases. For the input set of 20,000, the disparity is highest. Serial processing requires 159.099 s as against 1.4994 s of parallel processing.

5 Conclusion

This paper shows how parallelization and compaction of DFA helps in reducing the response time. Parallelization is obtained without any stalling or any requirement to wait until some other activity gets completed. This helps us to obtain an improvement in response time. How the use of parallelization helps in improvement of string matching algorithms efficiency has been illustrated in this paper. For processing a pattern set as large as 50,000, 100,000, only 3.67 s is taken. Compaction leads to further reduction in time. Lesser the size of memory, more it is fast. Hence, reducing the rules for string matching process also leads to improved results. The approach proposed could be used in all the applications which cannot bear the highly costly TCAM chips, and the application which require power saving facilities could also go with this approach.

References

1. V. Aho and Margaret J. Corasick, Efficient String Matching: An Aid to Bibliographic Search, communications of the ACM June 1975, Volume 18 Number 6.
2. Anat Bremler-Barr, Member, IEEE, David Hay, Member, IEEE, and Yaron Koral, Student Member, IEEE, CompactDFA: Scalable Pattern Matching Using Longest Pre x Match Solutions, IEEE/ACM TRANSACTIONS ON NETWORKING, VOL. 22, NO. 2, APRIL 2014.
3. Sang Kyun Yun Dept. of Comput. Telecommun. Eng., Yonsei Univ., Wonju, South Korea, An Efficient TCAM-Based Implementation of Multipattern Matching using Covered State Encoding, IEEE TRANSACTIONS ON COMPUTERS, VOL. 61, NO. 2, FEBRUARY 2012.
4. M. Becchi and P. Crowley, An improved algorithm to accelerate regular expression evaluation, in Proc. ACM/IEEE ANCS, 2007, pp. 145–154.
5. A. Bremler-Barr, D. Hay, and Y. Koral, CompactDFA: Generic state machine compression for scalable pattern matching, in Proc. IEEE INFOCOM, 2010, pp. 657–667.
6. Tzu-Fang Sheu Inst. of Commun. Eng., Nat. Tsing-Hua Univ., Hsinchu Nen-Fu Huang; Hsiao-Ping Lee, A Time- and Memory- Efficient String Matching Algorithm for Intrusion Detection Systems, Global Telecommunications Conference, 2006. GLOBECOM '06. IEEE.
7. S. Kumar, J. Turner, P. Crowley, and M. Mitzenmacher, HEXA: Compact data structures for faster packet processing, in Proc. IEEE ICNP, 2007, pp. 246–255.
8. R. Sidhu and V.K. Prasanna, Fast Regular Expression Matching Using FP-GAs, Proc. Ninth Ann. IEEE Symp. Field-Programmable Custom Computing Machines (FCCM 01), pp. 227–238, 2001.
9. Y. Weinsberg, S. Tzur-David, D. Dolev, and T. Anker, High performance string matching algorithm for a network intrusion prevention system (NIPS), in Proc. IEEE HPSR, 2006, pp. 147–153.
10. J. van Lunteren, High-performance pattern-matching for intrusion detection, in Proc. IEEE INFOCOM, Apr. 2006, pp. 113.
11. Cheng-Hung Lin, Member, IEEE, Chen-Hsiung Liu, Lung-Sheng Chien, and Shih-Chieh Chang, Member, IEEE, Accelerating Pattern Matching Using a Novel Parallel Algorithm on GPUs, IEEE TRANSACTIONS ON COMPUTERS, VOL. 62, NO. 10, OCTOBER 2013.

A Novel Speckle Reducing Scan Conversion in Ultrasound Imaging System

Dipannita Ghosh, Debashis Nandi, Palash Ghosal and Amish Kumar

Abstract Quality of ultrasound image is dominantly limited by two major issues such as low resolution and speckle noise. The existing speckle reduction techniques are mostly applied either before or after scan conversion. Filtering before scan conversion results in huge computational load since the amount of data handled is quite large while filtering after scan conversion provides poor image quality. In this paper, a novel and computationally efficient filtering technique has been proposed where filtering is performed along with scan conversion using spatial linear adaptive and nonlinear filters in two directions of scan conversion geometry. The proposed framework is found suitable for the real-time applications and improves the visual quality of the image. Quality metrics for the proposed method have been compared to other existing methods to show the novelty of the work.

Keywords Ultrasound image · Scan conversion · Speckle reduction

1 Introduction

Ultrasound imaging modality is predominantly used as a diagnostic tool in modern medicine. It is a noninvasive means of examining body's internal organs and is practically risk-free to human body. The quality of the ultrasound image is largely affected by a prominent factor known as speckle noise. Speckle is an inherently generated

D. Ghosh (✉) · D. Nandi · P. Ghosal · A. Kumar
Department of Information Technology, National Institute of Technology,
Durgapur 713209, West Bengal, India
e-mail: dipannitaghosh21@gmail.com

D. Nandi
e-mail: debashisn2@gmail.com

P. Ghosal
e-mail: ghosalpalash@gmail.com

A. Kumar
e-mail: amishkumar562@gmail.com

© Springer Nature Singapore Pte Ltd. 2018 335
P.K. Sa et al. (eds.), *Progress in Intelligent Computing Techniques: Theory,*
Practice, and Applications, Advances in Intelligent Systems and Computing 518,
DOI 10.1007/978-981-10-3373-5_34

noise by the ultrasound imaging acquisition system degrading resolution and contrast of the image. Speckle appears as a granular pattern [1] originating from a waveform with various independent scattered components. It seems as bright and dark spots over the surface of the image and brings difficulties to experts in medical diagnosis. Thus, speckle reduction becomes a rising area of research in order to make the ultrasound imaging modality comparable to the other medical imaging practices in terms of image excellence. In ultrasound imaging system, speckle noise is generally suppressed at the preprocessing stage or at the post-processing stage. The existing noise reduction filtering techniques [2–11] are, therefore, mostly applied either on the raw scan data (i.e., before scan conversion) or on the scan-converted images (i.e., after scan conversion). Filtering on raw scan data generates huge computational load whereas filtering after scan conversion lacks in image quality as information content in image hampers largely during scan conversion. Consequently, a new efficient speckle reduction technique is proposed which unifies filtering and scan conversion simultaneously, i.e., filtering along with scan conversion. The resultant image of filtering scan conversion is the outcome of filtering separately in the two directions of scan conversion geometry: radial direction and horizontal direction. The choice of filter can be made different in the two directions of scan conversion filtering. The present article, therefore, has investigated the performance of the speckle filtering scan conversion using either same or different combinations of linear and nonlinear filters imposed in two different phases of scan conversion.

2 Noise Model

The noise embedded to the ultrasound scan lines is the combined form of speckle and Gaussian noise. Speckle is multiplicative in nature and generally assumed as Rayleigh distribution. This speckle noise is multiplied with the signal and then log-compressed by the logarithmic amplifier. As a result, the multiplied speckle noise becomes additive to the log-compressed signal at the output of the logarithmic amplifier. Afterward, the Gaussian noise is added to the logarithmic amplifier output for the simulation of the ultrasound scan lines that are made corrupted by speckle and Gaussian noise. The noisy signal is then scan-converted. The noisy scan data is thus modeled as

$$S_0(i,j) = S(i,j) + n_G(i,j) + n_s p(i,j) . \tag{1}$$

where $S(i,j)$ is the log-compressed signal, $n_G(i,j)$ is the additive Gaussian noise, and $n_s p(i,j)$ is the log-compressed speckle noise.

3   Speckle Reduction Methods

The existing speckle reduction methods can be broadly categorized as compounding techniques [12, 13], spatial linear adaptive and nonlinear filtering [3–7], multi-scale method [8–11], non-local means denoising [14, 15], and sparse representation-based denoising [16]. Lee filter [3], Kuan filter [5], Dutt and Greenleaf [7], Bamber and Daft [4] are the examples of linear spatial adaptive filters. These filters are based on local statistics and perform filtering within the fixed size window centering the pixel under consideration increasing smoothness in homogeneous region of the image. Median, weighted median, adaptive weighted median [6], and directional median [17] are some of the nonlinear filters useful for preserving edges in an image while reducing random noise. The homogeneity map method (HMM) is introduced [18] for speckle reduction based on the mapping of homogeneous and non-homogeneous regions of the speckled image. Again, optimized Bayesian NL-means with block selection (OBNLM) [15] uses the adaptation of non-local (NL) means using Bayesian formulation. It uses the Pearson distance for patch comparison for speckle reduction generating a competitive performance.

Scan conversion in ultrasound imaging system is conventionally done by interpolation method. In the proposed method, the interpolation is restored by filtering using some of the spatial linear and nonlinear speckle reduction filters and is also compared with the two competitive state-of-the-art methods such as HMM and OBNLM.

4   Proposed Method for Speckle Reduction Through Filtering Scan Conversion

The simplified schematic block diagram of a typical diagnostic B-mode ultrasound imaging system that employs filtering scan conversion is shown in Fig. 1. In filtering scan conversion framework, the reconstruction of image from noisy scan data is performed along two directions in two successive phases, respectively: (i) filtering scan conversion along radial direction (phase 1) and (ii) filtering scan conversion along horizontal direction (phase 2). The essential steps of the proposed filtering scan conversion algorithm for obtaining the rectangular grids have been discussed in detail below.

Phase 1

Aim: Computation of the values at the points where radial lines intersect with the horizontal grid lines marked with solid samples in Fig. 2a.
Assumption: Solid triangular points in scan conversion geometry (Fig. 2a) are the sample points.

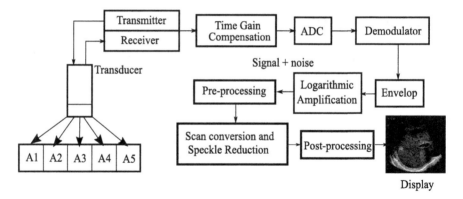

Fig. 1 Block diagram of ultrasound imaging system (speckle reduction is done with scan conversion)

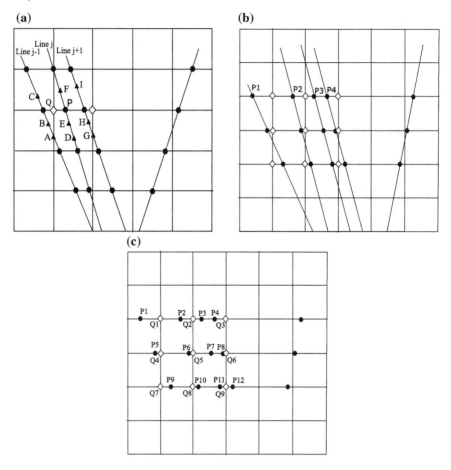

Fig. 2 **a** Scan conversion geometry **b** Geometry of first-stage computation **c** Pixel geometry for raster grid points computation

Steps

1. Consider three successive lines, Line $j - 1$, Line j, and Line $j + 1$.
2. P is considered as a point where the radial line, i.e., Line j cuts the horizontal grid line.
3. Find the nine surrounding nearest points around P.
4. A, B, C, D, E, F, G, H, and I are the nearest points around P along Line $j - 1$, Line j, and Line $j + 1$. E is the nearest along Line j termed as s(nearest,j).
5. To calculate the pixel value at P, these nine samples around point P are considered as a member of local window.
6. Use any spatial linear or nonlinear filters to find P. To adapt Kuan filter [5], the following equation is used

$$p = \bar{s} + k[\theta_r - \bar{s}] . \tag{2}$$

where $\bar{s}$ is the average value of the pixels within the local window and θ_r is the interpolated value at the desired point P on the jth scan line for rth-order interpolation. The parameter k can be determined from the equation below

$$k = \frac{Var(x)}{Var(x) + \bar{x}^2 + Var(x)} . \tag{3}$$

where $\bar{x}$ and $Var(x)$ represent the mean and variance of local window, respectively. The interpolated value can be taken as $\theta_r = s(nearest, j)$, i.e., nearest neighbor (*or zeroth − order interpolated value*).

7. Find all the points where radial lines cut the horizontal grid lines by moving the window along the scan lines. P1, P2, and P3 are such points as shown in Fig. 2b.

The geometry will be then converted as shown in Fig. 2c. Now, with the help of available points P1, P2, and P3, the raster grid points of the raster scan are computed. It can be illustrated with the help of Fig. 2c.

Phase 2

Aim: Computation of the pixel values at the grid points of the Cartesian coordinates (i.e., where vertical grid line cuts the horizontal grid lines) from the computed values of the previous step. Pixel values at raster grid points Q1, Q2, and Q3 (Fig. 2c) are to be computed.

Steps

1. Consider the raster grid point Q5 in the ith row and jth column. Find three nearest points of Q5 along ith row.
2. P7, P6, and P8 are such three nearest points. P7 is the nearest one assigned as p(i,nearest).
3. Find three nearest points from previous and next rows, i.e., $(i − 1)$th row and $(i + 1)$th row.
4. Find the grid point Q2 of the same column and $(i − 1)$th row and search three points around Q2 along the row. In a similar way, three nearest points from next row, i.e., $(i + 1)$th row can be found out.

5. Finally, the pixel point at the grid point Q5 can be computed from these nine points by using similar type of Eq. 2

$$q = \bar{p} + k[\phi_r - \bar{p}] . \qquad (4)$$

where $\bar{p}$, ϕ_r indicate the same meaning as of $\bar{s}$, ϕ_r in Eq. 2. The final pixel values at the grid point are evaluated by moving the window along horizontal direction.

In the proposed method, filtering can be performed in two different manners: FSC (2 pass) and FSC (1 pass). When filtering is performed in both the radial and horizontal directions, the technique is known as speckle reduction through filtering scan conversion in two pass (FSC (2 pass)). After computation of the point P, the raster grid point can also be computed by simple linear interpolation. This technique is termed as speckle reduction through filtering scan conversion in single pass (or FSC (1 pass)).

5 Simulation and Results

The efficiency of the algorithm has been evaluated by generating phantom data from analytic function $f(x, y) = \frac{1}{4}[\sin(wx)]$, placing few holes of different sizes. The simulations were performed in MATLAB and were tested in Windows 7 Home Basic, Intel(R) Core(TM) i5-4690 CPU @ 3.50 GHz processor, 4 GB RAM, 32-bit OS, with 500 GB hard disk.

The scan lines of the raw scan data are corrupted with the three levels of speckle and Gaussian combined parameters, speckle ($\sigma = 1.0$) with 20 dB Gaussian noise (overall noise = 17.6 dB), speckle ($\sigma = 1.5$) with 15 dB Gaussian noise (overall noise = 12.6 dB), and speckle ($\sigma = 2.0$) with 10 dB Gaussian noise (overall noise = 7.6 dB).

A quantitative measure of the image quality is performed using four well-defined quality metrics such as MSE (mean squared error), PSNR (peak signal-to-noise ratio), MSSIM (mean structural similarity) [19], and BLUR [20]. Table 1 shows the performance of Lee filter in terms of quality metric in the proposed framework both

Table 1 Quality metrics of Lee filter output under combined speckle and Gaussian noise (overall SNR 17.6 dB)

Lee	Filtering before scan conversion	Filtering after scan conversion	Filtering with scan conversion (2 pass)	Filtering with scan conversion (1 pass)
MSE	1090.19	1363.12	652.47	1162.47
PSNR	17.79	16.82	20.02	17.51
MSSIM	0.8375	0.7911	0.8949	0.8285
BLUR	0.2541	0.2837	0.3492	0.2493

in pass1 and pass2 (Lee filter in both directions) over the two conventional filtering techniques. It is found that filtering scan conversion (2 pass) gives the best performance in terms of MSE, PSNR, MSSIM, and BLUR.

As compared to filtering before and after scan conversion, the proposed method achieves a PSNR improvement of 2.23 dB and 3.2 dB, respectively.

The proposed algorithm also employs four combinations of spatial linear and nonlinear filters in the two paradigms of speckle filtering scan conversion such as Lee-Kuan, Lee-Med, Med-Lee, and Med-Med. 'Lee-Kuan' indicates Lee filtering in radial direction and Kuan filtering in horizontal direction. The same notation is followed for other filter combinations also. Med is indicating median filter. Performance of these filtering combinations in the proposed framework is also compared with two prominent speckle reduction filters, HMM and OBNLM, as shown in Table 2. In terms of MSE and PSNR, 'Lee-Kuan' achieves the best performance with an improvement of 7.75 in MSE and a significant improvement in PSNR can be achieved with respect to 'Lee-Med' that attains the second best performance. The PSNR for HMM and OBNLM is as low as compared to 'Lee-Kuan,' 'Lee-Med,'

Table 2 Quality metrics of different filtering combinations for the noise levels

MSE						
Noise level (dB)	Lee-Kuan	Lee-Med	Med-Lee	Med-Med	HMM	OBNLM
17.6	650.0608	657.8171	678.9877	781.004	1287.915	1253.893
12.6	748.2261	820.9659	1064.882	1350.91	2007.842	2016.524
7.6	806.7585	968.2591	1661.91	2277.55	3067.872	3092.939

PSNR						
Noise level (dB)	Lee-Kuan	Lee-Med	Med-Lee	Med-Med	HMM	OBNLM
17.6	20.0329	20.0101	19.837	19.2351	17.0822	17.1965
12.6	19.4772	19.1384	17.9217	16.9198	15.1552	15.132
7.6	19.1112	18.32	15.9784	14.5926	13.2762	13.244

MSSIM						
Noise level (dB)	Lee-Kuan	Lee-Med	Med-Lee	Med-Med	HMM	OBNLM
17.6	0.8849	0.8636	0.8708	0.8399	0.847	0.847
12.6	0.8246	0.797	0.7923	0.7509	0.7048	0.7088
7.6	0.7324	0.6996	0.6686	0.625	0.5782	0.5743

BLUR						
Noise level (dB)	Lee-Kuan	Lee-Med	Med-Lee	Med-Med	HMM	OBNLM
17.6	0.384	0.2782	0.3767	0.2553	0.2959	0.3142
12.6	0.349	0.2521	0.337	0.2305	0.247	0.2547
7.6	0.3086	0.2249	0.2985	0.2068	0.223	0.2224

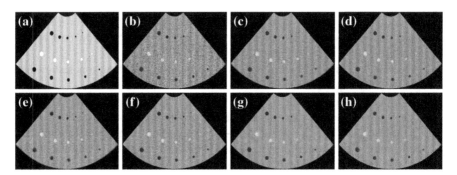

Fig. 3 **a** Original image **b** noisy image, overall noise 17.6 dB (speckle $\sigma = 1.0$, Gaussian $= 20$ dB) **c** Lee-Kuan **d** Lee-Med **e** Med-Lee **f** Med-Med **g** HMM **h** OBNLM filter output for speckle parameter $\sigma = 1.0$ combined with 20 dB Gaussian noise (overall noise 17.6 dB)

'Med-Lee,' and 'Med-Med.' The output images (for 17.6 dB overall noise) for different filtering combinations are represented in Fig. 3. Figure 4 shows the plot of the quality metric for three levels of noise. Table 3 depicts the real-time application of the proposed framework in terms of normalized mean time and variance, and Table 4 shows the running time of HMM and OBNLM on the scan-converted image. Filtering scan conversion examined with real scan data of ultrasound machine is given in Fig. 5.

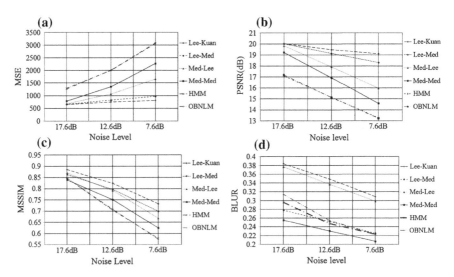

Fig. 4 **a** MSE **b** PSNR **c** MSSIM **d** BLUR at overall input SNR 17.6 dB (speckle noise $\sigma = 1.0$ added with 20 dB Gaussian noise), 12.6 dB (speckle noise $\sigma = 1.5$ added with 15 dB Gaussian noise), and 7.6 dB (speckle noise $\sigma = 2.0$ added with 10 dB Gaussian noise.)

Table 3 Normalized mean time (s) and variance of the filtering techniques in different paradigms

	Filtering before scan conversion		Filtering after scan conversion		Filtering with scan conversion	
	Normalized mean time	Variance	Normalized mean time	Variance	Normalized mean time	Variance
Lee-Kuan	0.989483	0.004642	0.695837	0.030766	0.826098	0.007313
Lee-Med	0.9892935	0.004867	0.721017	0.029154	0.839843	0.010799
Med-Lee	0.976750	0.004573	0.652359	0.017553	0.825114	0.004467
Med-Med	0.979409	0.010550	0.678938	0.022477	0.8522706	0.007719

Table 4 Normalized mean time (s) and variance of HMM and OBNLM filtering techniques

	Normalize mean time	Variance
HMM	0.56408	0.005993
OBNLM	0.41979	0.001993

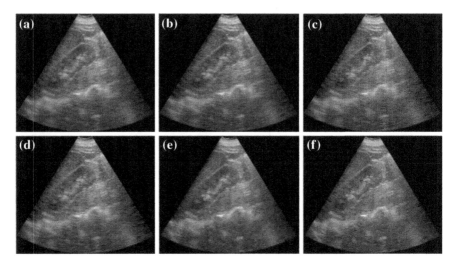

Fig. 5 **a** Lee-Kuan **b** Lee-Med **c** Med-Lee **d** Med-Med **d** HMM **e** OBNLM Filter output obtained from scan data of real ultrasound machine

6　Conclusion

Despite speckle filtering in pre- or post-processing stages, the article presents a novel framework of speckle suppression where filtering and scan conversion are unified as a single operation. The image quality metric justifies the proposed methods' efficiency when compared with other two techniques. The proposed algorithm is also compared with the two speckle reducing techniques lying in the state of the art: HMM and OBNLM. Though the image output of HMM and OBNLM looks less noisy but the fine details are found absent. On concluding the results in terms of running time of the algorithm, filtering with scan conversion takes comparable time with respect to other frameworks. Though HMM and OBNLM takes less time than the proposed technique but there is a high prospect of object loss as it is visualized from Figs. 3 and 5. The proposed algorithm is therefore well-suited with the real-time application of ultrasound imaging system with enhanced output images.

References

1. Goodman, J.W.: Some fundamental properties of speckle. J. Opt. Soc. Am. 66, 1145–1150 (1976)
2. Michailovich, O.V., Tannenbaum, A.: Despeckling of Medical Ultrasound Images. IEEE Transactions on Ultrasonics Ferroelectrics and Frequency Control. 53(1), 64–78 (2006)
3. Lee, J.S.: Refined Filtering of Image Noise Using Local Statistics. Computer Graphics and Image Processing. 15(1), 380–389 (1981)
4. Bamber, J.C., Daft, C.: Adaptive filtering for reduction of speckle in ultrasonic pulse-echo images. Ultrasonics. 24(1), 41–43 (1986)
5. Kuan, D.T., Sawchuk, A.A., Strand, T.C., Chavel, P.: Adaptive restoration of images with speckle. IEEE Transactions Acoustics. Speech and Signal Processing. 35(3), 373–383 (1987)
6. Loupas, T., McDicken, W.N., Allan, P.L.: An Adaptive Weighted Median Filter for Speckle Suppression in Medical Ultrasonic Images. IEEE Transactions on Circuits and Systems. 36(1), 129–135 (1989)
7. Dutt, V., Greenleaf, J.F.: Adaptive speckle reduction filter for log compressed B-scan images. IEEE Transactions on Medical Imaging. 15(6), 802–813 (1996)
8. Perona, P., Malik, J.: Scale-Space and Edge Detection Using Anisotropic Diffusion. IEEE Transactions on Pattern Analysis and Machine Intelligence. 4(7), 629–639 (1990)
9. Krissian, K., Fedrij, C.: Oriented Speckle reducing aniosotropic diffusion. IEEE Transactions on Image Processing. 16(5), 1412–1424 (2007)
10. Donoho, D.L.: De-Noising by Soft-Thresholding. IEEE Transactions on Information Theory. 41(3), 613–627 (1995)
11. Gleich, D., Datcu, M.: Wavelet-Based SAR Image Despeckling and Information Extraction Using Particle Filter. IEEE Transactions on Image Processing. 18(10), 2167–2184 (2009)
12. Behar, V., Adam, D., Friedman, Z.: A new method of spatial compounding imaging. Ultrasonics. 41(5), 377–384 (2003)
13. Li, P.C., Chen, M.J.: Strain Compounding: A New Approach for speckle reduction. IEEE Transactions on Ultrasonics Ferroelectrics and Frequency Control. 49(1), 39–46 (2002)
14. Buades, A., Coll, B., Morel, J.M.: A non-local algorithm for image denoising. Computer Vision and Pattern Recognition. CVPR. IEEE computer Society Conference. 2, 60–65 (2005)
15. Coupe, P., Hellier, P., Kervrann, C., Barillot, C.: Nonlocal Means-Based Speckle Filtering for Ultrasound Images. IEEE Transactions on Image Processing. 18, 2221–2229 (2009)
16. Foucher, S.: SAR image Filtering via learned dictionaries and sparse representations. Geoscience and Remote sensing symposium. IGARSS. IEEE International. 1, I-229–I-232 (2008)
17. Czerwinski, R. N., Jones, D. L., OBrien, Jr. W. D.: Ultrasound Speckle Reduction by Directional Median Filtering'. IEEE Proceedings. International Conference on Image Processing. 1, 358–361 (1995)
18. Gungor, M.A., Karagoz, I.: The homogeneity map method for speckle reduction in diagnostic ultrasound images. Measurement: Journal of the International Measurement Confederations. 68, 100–110 (2015)
19. Wang, Z. and Bovik, A. C.: Image quality Assessment: From Error Visibility to Structure Similarity, IEEE Trans. on Image Processing. 13(4), 600–612 (2004)
20. Crete, F., Dolmiere, T., Ladret P., Nicolas M.: The Blur Effect: Perception and Estimation with a New No-Reference Perceptual Blur Metric. SPIE Electronic Imaging Symposium Conf Human Vision and Electronic Imaging. San Jose:tats-Unisd'Amrique (2007)

Hierarchical Clustering Approach to Text Compression

C. Oswald, V. Akshay Vyas, K. Arun Kumar, L. Vijay Sri
and B. Sivaselvan

Abstract A novel data compression perspective is explored in this paper and focus is given on a new text compression algorithm based on clustering technique in Data Mining. Huffman encoding is enhanced through clustering, a non-trivial phase in the field of Data Mining for lossless text compression. The seminal hierarchical clustering technique has been modified in such a way that optimal number of words (patterns which are sequence of characters with a space as suffix) are obtained. These patterns are employed in the encoding process of our algorithm instead of single character-based code assignment approach of conventional Huffman encoding. Our approach is built on an efficient cosine similarity measure, which maximizes the compression ratio. Simulation of our proposed technique over benchmark corpus clearly shows the gain in compression ratio and time of our proposed work in relation to conventional Huffman encoding.

Keywords Hierarchical clustering · Compression ratio · Cosine similarity measure · Huffman encoding · Lossless compression

C. Oswald (✉) · L. Vijay Sri · B. Sivaselvan
Department of Computer Engineering, Indian Institute of Information Technology,
Design and Manufacturing Kancheepuram, Chennai, Tamil Nadu, India
e-mail: coe13d003@iiitdm.ac.in

L. Vijay Sri
e-mail: coe13b014@iiitdm.ac.in

B. Sivaselvan
e-mail: sivaselvanb@iiitdm.ac.in

V. Akshay Vyas · K. Arun Kumar
Department of Computer Science and Engineering, Department of Information
Technology, Sona College of Technology, Salem, Tamil Nadu, India
e-mail: vyasakshay4@gmail.com

K. Arun Kumar
e-mail: arunk8517@gmail.com

© Springer Nature Singapore Pte Ltd. 2018   347
P.K. Sa et al. (eds.), *Progress in Intelligent Computing Techniques: Theory,*
Practice, and Applications, Advances in Intelligent Systems and Computing 518,
DOI 10.1007/978-981-10-3373-5_35

1 Introduction

With the advent of the WWW, the need for transmission and storage of large amount of data has increased. To transmit large data in the form of text, images, videos, etc. over network channels and to reduce the storage space occupied, data compression techniques are of immense need. Since data in the machine is stored and transmitted in the form of bits and the technique where the number of bits is reduced to store the data is termed as compression, types of compression include lossy or lossless [1]. Lossy compression reduces the size by removing irrelevancy in addition to redundancy. It produces better compression than lossless technique at the cost of reduction in quality to level which is not visually perceptible. Some of them are JPEG, MP3, MPEG, PGF, etc. The process of reconstructing original data from compressed data is termed as lossless compression, and some of them are discussed in Sect. 2 [1].

This paper concentrates on bringing an efficient version of Huffman encoding which is one of the seminal algorithm and is a lossless compression technique [2]. Some commercial solutions use it as an intermediary phase and the codes are prefix free. For lossless compression techniques, many models based on statistical, sliding window, and dictionary have been proposed. A huge memory space to store the dictionary data structure, where a large static collection of words is involved, forms the major disadvantage of these models. Almost all of these algorithms employ character/pattern (sequence of characters)-based encoding [1].

Induction, compression, approximation, search, and querying were the five perspectives of Data Mining identified by Naren Ramakrishnan et al. [3]. The scope of Data Mining in the domain of data compression is explored in our work. Data Mining is the process of extracting hidden and useful information from large DB's [4]. The condensed/compressed representation of the original large data is the result of the knowledge (pattern base) represented, when Data Mining is viewed as a compression technique. Data Mining techniques include ARM, clustering, outlier analysis, classification. [4, 5]. They are widely applied in personal recommendation systems such as Amazon and Priority Inbox (Gmail), and Medical Diagnosis. To the best of our knowledge, no literature exists which uses hierarchical clustering approach blended with cosine similarity measure to perform text compression. We have used the concept of hierarchical clustering, an important technique in Data Mining in combination with lossless compression to group similar words (patterns) of a text for efficient compression. This work exploits the principle of assigning shorter codes to frequently occurring words in relation to single character-based approach of Huffman encoding. Moreover, we concentrate on employing an efficient similarity measure to improve the intracluster similarity by grouping frequently occurring words for an efficient compression.

The technique of grouping/clustering a set of data points such that points in the same group/cluster are highly similar to each other than to those in other clusters is referred as clustering. Clustering results in high intrasimilarity within the clusters and less intersimilarity with other clusters. Several similarity measures are used to cluster the data in the literature [4, 6, 7]. We felt it is better to use cosine similarity measure to cluster data and have used agglomerative clustering. The cosine of the angle AB is the dot product of A and B divided by the product of the lengths of the vectors A and B [6]. That is, the cosine is,

$$\cos \theta = \frac{\sum_{i=1}^{n}(A_i B_i)}{\sqrt{\sum_{i=1}^{n} A_i^2} \sqrt{\sum_{i=1}^{n} B_i^2}}$$

The similarity obtained, ranges from -1 meaning exactly opposite to 1 meaning exactly the same where 0 indicates orthogonality (De-correlation) and s indicates intermediate similarity or dissimilarity where $0 < s < 1$. In Sect. 2, we present the related literature of text compression and hierarchical clustering. We propose the design of our clustering-based Huffman algorithm approach for text compression in Sect. 3. In Sect. 4, simulation results are shown. The summary and future work in Sect. 5 is presented finally.

2 Related Work

The seminal work on text compression was proposed by Shannon Fano and David A. Huffman in the year 1948 and 1952 [2, 8]. Those were entropy encodings, giving prefix codes by assigning short codes to frequent characters. Arithmetic encoding, run-length encoding, and adaptive Huffman encoding are a few and they have their own demerits [9–11]. In sliding window-based techniques like LZ77 family of algorithms, by employing a dictionary and selecting strings from input data, each string is encoded as a token. It is not true in all cases that the patterns in the text occur close together and this assumption is a disadvantage of these methods. Some of them are LZR, LZHuffman, LZPP, LZX, etc. [1]. Dictionary-based algorithms such as LZ78, LZW, UNIX Compress, ZIP, and RAR suffer from the limitations of non-optimal codes. In these algorithms, large dictionaries are created, and hence, it costs time and memory heavily. In all the above given methods, focus was given mostly to character-based encoding/sequence of character-based encoding. [12] has shown that text compression by frequent pattern mining (FPM) technique is better than conventional Huffman, but time taken to compress is more. Our algorithm takes lesser time than [12] using efficient clustering mechanism.

A short discussion on data clustering follows. The most well-known partitioning algorithm is the k-means approach which is good in terms of its simplicity and ease of implementation. Some of the partitioning algorithms are k-medoids, PAM, and CLARA [4]. In Hierarchical clustering, each point is taken as its own cluster. Using one of many definitions of close, clusters are combined based on their closeness [6]. Until all the clusters are merged into one, or a termination condition holds, it successively merges the clusters close to one another [4]. The advantages of these methods is the smaller computation costs where combinatorial number of different choices is not of major concern. AGNES (AGglomerative NESting), BIRCH, and DIANA (DIvisive ANAlysis) were to name a few [4, 7, 13–15]. A detailed survey can be seen in [7].

3 Proposed Clustering-Based Huffman Algorithm (CBH)

The input file T with total words W is set to find the number of unique words $w(w \leq W)$ and assign a unique Word ID for each unique word. The words are tokenized based on the character *space*. This method of tokenizing has an advantage of reducing computation that includes *newline* being computed along with the other characters. The text file T containing W words is partitioned equally into x partitions. For every partition $p_i[1 \leq i \leq x]$, the unique words contained in it are found. A partition frequency vector matrix f_p of size $x \times w$ is constructed where p_i is the partition and $p_y[1 \leq y \leq w]$ is a unique word *id* with word y contained in partition p_x. Let us consider an example given below.

Text T: *where there is a will there is a way*
Let us assume x: 4, W: 9, and w: 6 where q_1 = where $''$ ($''$ denotes space), q_2 = there $''$, q_3 = is $''$, q_4 = a $''$, q_5 = will $''$, and q_6 = way and number of unique words in every partition are p_1 : 2, p_2 : 2, p_3 : 2, and p_4 : 3.

$$
\text{Partition frequency vector matrix } f_p = \begin{array}{c} \\ p_1 \\ p_2 \\ p_3 \\ p_4 \end{array} \overset{\displaystyle q_1 \ q_2 \ q_3 \ q_4 \ q_5 \ q_6}{\begin{bmatrix} 1 & 1 & 0 & 0 & 0 & 0 \\ 0 & 0 & 1 & 1 & 0 & 0 \\ 0 & 1 & 0 & 0 & 1 & 0 \\ 0 & 0 & 1 & 1 & 0 & 1 \end{bmatrix}}
$$

These partitions are clustered based on the cosine similarity measure. Initially, x partitions are the x clusters. The cosine similarity is found between all pairs of cluster frequency vectors and the pair having maximum similarity is grouped into a single cluster. In the example, $cos(p_2, p_4) = 0.816497$ and $cos(p_1, p_3) = 0.5$ and so *Cluster*1 contains p_1 and p_3 and *Cluster*2 contains p_2 and p_4. After clustering the pair, the newly formed vector's frequency is the sum of frequencies of both the vectors and this process terminates until 2 clusters are formed. Each of the 2 clusters that are formed have the frequencies of the unique words w, so that the respective Huffman codes can be generated individually for each cluster. A cluster frequency vector f_c (now f_p is modified to f_c, once the clusters are finalized) eliminates those q_y's where frequency of $q_y = 0$. This forms a modified cluster frequency vector matrix $f_{c_m}[m \in \{1, 2\}]$ with only nonzero frequency in the vector space. The Huffman codes are now generated for f_{c_m}. The entire algorithm is explained in Algorithm 1 and Procedure 2.

The purpose of generating Huffman codes for f_{cm} is to minimize the code length being generated in Huffman algorithm. The Huffman codes for words in cluster c_1 are $q_1 : 10$, $q_2 : 0$, and $q_5 : 11$. The Huffman codes for words in Cluster c_2 are $q_3 : 11$, $q_4 : 0$, and $q_6 : 10$. The procedure for generating Huffman codes is same as conventional Huffman algorithm with the only difference being encoding words instead of characters. A unique cluster ID corresponding to the cluster is appended, before encoding every partition to differentiate them from which cluster they come from. The decoding is done similar to the conventional Huffman decoding with the difference being, the pattern for the code is taken from the corresponding cluster ID it is appended to.

$$f_c = \begin{matrix} \\ c_1 \\ c_{24} \\ c_3 \end{matrix} \begin{matrix} q_1\ q_2\ q_3\ q_4\ q_5\ q_6 \\ \begin{bmatrix} 1\ 1\ 0\ 0\ 0\ 0 \\ 0\ 0\ 2\ 2\ 0\ 1 \\ 0\ 1\ 0\ 0\ 1\ 0 \end{bmatrix} \end{matrix} \quad \text{Final } f_c = \begin{matrix} \\ c_{13} \\ c_{24} \end{matrix} \begin{matrix} q_1\ q_2\ q_3\ q_4\ q_5\ q_6 \\ \begin{bmatrix} 1\ 2\ 0\ 0\ 1\ 0 \\ 0\ 0\ 2\ 2\ 0\ 1 \end{bmatrix} \end{matrix} \quad f_{c_1} = c_{13} \begin{matrix} q_1\ q_2\ q_5 \\ \begin{bmatrix} 1\ 2\ 1 \end{bmatrix} \end{matrix} \quad f_{c_2} = c_{24} \begin{matrix} q_3\ q_4\ q_6 \\ \begin{bmatrix} 2\ 2\ 1 \end{bmatrix} \end{matrix}$$

f_c is the summed up cluster frequency vector matrix; final f_c is the final cluster frequency vector matrix; and f_{c_1} and f_{c_2} are modified cluster frequency vector matrices from which the Huffman codes are generated.

4 Experimental Results

We tested our CBH algorithm over Calgary compression corpus datasets, which is a benchmark data which involves files in the range 76 kB–10 MB [16]. The method for CBH algorithm is written in C and executed in Ubuntu on 2:26 GHz machine with 4 GB memory. Table 1 reports the compression ratio and execution time of the tested algorithms. Figure 1 shows the efficiency of our algorithm in terms of compression ratio (C_r) of the proposed clustering-based Huffman technique in relation to conventional Huffman encoding, at varying partition size (x) for census and bible dataset while partition size is fixed from 2 to 9. The compression ratio C_r is defined as follows:

$$C_r = \frac{\text{Uncompressed size of text}}{\text{Compressed size of text}}$$

The code table's size along with the encoded text size post CBH is denoted by the compressed size. Our algorithm significantly outperforms the seminal conventional Huffman in compression ratio, with lesser partition size ($2 \leq x \leq 9$). In the bible and census corpora, the C_r obtained at $x = 9$ is 2.82932 and 6.40910 whereas conventional Huffman gave a C_r of 1.82434 and 1.75479 only. This is because, when partitions in T with high cosine similarity form clusters, it leads to clusters of high inter similarity measure. This implies that more non-unique words (less % of unique words) with high frequency exist across clusters, yielding less code table size. Moreover, total number of words W across the final two clusters are unequally distributed. This leads to almost same code length for words. In our approach, encoded size is

Table 1 C_r for various benchmark datasets

File name	Size (bytes)	W	% of unique words	Code table size (bytes)	Conv. Huffman C_r	CBH C_r	C_r Efficiency (%)	Time for CBH (s)	Conv. Huffman time (s)
Bible	4,047,392	766,112	3.88	522,221	1.8243	**2.82932**	35.52	5931.74	1.82
Census	10,485,371	1,690,219	1.14	329,008	1.7547	**6.4091**	72.62	2213.07	1.75
World192	2,473,400	303,739	16.80	966,027	1.6009	**1.73716**	7.83	1808.21	1.60

Fig. 1 Compression ratio versus x for *bible* and *census*

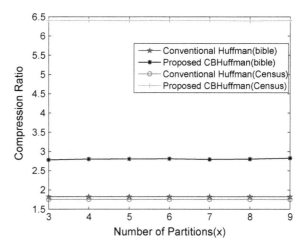

more than the code table size and this is because of the presence of more non-unique words, and every word q_y can get two different codes and the length of the them differs. In most of the datasets, in the final 2 clusters, more number of partitions fall into a single cluster leading to a biased situation. Since we encode words instead of characters, encoded size of CBH is much lesser (34% reduction) than conventional Huffman, even though code table size of CBH is more.

Algorithm 1 Clustering based Huffman Encoding(CBH)

Input: Text File: T, Number of partitions: x
Output: Encoded File: T' ▷

1: Scan T once to find unique words w from the total no. of words W; $w \leq W$
2: Divide T such that number of words contained in p_i [$1 \leq i \leq x$] $= \lfloor W/x \rfloor$
3: **for** each p_i **do**
4:     Use w to form Partition Frequency Vector f_{p_i} where $f_{p_i} = c_1 q_1 + c_2 q_2 + c_3 q_3 + \ldots + c_w q_w$,
    $c_1, c_2, c_3, \ldots, c_w$ are the frequencies of $q_1, q_2, q_3, \ldots, q_w$.
5: **end for**
6: $f_c = form\_Cluster(f_p)$ where $f_c = \begin{pmatrix} f_{c_1} \\ f_{c_2} \end{pmatrix}$, $c = \{c_1, c_2\}$, $p = \{p_1, p_2, \ldots, p_i\}$ and $f_p = \begin{pmatrix} f_{p_1} \\ f_{p_2} \\ \vdots \\ f_{p_i} \end{pmatrix}$
7: **for** each c_m **do**
8:     **for** each q_y in c_m **do**
9:         Eliminate q_y from f_{c_m} where frequency of $q_y = 0$ to get modified cluster frequency matrix f_{c_m}
10:     **end for**
11:     $gen\_HuffmanCodes(f_{c_m})$ //Generates Huffman codes using conventional method and stores in cod table c_1 and c_2
12: **end for**
13: **for** each p_i in T **do**
14:     **for** each word q_y in p_i **do**
15:         **if** (p_i is in c_0) **then**
16:             Append 0 as prefix to the encoded text and scan from c_0 code table and replace it with its respective Huffman codes
17:         **end if**
18:         **if** (p_i is in c_1) **then**
19:             append 1 as prefix to the encoded text and scan from c_1 code table and replace it with its respective Huffman codes
20:         **end if**
21:     **end for**
22: **end for**

Procedure 1 $form\_Cluster(f_p)$

Input: Partition frequency vector matrix $f_{p_{i} * w}$
Output: Cluster frequency vector matrix $f_{c_{2} * w}$

▷

1: $cnt = 0$
2: **for** each f_i **do**
3:     $f_{c_i} \leftarrow f_{p_i}$
4: **end for**
5: **while** $(x - cnt) > 2$ **do**
6:     **for** each pair of cluster frequency vectors $f_{c_i}, f_{c'_i}$ out of $\binom{x-cnt}{2}$ **do**
7:         Calculate Cosine Similarity measure, $\cos \theta = \frac{\sum_{i=1}^{x} f_{c_i} f_{c'_i}}{\sqrt{\sum_{i=1}^{x} f_{c_i}^2} \sqrt{\sum_{i=1}^{x} f_{c'_i}^2}}$ where θ denotes the angle be-
        tween the vectors c_i and c'_i and store the $\cos \theta$ value in a cosine array against the cluster pair $(f_{c_i}, f_{c'_i})$.
8:     **end for**
9:     Find the pair $(f_{c_i}, f_{c'_i})$ with the maximum $\cos \theta$ value.
10:    Form $f_{c_i c'_i}$ and update its cluster frequency vector as $c_{ii} \leftarrow c_i + c'_i$
11:    Delete the entries in the table with atleast one element of the pair $(f_{c_i}, f_{c'_i})$ except $f_{c_i c'_i}$.
12:    $cnt + +;$
13: **end while**
14: **return** Cluster frequency vector f_c

Procedure 2 Clustering based Huffman Decoding Algorithm

Input: Encoded File T
Output: Original Text File T' ▷
1: **while** end of file **do**
2:     **while** end of partition p_i **do**
3:         **if** (bit $b == 0$) **then**
4:             **for** all words q_y in p_i **do**
5:                 Read from c_0 code table and the corresponding word for the code is replaced
6:             **end for**
7:         **else**
8:             **for** all words q_y in p_i **do**
9:                 Read from c_1 code table and the corresponding word for the code is replaced
10:            **end for**
11:        **end if**
12:    **end while**
13: **end while**

For any corpus, C_r does not increase significantly while increasing x, because individually code table size and encoded size are almost same for any x. This is due to the fact that w is same for all x and the number of words in every partition is constant, w words in partition are distributed almost equally across clusters in the clustering process, getting almost same code length for words. Moreover, encoded size remains almost same. The similar behavior was observed for any input set of corpus (Fig. 2).

The percentage of unique words play a major role along with the partition size in achieving a better C_r as shown in Fig. 2. With the decrease in the percentage, the compression ratio increases since more number of frequent words generate less number of codes leading to better compression. Since all partitions go into any of the two clusters, it has been observed that when the clusters does not have much equal number of partitions (even ranging from 11% in *cluster*1 to 89% in *cluster*2) in some cases, it leads to higher compression ratio in all corpus. This is because, when partitions contain less words with high frequency, it leads to high cosine similarity (high inter similarity measure), and hence, their code size is less leading to lesser

Fig. 2 % of Unique words versus C_r for all datasets

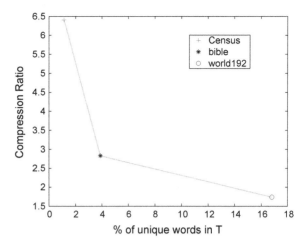

Fig. 3 Execution time versus x for *bible* and *census*

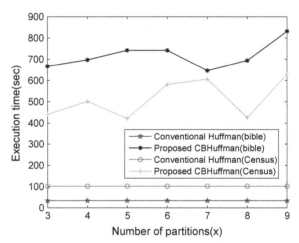

encoded size as well. A high compression ratio by CBH was observed for census dataset because of its dense occurrence of patterns (words).

Figure 3 shows the partition size x and the encoding time for bible and census corpora. The time taken to compress the corpus by CBH is more than conventional Huffman. This primarily depends on the size of the input file, x and less likely dependent on frequency of words in T. If the number of partitions are more, the time taken to find the cosine measure for all combinations of clusters adds up the cost. The time taken to encode more number of partitions also adds up the cost, because of locating the clusters x times and to traverse the tree to find the codes. Since the time taken to locate the cluster for W codes and to traverse the trees to retrieve the word is more, decoding time is more than the time to encode. Figure 4 denotes the % of unique words and the time for 3 different corpora. As the % of unique word increases, time

Fig. 4 % of Unique words
versus time for all datasets

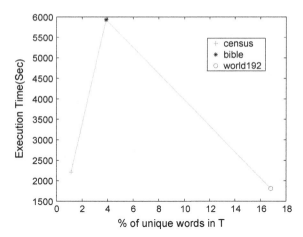

taken to encode increases for bible and census corpora. This is because, if w is more, the size of the clustering-based Huffman tree is large, generating more codes. The similar behavior was observed for any input set of corpus.

5 Conclusion

We explored a novel text compression algorithm in this paper by employing clustering in conventional Huffman encoding. We show that the proposed CBH algorithm achieves efficient compression ratio, encoded size, and execution time. However, we still consider our CBH technique to be a prototype that needs further improvement. As a part of future work, we want to investigate the scope of the CBH algorithm in other lossless word-based compression and lossy compression techniques. In addition, we want to focus on improving the method of clustering using other efficient hierarchical algorithms.

References

1. David, S.: Data Compression: The Complete Reference. Second edn. (2004)
2. A, H.D.: A method for the construction of minimum redundancy codes. proc. IRE **40**(9) (1952) 1098–1101
3. Ramakrishnan, N., Grama, A.: Data mining: From serendipity to science - guest editors' introduction. IEEE Computer **32**(8) (1999) 34–37
4. Han, J., Kamber, M.: Data Mining: Concepts and Techniques. Morgan Kaufmann (2000)
5. Agarwal, R., Srikant, R.: Fast algorithms for mining association rules in large databases. In Bocca, J.B., Jarke, M., Zaniolo, C., eds.: VLDB'94, Proceedings of 20th International Conference on Very Large Data Bases, September 12-15, 1994, Santiago de Chile, Chile, Morgan Kaufmann (1994) 487–499

6. Rajaraman, A., Ullman, J.D., Ullman, J.D., Ullman, J.D.: Mining of massive datasets. Volume 1. Cambridge University Press Cambridge (2012)
7. Aggarwal, C.C., Reddy, C.K.: Data clustering: algorithms and applications. CRC Press (2013)
8. Shannon, C.E.: A mathematical theory of communication. ACM SIGMOBILE Mobile Computing and Communications Review **5**(1) (2001) 3–55
9. Pountain, D.: Run-length encoding. Byte **12**(6) (1987) 317–319
10. Witten, I.H., Neal, R.M., Cleary, J.G.: Arithmetic coding for data compression. Communications of the ACM **30**(6) (1987) 520–540
11. Vitter, J.S.: Design and analysis of dynamic huffman codes. Journal of the ACM (JACM) **34**(4) (1987) 825–845
12. Oswald, C., Ghosh, A.I., Sivaselvan, B.: An efficient text compression algorithm-data mining perspective. In: Mining Intelligence and Knowledge Exploration. Springer (2015) 563–575
13. Kaufman, L., Rousseeuw, P.J.: Agglomerative nesting (program agnes). Finding Groups in Data: An Introduction to Cluster Analysis (2008) 199–252
14. Zhang, T., Ramakrishnan, R., Livny, M.: Birch: An efficient data clustering method for very large databases. In: Proceedings of the 1996 ACM SIGMOD International Conference on Management of Data. SIGMOD '96, New York, NY, USA, ACM (1996) 103–114
15. Jain, A.K.: Data clustering: 50 years beyond k-means. Pattern Recognition Letters **31**(8) (2010) 651–666 Award winning papers from the 19th International Conference on Pattern Recognition (ICPR)19th International Conference in Pattern Recognition (ICPR)
16. Calgary compression corpus datasets. http://www.corpus.canterbury.ac.nz/descriptions/ Accessed: 2015-07-23

Design of Cerebrovascular Phantoms Using Fuzzy Distance Transform-based Geodesic Paths

Indranil Guha, Nirmal Das, Pranati Rakshit, Mita Nasipuri, Punam K. Saha and Subhadip Basu

Abstract Generation of digital phantoms specific to the patients' carotid vasculature is a complicated and challenging task because of its complex geometrical structure and interconnections. Such digital phantoms are extremely useful in quick analysis of the vascular geometry and modelling blood flows in the cerebrovasculature. All these analyses lead to effective diagnosis and detection/localization of the diseased arterial segment in the cerebrovasculature. In this work, we have proposed a semi-automatic geodesic path propagation algorithm based on fuzzy distance transform to generate digital cerebrovascular phantoms from the patients' CT angiogram (CTA) images. We have also custom-developed a 2-D/3-D user interface for accurate placement of user specified seeds on the input images. The proposed method effectively separates the artery/vein regions from the soft bones in the overlapping intensity regions using minimal human interaction. Qualitative results along with 3-D rendition of the cerebrovascular phantoms on four patients CTA images are presented here.

Keywords Digital phantom design · Carotid vasculature · Fuzzy distance transformation · 3-D rendering · Geodesic paths

I. Guha · N. Das · P. Rakshit · M. Nasipuri · S. Basu (✉)
Department of Computer Science and Engineering, Jadavpur University,
Kolkata 700032, West Bengal, India
e-mail: subhadip@cse.jdvu.ac.in

P.K. Saha
Department of Electrical and Computer Engineering, University of Iowa,
Iowa City, IA 52242, USA

© Springer Nature Singapore Pte Ltd. 2018
P.K. Sa et al. (eds.), *Progress in Intelligent Computing Techniques: Theory,
Practice, and Applications*, Advances in Intelligent Systems and Computing 518,
DOI 10.1007/978-981-10-3373-5_36

359

1   Introduction

Reconstruction of accurate digital phantoms of human cerebrovasculature is an active area of research in the field of medical image analysis. It helps researchers to explore the underlying hemodynamics, which is one of the major factors in determining one's cerebrovascular health. In general, human cerebrovasculature refers to the vessel network circulating blood throughout the entire brain. Arteries and veins are the building blocks of this network. In this work, we are mainly interested in the carotid arteries, which are supplying oxygenated and nutrient filled blood to the various parts of the cerebral system. Anatomical analysis of carotid arteries is the key to elucidate the blood flow patterns in the vasculature, determination of irregular dilation of vessel wall, detection of possible obstruction in the flow, etc. To unveil the anatomy of carotid arteries properly, we should have the knowledge about the anatomy of *Circle of Willis*. The left and right Internal Carotid Arteries, Anterior and Posterior Cerebral Arteries (left and right), Anterior and Posterior communicating artery form this circular vasculature. The basilar and middle cerebral arteries are also part of this circle.

Generally, there are two ways to analyse the carotid arteries (1) using physical vascular phantom, which is a replica of the original one (2) digital modelling of the vessel network using mathematical model. Both of them have individual advantages and disadvantages, and researchers sometimes acquire both models to verify the experiments as in work [1–4]. In this paper, we have focused only on construction of digital vascular phantoms using mathematical modelling from human cerebral CT angiogram (CTA) images. There are many existing works on generation of carotid vasculature [5, 6], but all these methods need active human participation in the reconstruction process. In our work, user interaction with the system has been minimized by the notion of fuzzy distance transform (FDT)-based geodesic path propagation approach [7]. Please note that the concept FDT has been widely used before in various vessel reconstruction algorithms [8–10].

Separation of vessel from bone and soft tissues is most critical in the cerebrovascular system, especially in the regions with high overlapping of vessels, bones and tissues. Therefore, construction of accurate cerebrovascular phantoms in complex regions of human brains and capability to independently execute them has a multiple potential applications in the field of medical science. Moreover, Construction of phantoms from original CTA image helps to detect diseased vascular segments or disorders.

In the subsequent discussions, first we introduce the theory and notations used in the mathematical simulation of the digital phantoms of cerebrovasculature, followed by the methodology of our proposed algorithm and experimental results.

2　Theory and Methods

A three-dimensional cubic grid is expressed by $\{Z^3 | Z$ is the set of positive integers$\}$. A point on the grid, often referred to as a voxel, is a member of Z^3 and is denoted by a triplet of integer coordinates. Each voxel has 26 adjacent voxels, i.e. two voxels $P = (x_1, x_2, x_3)$ and $Q = (y_1, y_2, y_3) \in Z^3$ are adjacent if and only if

$$\{max(|x_i - y_i|) \leq 1 | 1 \leq i \leq 3\},$$

where $|\cdot|$ means the absolute value. Two adjacent points in a cubic grid are often referred to as neighbours of each other. Twenty-six neighbours of a voxel P omitting itself are symbolized as $N^*(P)$.

CTA is a 3-D grey-scale image where each voxel is represented as 8 bit character or 16 bit unsigned short value. Numeric value of a voxel implies the intensity of the voxel. Artery, veins and soft tissues occupy small intensity value, where bones take high intensity values. We will denote artery and veins together as vessels. Intensity of vessels and soft tissues is highly overlapping. In this paper, we are mainly interested in reconstruction of arterial tree of human cerebrovasculature.

It has been observed that in general intensity of vessels lie between 130 and 450 Hu (Hounsfield unit) [2], but there is almost zero overlapping around the middle point of this intensity scale. Hence, voxels within this intensity range are considered as object voxel and rest of the intensity regions are taken as background.

The distance transform (DT) is an algorithm generally applied to binary images. The output of this algorithm is same as the input image except that the values of each foreground points of the image are changed to the distance to the nearest background from that point. Over the years several distance transformation algorithm have been developed both in 2-D and 3-D [11, 12]. If $P = (x_1, x_2, x_3)$ is a point in a 3-D image, then DT value of that point will be,

$$DT(P) = \begin{cases} DT(Q_i) + d_k, & |DT(Q_i) + d_k < DT(P) \\ DT(P), & |otherwise \end{cases} \tag{1}$$

where, Q_i is the neighbour of P, $i = 1, 2, \ldots, 26$. $d_{k=1,2,3}$ is the approximate Euclidean distance from three different kinds of neighbour. Distance transform performs very well in case of binary images. But in case of digital phantom generation, distinction between vessel intensity and other objects intensity is crucial which cannot be done in DT image. Hence, fuzzy distance transform is more suitable. In this work, we have used triangular fuzzy membership function to convert a DT image into a FDT image [13].

$$FDT(P) = DT(P) * \mu(P) \tag{2.a}$$

$$\mu(P) = \begin{cases} \frac{I - I_A}{I_M - I_A}, & |I_A \leq I \leq I_M \\ \frac{I_B - I}{I_B - I_M}, & |I_M < I \leq I_B \end{cases} \tag{2.b}$$

where, I denotes the grey-scale value of voxel P. I_A, I_B denotes the max and min grey-scale intensity, respectively. I_M is the median of I_A and I_B,

$$I_M = \frac{I_A + I_B}{2} \tag{2.c}$$

We represent whole image as an undirected graph $G = (V, E)$ where V is the vertex set denoted by $\{P|P$ is a voxel in the 3-D image$\}$, E is the set of edges denoted by $E = \{(P_1, P_2)|P_1$ and P_2 are adjacent$\}$.

Methodology used here is to find the centre point of the presumed artery between input seed points and draw spheres in these points with radius equal to the FDT value of that point. We may define a sphere $S(P, r)$ with centre P having coordinates (x_c, y_c, z_c), and radius r is the locus of all points (x, y, z) in z^3 such that

$$(x - x_c)^2 + (y - y_c)^2 + (z - z_c)^2 = r^2. \tag{3}$$

A point Q is called local maxima if,

$$\{Q|Q\epsilon Z^3, FDT(Q) = max(FDT(Q_i)),$$

$$1 \leq i \leq 26, \text{ all } Q_i's \text{ have same Eucledian distance to } P, \} \tag{4}$$

Therefore, a point may have several local maxima points in its neighbourhood or it may not have any local maxima point in its neighbourhood. Geodesic path is defined as minimum cost shortest path between two points in two or three-dimensional space. Dijkstra's shortest path algorithm has been used for geodesic path propagation with modification so that shortest path will always pass through the nearest local maxima point if it exists; otherwise, it will pass through the maximum FDT value point in its neighbourhood. To force the geodesic path between two seed points to pass through nearest local maxima points or maximum FDT value point, edge weight between any point and corresponding nearest local maxima point or maximum FDT value point should be minimum. Here we introduce a cost function β that computes edge weights between two adjacent voxel $P = (x_1, x_2, x_3)$ and $Q = (y_1, y_2, y_3)\epsilon Z^3$ as follows:

$$\beta(P, Q) = \left(\frac{2}{(DT(P) + DT(Q))}\right) \times DIST(P, Q). \tag{5}$$

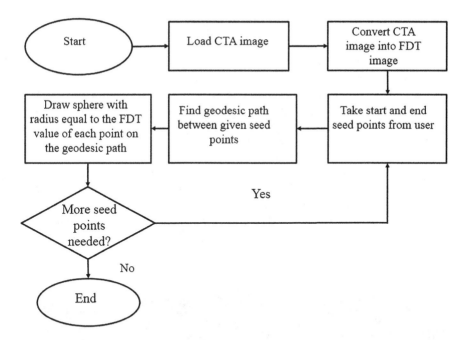

Fig. 1 Modular representation of the algorithm used to generate phantoms

where

$$DIST(P,\ Q) = \sqrt{(x_1 - y_1)^2 + (x_2 - y_2)^2 + (x_3 - y_3)^2}. \tag{6}$$

Dijkstra's algorithm is a greedy method where from current point algorithm chooses a neighbour connected through least cost edge for next iteration. In the proposed method due to cost function β, it will always select the nearest local maxima point or maximum FDT value point in the neighbourhood of the current point. After termination of this, algorithm will return connected centre points of the presumed artery. If we draw sphere taking these points as centre and radius equal to the FDT values of the point, we will get the desired digital phantom.

As the proposed method is a semi-automatic algorithm, hence user may not get accurate result at first; hence, trial and error method should be used to generate accurate phantoms. In this method, user can modify the generated phantom by giving extra seed points. We can summarize the whole method in following steps Fig. 1.

3 Experimental Methods and Results

To facilitate the experimental methods, an integrated custom designed 2-D/3-D graphical user interface was developed in our laboratory allowing axial, coronal and sagittal view of segmented data. Facilities of selecting and editing different seed points are supported within the graphical user interface. Using developed GUI, we have taken seed points as input and visualized generated digital phantoms in axial, coronal, sagittal and 3-D views. In Fig. 2, 3-D view of different phases of generation of a digital phantom from CTA image is shown.

In Fig. 3, we have shown 3-D view of four digital phantoms constructed from four different CTA images. Blue coloured portion in Fig. 3a shows aneurysm present in the cerebrovasculature.

Generated phantoms are overlaid with original CTA image to analyse the accuracy of the phantom. Figure 4 shows the overlay of the digital phantom over original CTA image.

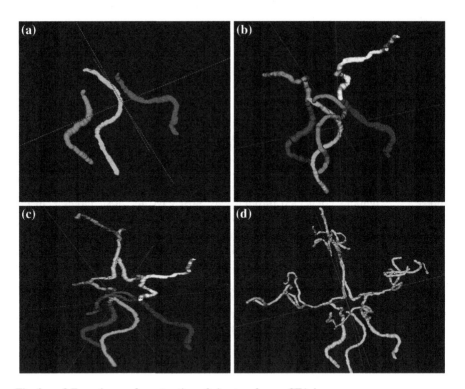

Fig. 2 a–d Four phases of construction of phantom from a CTA image

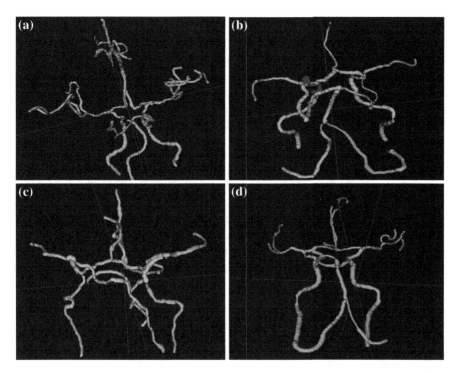

Fig. 3 a–d Digital phantoms generated around *circle of Willis* from four different CTA images

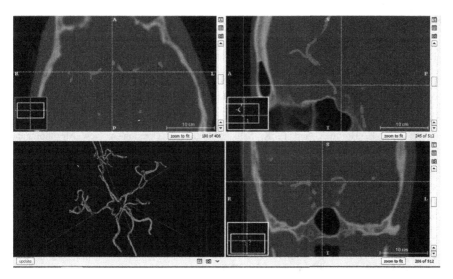

Fig. 4 A segmented phantom image is shown with respect to the original CTA image in axial, coronal, sagittal and 3-D views

4 Conclusion

In this present work, we have shown an application of fuzzy distance transformation-based geodesic path propagation approach to generate accurate phantoms from human CTA images. Local maxima points and distance transformation are used to find the radius of the arterial tube. The proposed process is semi-automatic, and the user can modify the generated digital phantom structures to make it more accurate. We argue that the proposed algorithm is both efficient and precise. Digital phantoms generated through this algorithm can be helpful in studying the arterial bends, bifurcated regions, joins and possible modelling of digital fluid flows in human cerebrovasculature. We have used the ITK-SNAP [14] open-source software to overlay generated phantom structures over original CTA images. In future, we may attempt to use the generated digital phantoms for homodynamic analysis in human cerebrovasculature. Digital phantoms are also useful in various other bio-imaging applications, and we propose to develop similar synthetic structures for analysis of structural/plastic changes of hippocampal dendritic spines [15].

Acknowledgements Authors are grateful to Dr. Robert E. Harbaugh, Penn State Hershey Medical Center and Prof. Madhavan L. Raghavan, Department of Biomedical Engineering, University of Iowa, for sharing the CTA datasets used in this study. This work is partially supported by the FASTTRACK grant (SR/FTP/ETA-04/2012) by DST, Government of India and Research Award (No.F.30-31/2016(SA-II)) by UGC, Government of India.

References

1. P. K. Saha, J. K. Udupa, and D. Odhner, "Scale-Based Fuzzy Connected Image Segmentation: Theory, Algorithms, and Validation," *Comput. Vis. Image Underst.*, vol. 77, no. 2, pp. 145–174, 2000.
2. S. Basu, M. L. Raghavan, E. A. Hoffman, and P. K. Saha, "Multi-scale opening of conjoined structures with shared intensities: methods and applications," in *2011 International Conference on Intelligent Computation and BioMedical Instrumentation*, 2011, pp. 128–131.
3. S. Basu, E. Hoffman, and P. K. Saha, "Multi-scale Opening–A New Morphological Operator," in *Image Analysis and Processing—ICIAP 2015*, Springer, 2015, pp. 417–427.
4. P. Saha, S. Basu, and E. Hoffman, "Multi-Scale Opening of Conjoined Fuzzy Objects: Theory and Applications," *IEEE Trans. Fuzzy Syst.*, vol. in press, 2016.
5. A. Banerjee, S. Dey, S. Parui, M. Nasipuri, and S. Basu, "Design of 3-D Phantoms for Human Carotid Vasculature," in *ICACC*, 2013, pp. 347–350.
6. A. Banerjee, S. Dey, S. Parui, M. Nasipuri, and S. Basu, "Synthetic reconstruction of human carotid vasculature using a 2-D/3-D interface," in *ICACCI*, 2013, pp. 60–65.
7. A. F. Frangi, S. Member, W. J. Niessen, R. M. Hoogeveen, T. Van Walsum, and M. A. Viergever, "Model-Based Quantitation of 3-D Magnetic Resonance Angiographic Images," vol. 18, no. 10, pp. 946–956, 1999.
8. Y. Xu, P. K. Saha, G. Hu, G. Liang, Y. Yang, and J. Geng, "Quantification of stenosis in coronary artery via CTA using fuzzy distance transform," *Proc. SPIE*, vol. 7262, p. 72620K–72620K–12, 2009.

9. S. Svensson, "A decomposition scheme for 3D fuzzy objects based on fuzzy distance information," *Pattern Recognit. Lett.*, vol. 28, no. 2, pp. 224–232, 2007.

10. B. Aubert-Broche, M. Griffin, G. B. Pike, A. C. Evans, and D. L. Collins, "Twenty new digital brain phantoms for creation of validation image data bases," *IEEE Trans. Med. Imaging*, vol. 25, no. 11, pp. 1410–1416, 2006.

11. C. Vision, I. Understanding, and A. No, "On Digital Distance Transforms in Three Dimensions," vol. 64, no. 3, pp. 368–376, 1996.

12. Borgefors, Gunilla. "Distance transformations in arbitrary dimensions." Computer vision, graphics, and image processing 27, no. 3 (1984): 321-345.

13. Hong, Tzung-Pei, and Chai-Ying Lee. "Induction of fuzzy rules and membership functions from training examples." Fuzzy sets and Systems 84, no. 1 (1996): 33-47.

14. Paul A. Yushkevich, Joseph Piven, Heather Cody Hazlett, Rachel Gimpel Smith, Sean Ho, James C. Gee, and Guido Gerig. User-guided 3D active contour segmentation of anatomical structures: Significantly improved efficiency and reliability. Neuroimage 2006 Jul 1; 31 (3):1116-28.

15. Basu, Subhadip, Dariusz Plewczynski, Satadal Saha, Matylda Roszkowska, Marta Magnowska, Ewa Baczynska, and Jakub Wlodarczyk. "2dSpAn: semiautomated 2-d segmentation, classification and analysis of hippocampal dendritic spine plasticity." Bioinformatics (2016): btw172.

High-Performance Linguistics Scheme for Cognitive Information Processing

D. Suryanarayana, Prathyusha Kanakam, S. Mahaboob Hussain
and Sumit Gupta

Abstract Natural language understanding is a principal segment of natural language processing in semantic analysis to the use of pragmatics to originate meaning from context. Information retrieval (IR) is one of the emerging areas to deal with enormous amounts of data, which are in the form of natural language. Content of the query posed will affect both volume of data and design of IR applications. This paper presents a cognition-applied methodology termed as High-Performance Linguistics (HPL), which is a question-answering system for interpreting a natural language sentence/query. It constitutes three phases of computations: parsing, triplet generation and triplet mapping/matching. The generation of the triplets for the knowledge base is to create new data and compare them with that of stored triplets in the database. Thus, the generation of the cognitive question-answering system can make easy using this machine learning techniques on the generated triplet database.

Keywords Pragmatics · RDF · Triplets · Ontology · Information retrieval · Linguistics · Semantics · Indexing

D. Suryanarayana (✉) · P. Kanakam · S.M. Hussain · S. Gupta
Department of Computer Science & Engineering, Vishnu Institute of Technology,
Vishnupur, Bhimavaram, Andhra Pradesh, India
e-mail: suryanarayanadasika@gmail.com

P. Kanakam
e-mail: prathyusha.kanakam@gmail.com

S.M. Hussain
e-mail: mahaboobhussain.smh@gmail.com

S. Gupta
e-mail: sumit108@hotmail.com

© Springer Nature Singapore Pte Ltd. 2018   369
P.K. Sa et al. (eds.), *Progress in Intelligent Computing Techniques: Theory,
Practice, and Applications*, Advances in Intelligent Systems and Computing 518,
DOI 10.1007/978-981-10-3373-5_37

1 Introduction

Nowadays, computation on the Web is a critical task to retrieve the accurate information because every minute the World Wide Web (WWW) is becoming big and big with lot of information and resources. Web crawlers handle this critical job to retrieve related information from the Web documents. Knowledge engineering is the major task to execute to achieve the information semantically from the current semantic Web. Semantic Web technologies initiate a huge impression to work on the semantic Web and semantic search make promising. To step into enhanced progress of information retrieval from the semantic web documents, some of the authors offered various practices such as Probabilistic Model and Vector Space Model [1]. For the advancement of semantic search and retrieving process, a variety of implements is put into practice and broadens by means of latent semantic indexing [2], machine learning [3] and probabilistic latent semantic analysis [4]. Semantic information helps computers to understand what we put on the web, and it was the current research issue of World Wide Web (WWW) to provide semantic data according to the query. The intent of the users to query the search engine in natural language interrelated to the human cognition. Thus, semantics are related to the intent and meaning of the users query. Most of the search engines try to provide the results as per the query posted from the huge repository of databases depending/according to the terms located in the query even though for the direct questions/query some search engines failed to answer. Some of the current search engines, especially semantic search engines, are trying to understand the intent/semantics of the user and their queries. They can provide the better results for any type of natural language queries. [5]. Interpreting the formal languages in the web content is more effortless using ontologies. Resource Description Framework (RDF) imparts to add semantic information to web pages.

2 Related Work

Estimating the cognition of user or their natural language queries (NLQ) is a tricky task for the system to retrieve the expected results. Most users are not satisfying with the results retrieved by a question-answering system. To facilitate the best results, every system needs to undergo a technique for understanding the content and semantics of the query.

2.1 Question-Answering System

In 1993, Boris Katz and his team developed the web-based question-answering system called START. It is not just like a search engines to retrieve the information

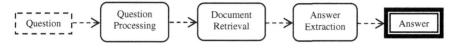

Fig. 1 Conventional question-answering system

depending upon the hits, but it simply supply the right information from the default database [6]. Richard J Cooper introduces a question-answering system in Perl that uses CMU Link Parser [7]. The main goal of the proposed question-answering system is to provide accurate content for the natural language query by the user to the system. Depending upon the semantic structure of the data, it is easy to build up a question-answering system with precise information for the posed queries and satisfies the user needs. The proposed system processes the database for the query and searches for the relative information, which is in triplet form constructed by the resource description framework. Question processing performs for the question classification of the trained machine learning algorithms. Figure 1 shows the three main segments of the question-answering system query processing, retrieval of documents and extraction of answers.

Question processing involves query posed by user through making it ready by changing its form to triplet for interpreting semantically in order to give better results by the machine. Document retrieval includes inquiring ontology database to fetch corresponding ontology for that query. Answer extraction implicates is to finding the property value of that particular triplet generated during question processing phase.

2.2 Machine Learning Using RDF and SPARQL

The general issue of machine learning is to look at a typically extensive space of potential theories to decide the one that will best fit the information and any earlier knowledge. The learner makes an expectation of the property of best, the right answer exhibited and the learner changes its theory appropriately. In supervised realizing, there is essentially the supposition that the descriptors accessible are identified with an amount of significance. The machine that learns to infer a function using trained data is supervised machine learning. Cognition applied to experience or to study whether result retrieved suits the query posed or not. Many training examples will be in the training data, and each example classified into pairs of input vector object and a supervisory signal as an output that actually desired.

Parsing of the sentence involves two main process-text lemmatization and text categorization. Semantic interpretation of natural language query obtained after text categorization. To retrieve the actual content of a Natural Language Query given by a user to a search engine, it must change its query form to a variety of forms, i.e., the search process will be done in the microlevel of the database. To grab semantics from the natural language sentence, Lambda Calculus need to apply. Always prefer disambiguated language (a language without ambiguity) while dealing with the semantics of the sentence in order to avoid ambiguities. There should be a compositional relation between syntax and semantics from the side of formal semantics. Principle of compositionality is defined as the significance of the sentence is a set of semantics of its elements and the process of that way of syntactically united [8]. In order to provide syntax and semantics for language L, every well-formed sentence in it must represent in a compositional way.

To retrieve exact results from the semantic web documents which are in the form of RDF format, a unique query language is used called SPARQL. This specification defines the semantics as well as syntax of the SPARQL to RDF. Finally, the outcome of the queries in SPARQL syntax will be in triplet or in graphical representation called RDF graphs. Mostly, the syntax of the query of SPARQL represents conjunctions, disjunctions and some optional patterns. Therefore, the entire semantic web documents are in <*subject, object, predicate*> triples [9]. SPARQL endpoint is RDF triple database on server usually, which is available on web and top of web transfer protocol, there is a SPARQL protocol layer means via http SPARQL query transfers to server and server gives its results to client. It is like SQL but works on RDF graphs not on tables. Graph pattern is RDF triple that contains some patterns of RDF variables. These patterns combined to get different patterns of more complex results.

3 High-Performance Linguistic Scheme

Information retrieval is one of the emerging areas to deal with massive amounts of data that presented in natural language. Content of the query posed will affects both volume of data and design of IR applications. Text Lemmatization is critical process involved in question-answering system, which is the process of finding lemmas from the natural language sentence as well to assign some categories to those particular lemmas. It gives the best solution to solve the problem of grasping enormous amounts of data and handle it more efficiently. High-Performance Linguistics, a question-answering system, is a cognition-applied machine to learn how to infer the content of the natural language sentence. The path to give output from input is a trivial task to done. For that, a systematic procedure will give a clue to machine to interpret natural language sentence. Most of the supervised machine learning algorithms uses a model to project known outputs from known inputs as shown in Fig. 2. However, applying cognition to machine to comprehend various categories of the text and mapping of text to document is a complicated task.

To overcome this, HPL algorithm applied to query to infer the content of query. Here, the forms of query/natural language sentence will changes. In earlier work, Palazzo Matrix Model (PMM) gives the occurrence of the term in the document and handle whether the document is appropriate to the query posed by the user [10]. At parsing level in Fig. 3, each natural language sentence is categorized as lemmas and allotted respective categories.

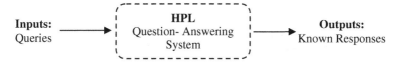

Fig. 2 Typical structure of a question-answering system

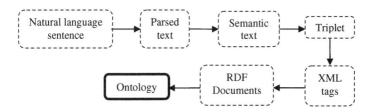

Fig. 3 Various states of a NLQ

Table 1 Category rules for semantic representations of NLP

Input trigger	Logical forms
A constant c	$NP : \lambda c$
Arity one predicate p	$N : \lambda x.p(x)$
Arity one predicate p	$S\backslash NP : \lambda x.p(x)$
Arity two predicate p	$(S\backslash NP)/NP : \lambda x.\lambda y.p(y, x)$
Arity two predicate p	$(S\backslash NP)/NP : \lambda x.\lambda y.p(x, y)$
Arity one predicate p	$N/N : \lambda g.\lambda x.p(x) \wedge g(x)$
Arity two predicate p and constant c	$N/N : \lambda g.\lambda x.p(x, c) \wedge g(x)$
Arity two predicate p	$(N\backslash N)/NP : \lambda x.\lambda g.\lambda y.p(y, x) \wedge g(x)$
Arity one function f	$NP/N : \lambda g.\text{argmax/min}(g(x), \lambda x.f(x))$
Arity one function f	$S/NP : \lambda x.\ f(x)$

There are *425* lexical categories such as noun, pronoun and determiner. For semantic representation, directionalities (forward/backward) are applied, and there by combinatory rules are generated. Finally, from XML to RDF, documents are used to create respective ontology. Using the rules and classical programming language (λ-Calculus), the sentence will be explicated as parts as shown in Table 1.

3.1 HPL Algorithm

This algorithm contains three fragments—parsing, triplet generation and triplet matching for accurate generation of results from the ontology database. Before applying the algorithm, NLQ undergo preprocessing (removal of stop words).

Begin

$Q <n_1, n_2, ..., n_p, ..., n_m>$ /* Query with 'm' words*/

Step 1: $L<Q>$ /* Apply Lemmatization by removing stop words*/

Step 2: Derive categories for each word

$$S = |n_1|, \quad |n_2|, \quad ... \quad |n_p|, ... \quad |n_m|$$

$$N/N \quad NP/N... \quad N/N \; ... \quad NP$$

Step 3: Apply parsing process.

/* applying forward/ backward directionalities along with λ-calculus*/

$$N/N \quad NP/N \; ... \quad N/NP \quad NP$$

$$N/NP \qquad N/NP$$

$$S$$

Apply λ-calculus then,

$\lambda[P(x)] <=> P(x)$, where P is predicate and x is subject

Step 4: Apply triplet generation process

Using λ- notation, obtain triplet form

$Q = <S> <O=?> <P>$

where S= subject, P= predicate, O=object

Step 5: Apply matching process

/*Compare this triplet set with that of stored triplet sets in database*/

$Q = \{<S>, <O=?>, <P>$

$Q_1 = \{<RS_1>, <RO_1>, <RP_1>\}$

where $Q_1 \rightarrow$ RDF Triplet for Query Q

$\quad\quad <S> = <RS_1>$ /* Resource*/

$\quad\quad <O> = <RO_1>$ /* Property Value*/

$\quad\quad <P> = <RP_1>$ /* Property*/

Here, $<RO_1>$ gives the value of $<P>$. Therefore, it will search in resource description table:

Resource	Property	Property value
<Subject>	<Predicate>	<Object>
$<RS_1>$	$<RP_1>$	$<RO_1=?>$

The value of $<RO_1>$ is answer to query 'Q'

Step 6: Ontology related to output value from matching process is derived.

End

The link between λ-notations and triplet generation as <Subject> <Object> <Predicate> is simple through single transformation and single function definition scheme. Here, the object value is unknown, and output is the object value or property value. Then, derived triplet compared with existed triplets in the database. K-Nearest Neighbor is applied to find the similarity between test triplet which is generated with that of training triplets stored in database by using <Subject> and <Predicate>. Euclidean distance measure along with K-NN is employed to find similarity distance D_s between nearest triplet (t_n) with that of k-triplet (derived triplet). Consider, triplet t_n = <$t_n(S)$> <$t_n(P)$> and triplet t_k = <$t_k(S)$> <$t_k(P)$>, then

$$D_s(t_n, t_k) = \sqrt{\sum_{w=1}^{m} \left((t_n(S)) - t_k(S) \right)^2} \tag{1}$$

if $D_s(t_n, t_k) = 0$, then t_n is the matching triplet for t_k otherwise not.

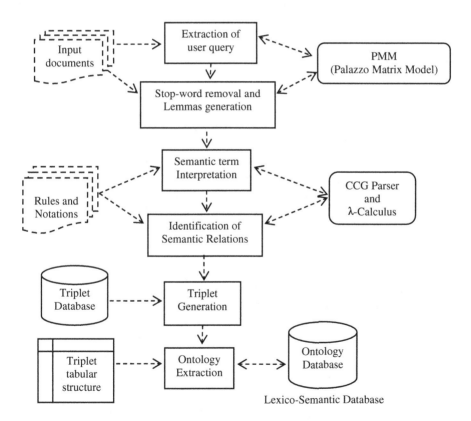

Fig. 4 Internal anatomy of HPL

Then, obtained triplet and associated ontology inferred from database by indexing it with from triplet tabular structures. Then, the machine projects the corresponding object or property value. It depicts a system that automatically learns ontologies. As shown in Fig. 4, the entire collection of ontology is termed as semantic knowledge bases enabling exciting applications such as question answering on open domain collections. This system automatically learns ontology from texts in a given domain. The domain-specific ontology that results from these knowledge acquisition methods incorporated into lexico-semantic database that various natural language processing systems may employ. This system helps to extract specific knowledge and for searching that knowledge from unstructured text on the web. It uses ontology-based domain knowledge base known as lexico-semantic database. Ontology conceptualizes a domain into a machine-readable format. Mostly, information on web represented as natural language documents. Knowledge extraction process involves reducing the documents into tabular structures (which indexed easily) for grabbing the context from the document i.e., answering to user's queries. HPL system mainly depends on triplet generated and domain ontology mapping to that triplet for extracting exact content or semantics from natural language queries posed by the user.

4 Experiments and Results

Observations were made on trial set of *10* queries in natural language, randomly collected from students. All the queries posed belong to single domain called educational ontology. Triplet generation for three queries among the ten is in Table 2.

Authors conducted several methods to find the exactness of HPL system on triplet database that contains 20 overall triplets on educational ontology and *10* sample natural language queries applied to that of triplets to match which are in the database. The correct value to give the exactness is precision, defined as,

Table 2 Triplet generation for the natural language queries

Query	Subject	Predicate	Object
What is the exam fee for JNTUK III–II semester external exams?	JNTUK III–II semester external exams	Has exam fee	# value (~650)
How to install python in PC?	Python in PC	Has install process	# value (URL)
Who is the author of social networks and the semantic web?	social networks and the semantic web	Has author	# value (Peter Mika)

Fig. 5 Triplet generated versus precision calculated

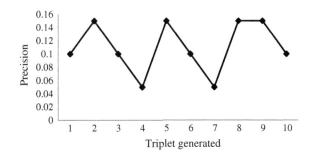

$$\text{Precision} = \frac{\text{Correct triplets matched}}{\text{total triplets in database}} \tag{2}$$

Thus, k (generated triplet) values ranging from *1* to *10* and the precision value for each k value is computed. Mostly, the precision value is range of *0–0.1* as shown in Fig. 5. The values produced are giving evidence to show the truthfulness of HPL system.

5 Conclusion

Most of the question-answering systems developed based on text retrieval or web documents retrieval methodology where users may retrieve embedded answers from the systems. The idea behind this paper is to create the database with the relationships among the subject, object and predicate and make it accurate to answer the questions.

By simply parse the natural language sentence using CCG and λ-Calculus, generation of triplet made easy. Thus, matching algorithm actively searches for the matching contents in the database and generates the accurate and coherent answer for the question in the same triplet form without any long and embedded sentences. This cognitive informative processing mechanism is helpful to the users for their desired information by providing relevance and factually correct data from the database. In future, this retrieval process employed images to retrieve the relevant information in the semantic manner.

Acknowledgements This work has been funded by the Department of Science and Technology (DST), Govt. of India, under the grants No. SRC/CSI/153/2011.

References

1. Baeza-Yates, R, A., Ribeiro-Neto, B, A.: Modern Information Retrieval. ACM Press/Addison-Wesley, (1999).
2. Deerwester, S, C,. et al.: Indexing by latent semantic analysis. Journal of the American Society for Information Science and Technology – JASIS, Vol. 41(6), 391–407 (1990).
3. H. Chen.: Machine learning for information retrieval: Neural Networks, Symbolic learning, and genetic algorithms. Journal of the American Society for Information Science and Technology – JASIS, Vol. 46(3), 194–216 (1995).
4. Thomas Hofmann.: Probabilistic latent semantic indexing. In: International conference SIGIR '99, ACM, New York, NY, USA, 50–57 (1999).
5. Dumais, Susan, Michele Banko, Eric Brill, Jimmy Lin, and Andrew Ng.: Web question answering: Is more always better?. In Proceedings of the 25th annual international ACM SIGIR conference on Research and development in information retrieval, ACM, 291–298 (2002).
6. Boris Katz.: START: Natural Language Question Answering System. (1993), http://start.csail. mit.edu/index.php.
7. Boris Katz, Gary Borchardt and Sue Felshin.: Natural Language Annotations for Question Answering. In: 19th International FLAIRS Conference (FLAIRS 2006), Melbourne Beach, FL, (2006).
8. Partee, B, H.: Introduction to Formal Semantics and Compositionality. (2013).
9. Suryanarayana, D., Hussain, S, M,. Kanakam, P., Gupta, S.: Stepping towards a semantic web search engine for accurate outcomes in favor of user queries: Using RDF and ontology technologies. 2015 IEEE International Conference on Computational Intelligence and Computing Research (ICCIC), Madurai, India, 1–6 (2015).
10. Mahaboob Hussain, S., Suryanarayana, D., Kanakam, P., Gupta, S.: Palazzo Matrix Model: An approach to simulate the efficient semantic results in search engines. In: IEEE International Conference on Electrical, Computer and Communication Technologies (ICECCT), 2015, Coimbatore, 1–6 (2015).

Introducing MIT Rule Toward Improvement of Adaptive Mechanical Prosthetic Arm Control Model

Biswarup Neogi, S.K. Sanaul Islam, Pritam Chakraborty, Swati Barui and Achintya Das

Abstract The article represents MRAC strategies for controlling a mechanical prosthetic arm utilizing gradient method MIT rule. Depending on the uncertainty parameters like mass of the mechanical arm, friction constant and spring constant, an adaptive controller is designed and outlined. MIT rule has been applied to this second-order system, and simulation is done in MATLAB-Simulink for various estimation of adaptation gain. With the changes in adaptation gain, the adaptation mechanisms are changed and results are analyzed. Variation in the parameters should be in the specified range for MIT rule. Further Lyapunov stability approach has been applied to the closed-loop dynamics to ensure global stability on variation of plant parameters.

Keywords Prosthetic arm · Model reference adaptive control · Gradient method MIT rule · Lyapunov rule

B. Neogi (✉) · S.K.S. Islam · P. Chakraborty
JIS College of Engineering, Kalyani, West Bengal, India
e-mail: drbneogi@gmail.com

S.K.S. Islam
e-mail: sksanaulislam@gmail.com

P. Chakraborty
e-mail: pritam29121988@gmail.com

S. Barui
Narula Institute of Technology, Sodepur, West Bengal, India
e-mail: engg2006@gmail.com

A. Das
Kalyani Government Engineering College, Kalyani, West Bengal, India

© Springer Nature Singapore Pte Ltd. 2018
P.K. Sa et al. (eds.), *Progress in Intelligent Computing Techniques: Theory, Practice, and Applications*, Advances in Intelligent Systems and Computing 518, DOI 10.1007/978-981-10-3373-5_38

379

1 Introduction

The modeling of prosthetic arm has been going through a significant research interest among the researchers. Rossini et al., contributed thought controlled prosthetic arm, to control a biomechanics arm connected with the nervous system by linking electrodes to human nerves [1], but designing is much expensive. Dynamic model of the prosthetic arm of a upper limb amputated patient with transfer functions of the arm model implementing various tuning methods presented by Neogi et al. [2], as an extension of our previous work, this paper target is to control the movement of fingers of prosthetic arm by considering an adaptive controller with a desired reference model, presents design methodology using auto adjustable adaptive controller with adaptation mechanism utilizing MRAC for solving the problem of prosthetic arm which in relation to mass-damper-spring process. Controller parameters converge to ideal values that force the response to track the reference model, i.e., the quick response between fingers of prosthetic arm with little or no vibrations the step response should give the desired performance with less or no peak overshoot, optimum rise time, less settling time, optimum damping ratio and a zero steady-state error. Reference behavior is modeled using transfer function. The Lyapunov stability analysis for system closed-loop asymptotic tracking also is done in this paper.

2 Dynamic and Mathematical Modeling of Prosthetic Arm Using Gradient MRAC-MIT Adaptation Technique

A standard mechanical arm represented as a damped second-order differential equation as,

$$f(t) = m\frac{d^2x}{dt^2} + K_1\frac{dx}{dt} + K_2x(t) \tag{1}$$

where $f(t)$ = applied force to create displacement between fingers, m = mass of the mechanical arm, $x(t)$ = displacement between fingers, K_1 = friction constant, K_2 = spring constant [2].

The dynamic modeling and mathematical calculations for deriving the prosthetic arm model are taken from our previous work, Neogi et al. [2]; for the analysis we have considered only first set of transfer function defined as,

Set 1:

$$G(s) = \frac{1}{0.21s^2 + 0.4038s + 0.0411} \tag{2}$$

Model reference adaptive control is utilized to make quick movement of fingers of prosthetic arm with little or no vibration. A reference model is then chosen that could respond quickly to a step input with settling time = 3 s. A Perfect tracking of model reference is achieved by developing parameter adaptation laws for control algorithm utilizing MRAC-MIT rule. Prosthetic arm movement x_p has to follow desired trajectory x_m of the reference model.

Since, the plant is described by mass-damper-spring process as a second-order element and has input as applied force $f(t)$ to create displacement and output as displacement between fingers $x(t)$,

$$G_p(s) = \frac{X_P(s)}{F(s)} = \frac{\frac{1}{m}}{\left[s^2 + \frac{K_1}{m}s + \frac{K_2}{m}\right]} \tag{3}$$

$$X_P(s) = G_p(s)F(s) = \frac{c}{[s^2 + as + b]}F(s) \tag{4}$$

A controller has to be design to reach desired closed-loop dynamics as system gain c, time constant a, b are unknown or time varying.

The reference model, described by the standard second order under damped system with transfer function,

$$G_m(s) = \frac{c_m}{[s^2 + a_ms + b_m]} \tag{5}$$

Tracking error,

$$e = X_P - X_m \tag{6}$$

The necessary step is to search the adaptation law so that closed-loop system dynamics are bounded as well as it able to track the desired reference model asymptotically. Adaptation law of MRAC structure in time domain taken as in the following form,

$$f = \left[S_1 r - S_2 x_p - S_3 \dot{x}_p\right] \tag{7}$$

where S_1, S_2, S_3 the adaptive feedforward–feedback controller parameter can be written in vector form as $\theta = (S_1, S_2, S_3)$ and $\delta = [r, -x_p, -\frac{d}{dt}x_p]^T$. Here, adaptive gain θ will be adjusted to track the reference model. Substituting Eq. (7) in time domain representation of Eq. (4), then Eq. (8) will follow the desired reference model of Eq. (5),

$$\ddot{x}_p + (a + cS_3)\dot{x}_p + (b + cS_2)x_p = cS_1 r \tag{8}$$

If controller parameters taken as

$$S_1 = \frac{c_m}{c}, S_2 = \frac{b_m - b}{c}, \text{ and } S_2 = \frac{b_m - b}{c} \tag{9}$$

The values of plant parameter a, b and c are unknown and varying, so adaptation mechanism for each controller parameter S_1, S_2, S_3 has to be found that are uniquely based on measurable quantity. The mathematical analysis of update law using MIT rule are derived and experimentally evaluated utilizing MATLAB simulation.

Generalized MIT Rule conveys that

$$\frac{d\theta}{dt} = -\tilde{\gamma}\frac{\partial J}{\partial \theta} = -\tilde{\gamma}e\frac{\partial e}{\partial \theta} \tag{10}$$

Here, $\tilde{\gamma}$ is adaptation gain.

Hence, cost function can be defined as, $J(\theta) = \frac{1}{2}e^2(\theta)$, which has to be minimized.

Using $s(.) = \frac{d}{dt}(.)$ from differential Eq. (8), we get

$$X_P = \frac{cS_1}{s^2 + (a + cS_3)s + (b + cS_2)}R(s)$$

The sensitivity derivatives $\frac{\partial e}{\partial \theta}$ are given as

$$\frac{\partial e}{\partial S_1} = \frac{\partial}{\partial S_1}(X_P - X_m) = \frac{c}{s^2 + (a + cS_3)s + (b + cS_2)}R(s)$$

$$\frac{\partial e}{\partial S_2} = -\frac{c}{s^2 + (a + cS_3)s + (b + cS_2)}X_P \tag{11}$$

$$\frac{\partial e}{\partial S_3} = -\frac{sc}{s^2 + (a + cS_3)s + (b + cS_2)}X_P$$

Perfect model tracking can be possible if as from Eq. (8)

$$S_1c = c_m, a + cS_3 = a_m, b + cS_2 = b_m$$

From Eq. (9), the update law for $\theta = (S_1, S_2, S_3)$ is given by,

$$\frac{dS_1}{dt} = -\gamma e\frac{c_m}{(s^2 + a_ms + b_m)}R(s)$$

$$\frac{dS_2}{dt} = \gamma e\frac{c_m}{(s^2 + a_ms + b_m)}X_P \tag{12}$$

$$\frac{dS_3}{dt} = \gamma e\frac{c_ms}{(s^2 + a_ms + b_m)}X_P$$

Here, single adaptation gain γ is taken as $\gamma = \tilde{\gamma}\frac{c}{c_m}$.

3 Simulation Results and Analysis

The simulation for the closed-loop system of prosthetic arm is done using MATLAB software (Simulink). The transfer functions of mechanical prosthetic arm are considered as plant. The parameters relating to these transfer function taken as, m = prosthetic arm mass = 210 gm = 0.21 kg, K_1 = friction constant = 0.4038, K_2 = spring constant = 0.0411.

Figure 1a shows block diagram of prosthetic arm with adaptive controller and adjustment mechanism. Controller parameters are updated in the adaptive controller block. A step signal is employed, the equivalent time reaction is shown in Fig. 1b and the transfer function of reference model is chosen in such a manner that it could respond quickly to a step input as follows,

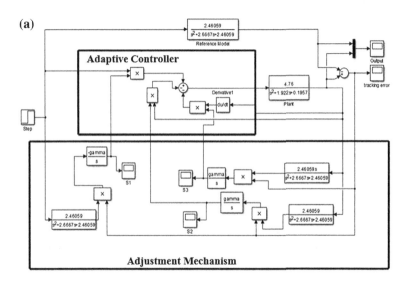

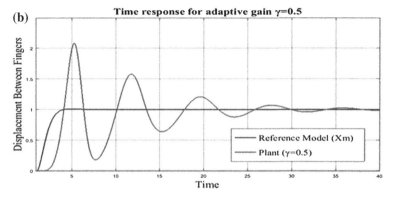

Fig. 1 **a** Block diagram of prosthetic arm and, **b** Time response corresponds to step input

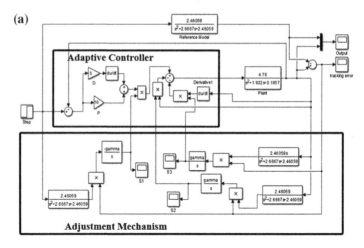

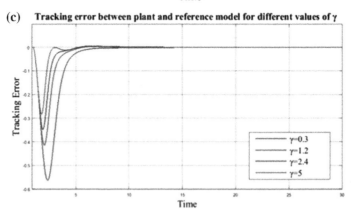

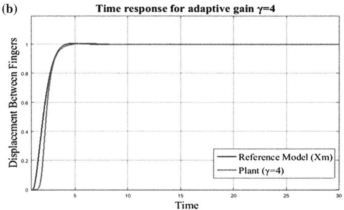

Fig. 2 a Modified block diagram of prosthetic arm with fixed gain PD control, **b** Time reaction corresponds to step input; **c** Tracking error and, **d–f** Adaptive controller parameters

(d)

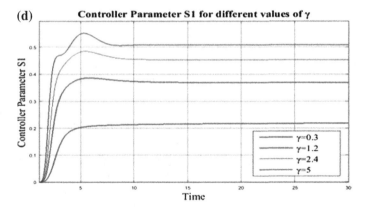

(e)

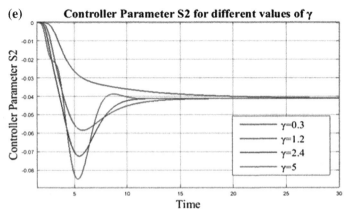

(f)

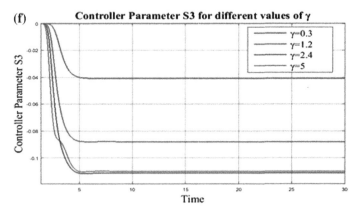

Fig. 2 (continued)

$$G_m(s) = \frac{2.460592}{s^2 + 2.6667s + 2.460592} \tag{13}$$

The reference model has settling time three sec, damping factor of 0.85 and percentage overshoot of 0.63%. With the simulation in Simulink utilizing above adaptation mechanism, it takes more time for the response of plant to be stable and shows much oscillations due to the fact of open-loop instability. A proportional and derivative (PD) control is applied across the plant to eradicate the oscillations due to instability. Figure 2a shows modified SIMULINK block diagram of prosthetic arm with added fixed gain PD Control in the adaptive controller block. A step signal is employed, and the closed-loop behavior is observed. The results of the simulation involve time response, tracking errors between plant and reference model, and controller parameters S_1, S_2, S_3 for adaptation gain 0.3, 1.2, 2.4, 5 are shown in Fig. 2b–f. Furthermore, time response characteristics for different values of adaptation gain for the plant and reference model are calculated and shown in Table 1.

4 Lyapunov Stability Analysis

The designing of adaptive controller based on MIT rule does not guarantee convergence or stability [3]. Lyapunov-based design can be effective to make the adaptive system stable. Subtracting Eq. (4) from Eq. (5) yields (in time domain representation);

$$(\ddot{x}_p - \ddot{x}_m) + (a + cS_3)\dot{x}_p - a_m \dot{x}_m + (b + cS_2)x_p - b_m x_m = cS_1 r - c_m r \tag{14}$$

Let,

$$(a + cS_3) - a_m = a_1; (b + cS_2) - b_m = b_1; cS_1 - c_m = c_1$$

Then; Eq. (13) can be represented as,

$$\ddot{e} + a_m \dot{e} + b_m e = c_1 r - b_1 x_p - a_1 \dot{x}_p \tag{15}$$

Table 1 Characteristic values for displacement between fingers

Specifications	Reference model	Values of gammas			
		$\gamma = 0.3$	$\gamma = 1.2$	$\gamma = 2.4$	$\gamma = 5$
Settling time (s)	3	19	4.156	3.875	3.7
% Overshoot or undershoot	0.63	−0.01	0.4	0.6	0.7

Lyapunov function candidate can be selected as,

$$V(e, \dot{e}, c_1, b_1, a_1) = \dot{e}^2 + b_m e^2 + \frac{1}{\gamma} c_1^2 + \frac{1}{\gamma} b_1^2 + \frac{1}{\gamma} a_1^2$$

Then, its derivative is,

$$\dot{V} = 2\dot{e}\ddot{e} + 2b_m e\dot{e} + \frac{2}{\gamma} c_1 \dot{c}_1 + \frac{2}{\gamma} b_1 \dot{b}_1 + \frac{2}{\gamma} a_1 \dot{a}_1$$

$$= 2\dot{e}\left(c_1 r - b_1 x_p - a_1 \dot{x}_p - a_m \dot{e} - b_m e\right) + 2b_m e\dot{e} + \frac{2}{\gamma} c_1 \dot{c}_1 + \frac{2}{\gamma} b_1 \dot{b}_1 + \frac{2}{\gamma} a_1 \dot{a}_1$$

$$= -2\dot{e}^2 a_m - 2b_m e\dot{e} + 2b_m e\dot{e} + 2c_1 \left(\frac{\dot{c}_1}{\gamma} + \dot{e}r\right) + 2b_1 \left(\frac{\dot{b}_1}{\gamma} - \dot{e}x_p\right) + 2a_1 \left(\frac{\dot{a}_1}{\gamma} - \dot{e}\dot{x}_p\right)$$

$$= -2\dot{e}^2 a_m + 2c_1 \left(\frac{\dot{c}_1}{\gamma} + \dot{e}r\right) + 2b_1 \left(\frac{\dot{b}_1}{\gamma} - \dot{e}x_p\right) + 2a_1 \left(\frac{\dot{a}_1}{\gamma} - \dot{e}\dot{x}_p\right)$$

The derivative of Lyapunov function will be negative if the terms in the brackets are zero.

This gives,

$$\dot{S}_1 = -\alpha\dot{e}r; \dot{S}_2 = \alpha\dot{e}x_p; \dot{S}_2 = \alpha\dot{e}\dot{x}_p$$

α is taken as, $\alpha = \frac{\gamma}{c}$. The above equation represents adaptation law of adaptive controller parameter which can be implemented. Also, the negativeness of the derivative indicates stable adaptive controller design.

5 Conclusion

From the simulation result, the design of adaptive controller for the prosthetic arm with fixed gain PD controller utilizing MRAC along proper adaptation law gives very well result. With the change in adaptation gain, the transient response performance of output prosthetic arm movement can be improved. The system will be stable for a range of adaptation gain between that transient responses will be improved. MIT rule gives the better performance, but it does not guarantee the convergence or stability. Design utilizing Lyapunov stability gives the stable adaptive controller.

Acknowledgements Authors are grateful to TEQIP-II Scheme of JIS College of Engineering for funding this research.

References

1. P.M. Rossini et.al., The Italian Tribune Newspaper, July 21, 2010. http://theitaliantribune.com/?p=119.
2. Neogi B., Ghosal S., Ghosh S., Bose T.K., Das. A., "Dynamic modeling and optimizations of mechanical prosthetic arm by simulation technique" Recent Advances in Information Technology (RAIT), IEEE Xplore, pp. 883–888, March 2012.
3. Astrom, K.J., and B. Wittenmark; Adaptive control; 2nd Edition: Prentice-Hall, 1994.
4. S. Coman, and Cr. Boldisor, "Adaptive PI controller design to control a mass-damper-spring process." Bulletin of the *Transilvania* University of Braşov • Series I • Vol. 7 (56) No. 2 – 2014.

Part V
Operating System, Databases, and Software Analysis

Framework for Dynamic Resource Allocation to Avoid Intercell Interference by Evidence Theory

Suneeta Budihal, R. Sandhya, S. Sneha, V.S. Saroja and M.B. Rajeshwari

Abstract The paper proposes a framework to allocate the radio resources dynamically to avoid intercell interference (ICI) using evidence theory in MIMO-OFDMA (Multiple Input Multiple Output–Orthogonal Frequency Division Multiple Access) downlink system. In order to utilize the available radio resources such as bandwidth and power for improving cell throughput, the neighboring cells need to share the common resources. Hence, user terminals (UTs) at cell boundary experience severe ICI and contribute to the reduced overall cell throughput. We propose evidence-based Dempster–Shafer Combination Rule (DSCR) to estimate the confidence factor *(CF)* based on evidence parameters. We estimate the confidence interval *(CI)* for the bifurcation of cell users based on either decision ratio or CF as cell center and cell edge. With DSCR, an uncertainty contained in the hypothesis is avoided. The obtained evidences are the decisive factors to allocate resources dynamically to avoid ICI. The simulation results demonstrate the improvement in cell edge performance and in turn enhance the overall cell capacity.

Keywords ICI avoidance · Resource allocation · DSCR · Capacity · Confidence factor · Confidence interval

S. Budihal (✉) · R. Sandhya · S. Sneha · V.S. Saroja · M.B. Rajeshwari
Department of ECE, BVBCET, Hubballi 580031, Karnataka, India
e-mail: suneeta_vb@bvb.edu
URL: http://www.bvb.edu

R. Sandhya
e-mail: sandhyaravi112@gmail.com

S. Sneha
e-mail: sneha.shiralikar19@gmail.com

V.S. Saroja
e-mail: sarojavs@bvb.edu

M.B. Rajeshwari
e-mail: banakar@bvb.edu

© Springer Nature Singapore Pte Ltd. 2018
P.K. Sa et al. (eds.), *Progress in Intelligent Computing Techniques: Theory,*
Practice, and Applications, Advances in Intelligent Systems and Computing 518,
DOI 10.1007/978-981-10-3373-5_39

391

1 Introduction

The frequency reuse techniques in cellular networks lead to reuse spectral resources, to significantly increase system capacity. We address the issue of interference in this paper. By the allocation of the OFDMA subcarriers, intracell interference can be minimized. However, the system behavior gets deteriorated from the interference caused by neighboring cells which use same resources for the users known as ICI.

Randomization, cancelation, and avoidance are the possible three ICI mitigation techniques. The first strategy distributes the interference to all users in a random manner after channel coding. Cancelation technique computes interference present in the received signal and removes it. ICI avoidance strategy deals with resource allocation and provides better throughput for edge users without compromising cell center throughput. This is considered the best technique to overcome ICI [1].

In dynamic allocation, edge users are allotted with higher preference compared to center user w.r.t. frequency and power allocation. Universal Frequency Reuse (UFR) implies the most basic frequency planning that results in over usage of frequency. To overcome this limitation, Soft Frequency Reuse (SFR) scheme was proposed where the available spectrum is divided into two fixed parts for edge and center users [2]. In paper [3], the author has proposed a unique interface which dynamically changes the size of edge bandwidths.

In [4], only one-third of available bandwidth is reserved for CEU, while the rest is reserved for CCU, but the power allocation is higher for the edge users. Author in [5] provided an relative analysis of UFR, FFR3, PFR, and SFR. Authors in [6] preferred to split the cell into different areas based on the received interference. To split the cell into regions, certain thresholds are used, but the CCUs are permitted to use the entire bandwidth, while the CEUs are restricted. In [7], the overall bandwidth is initially categorized into two parts for center users and edge users, but the edge is further sectioned into three sectors which guarantees intracell orthogonality, thereby reducing overall ICI for multi-cell OFDMA.

The power wastage is reduced by intelligently over allocating the power to some channels while under allocating power to other users. As per this, several methods have been devised to solve the channel assignment problem in a computational efficient way, such as fuzzy logic, game theory, genetic algorithms [8], integer programming [9], graph coloring [10], and water filling [11]. However, the discussed techniques do not provide the evidences for the decisions taken. In the proposed framework, we used the Dempster–Shafer Combination Rule (DSCR) that improves the decision-making capability. Theory of belief functions [12] is a tool for knowledge representation and uncertain reasoning in systems. This paper extends the work to develop a decision-making criterion to separate the cell edge users and cell center users and optimize the power allocation problem and achieve the approximate rate proportionality. The contributions of the paper are as follows:

1. We use evidence parameters and propose a methodology to compute confidence factors by DSCR.

2. We propose a criteria to determine confidence interval to estimate cell edge users based on either decision ratio or confidence factor.
3. We recommend a technique to partition the subcarriers and allocate to the cell edge and center user terminals based on confidence interval.
4. We intend a process to allocate the power for the respective subcarriers based on confidence factors.

The paper is structured as follows. In Sect. 2, the framework for dynamic resource allocation to avoid ICI using evidence theory is addressed. Section 3 highlights the results, and in Sect. 4, conclusion is provided.

2 Framework for Dynamic Resource Allocation to Avoid Intercell Interference Using Evidence Theory

Consider MIMO-OFDMA downlink system that consists of multiple transmit and a receive antennas. At the transmitter, the bits are transmitted to K different UTs, and each user k ranging from 1 to K is allotted with n ranging from 1 to N subcarriers and is assigned with a power $p_{k,n}$. The UTs at the cell boundary experience maximum ICI as the neighboring cells reuse the resources. In order to avoid ICI at the cell boundary, users need to be separated based on the confident factors obtained from evidence theory. For evidence theory, the selection of the evidence parameters are done such that it satisfies the Dempster–Shaffer rule of combination. The considered evidence parameters are EP_d, EP_r, and EP_s. The proposed framework can be explained in the following subsections.

2.1 Evidence Parameters to Compute Confidence Factors by Applying DSCR

Evidence theory is suited for making judgmental based decisions. In [13], the hypotheses are assigned with evidential weights. The evidential weights are the basic assignments and are represented as follows:

$$\Sigma_{X \subseteq (\Omega)} w(X) = 1 \tag{1}$$

where X is the hypothesis belonging to the set Ω known as the power set and $w(X)$ is its weight. The evidential weights are combined with DSCR to obtain CF as shown in Fig. 1. The CF provides the strength of belief function. According to DSCR, the masses of the evidences are combined to get the accumulated evidences that support the hypothesis Z known as combined hypothesis.

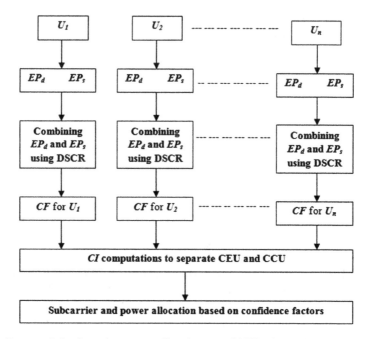

Fig. 1 Framework for dynamic resource allocation to avoid ICI using evidence theory. The evidence parameters are combined with DSCR to compute the confidence factors. These are used to set the confidence interval

$$w(Z) = \frac{\sum_{X \cap Y = Z \neq \phi} w(X).w(Y)}{\sum_{X \cap Y \neq \phi}(X).w(Y)} \qquad (2)$$

In Eq. (2), $w(X)$ and $w(Y)$ denote the weighted masses for the evidence parameters. The weighted masses are combined to provide $w(Z)$, the hypothesis toward the belief (h_1), disbelief (h_2), and ambiguity (h_3). In this section, the distances of all the UTs w.r.t. base station are computed by Euclidean formula and the *SINR* of the UTs is obtained using Eq. (3) as follows:

$$SINR_{k,n} = \frac{Ptr_n L_{k,n}}{\sum_{i=1}^{K} Ptr_n L_{k,n} + \sigma^2} \qquad (3)$$

where Ptr_n is the power transmitted from the base station on the RB, $L_{k,n}$ denotes the path loss, and σ^2 is the noise power spectral density. In evidence theory, the preferred evidences have a set of hypotheses known as h_1, h_2, and h_3. The confident factors are calculated by belief–disbelief hypothesis, without ambiguity and belief-ambiguity hypothesis, without disbelief. From Fig. 2, it is revealed that without the disbelief, linear relation is observed and without ambiguity, exponential relation is obtained between evidence parameters. The hypothesis which leads to exponential relation is

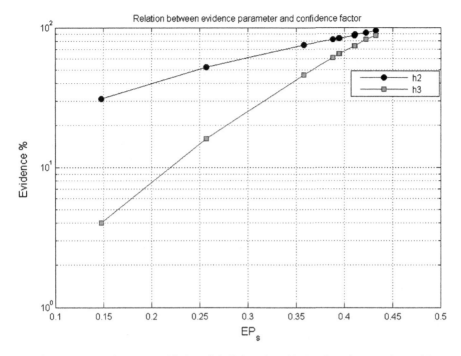

Fig. 2 Plot for evidences considering disbelief and ambiguity hypotheses using evidence parameters

best suited for the computation of *CF*. Hence, second combination of hypothesis is considered for further experimentation.

The normalized values of the hypothesis are the weights and are combined using DSCR as in Eq. (2) which provides a set of *CF*. Table 1 shows the overall combined hypotheses for EP_d and EP_s. The graph in Fig. 2 indicates the *CF* of each UT considering disbelief h_2 and ambiguity h_3. In case of h_2, consider the ambiguity to decide the UT position in the cell that leads to uncertainty in the belief. In case of h_3, there is no ambiguity to decide the position of the UT and provides strong evidence toward belief function. Hence, consider a hypothesis with ambiguity, for further decision of UT positions. UTs with higher *CF* are considered as CEU and vice versa. From Table 1, the belief of evidence parameter such as distance with belief (EP_{d1}),

Table 1 The combination table contains the combined hypothesis of EP_d and EP_s

$\cap$	$w(EP_{d1})$	$w(EP_{d2})$	$w(EP_{d3})$
$w(EP_{s1})$	$h_1 \leftarrow w(Z_{11})$	$\emptyset$	$h_1 \leftarrow w(Z_{13})$
$w(EP_{s2})$	$\emptyset$	$h_2 \leftarrow w(Z_{22})$	$h_2 \leftarrow w(Z_{23})$
$w(EP_{s3})$	$h_1 \leftarrow m(Z_{31})$	$h_2 \leftarrow w(Z_{32})$	$\Omega \leftarrow w(Z_{33})$

disbelief (EP_{d2}), and ambiguity (EP_{d3}) is combined with the evidence parameter such as $SINR$ with belief (EP_{s1}), disbelief (EP_{s2}), and ambiguity (EP_{s3}) to provide the belief function given by.

$$w(\{h_1\}) = \frac{\sum_{i=1}^{3} w(C_i)}{\sum_{i=1}^{7} w(C_i)}, \tag{4}$$

The mass $w(h_1)$ in Eq. (4) is the CF, where i in the numerator is the index for hypothesis providing belief and i in the denominator is the index for hypothesis providing belief, disbelief, and ambiguity. The measured CF helps to decide the UT as cell edge user. Similarly, the confidence of all the K UTs are calculated.

2.2 Criteria for Setting Confidence Interval to Identify Cell Edge Users Based on Dempster–Shaffer Combination Rule

Based on the CF, CI is defined which separates the UT as CCU and CEU. The CI is dynamically set based on either $CF_{threshold}$ or decision ratio. The decision ratio is defined by

$$DR = \frac{\sum_{v=1}^{V} U_l}{\sum_{k=1}^{K} U_k} \tag{5}$$

where $(1 \leq v \leq V) \subset (1 \leq k \leq K)$.

The ratio provides the criteria to separate the UTs such that the random grouping of total users in inner or outer region is avoided. The $CF_{threshold}$ is the criteria set to avoid the CI to go beyond the perimeter of the cell, given by, $10 < CI < CF_{threshold}$.

2.3 Partition the Subcarriers and Allocate to the Cell Edge and Center User Terminals Based on Confidence Interval

The frequency plan of Fractional Frequency Reuse of 3 is considered for a sector of seven cells. It is considered that every UT goes through individual fading. Channel gain of kth user on nth subcarrier is represented by $Ga_{k,n}$. The Gaussian noise is denoted by $\sigma^2 = N_0 \frac{B}{N}$ where N_0 is the power spectral density of noise. Therefore, the subsequent subchannel-to-noise power ratio is denoted by $H_{k,n} = \frac{Ga_{k,n}^2}{\sigma^2}$ which gives the signal-to-noise ratio of the kth UT on subcarrier n and is denoted by $\gamma_{k,n} = Ptr_{k,n}H_{k,n}$. The channel state information is assumed to be priorly known to the receiver in the cell. In order to meet the targeted BER, SNR has to be modified effectively. The bit error rate (BER) of *M-QAM* modulation, received *SNR*, $\gamma_{k,n}$, and number of bits $Rb_{k,n}$ can be approximated to be within 1 *dB* for $Rb_{k,n} \geq 4$ and $BER \leq 10^{-6}$.

$$BER_{M-QAM}(\gamma_{k,n}) \approx 0.2exp\left[\frac{-1.6\gamma_{k,n}}{2^{Rb_{k,n}} - 1}\right] \quad (6)$$

we have $Rb_{k,n}$, as

$$Rb_{k,n} = log_2\left(1 + \frac{\gamma_{k,n}}{\Gamma}\right) \quad (7)$$

where $\Gamma = -ln(5BER)/1.6$

The partitioning and managing of the subcarriers in order to avoid ICI is done by the evidential weights obtained from DSCR. The total available subcarriers N are divided into N_{inner} and N_{outer} depending on the *CI*. The N_{inner} and N_{outer} subcarriers are available to allocate for center and edge of the cell, respectively. Initially, UTs present in the exterior region of the cell are allotted with the subcarriers. The reasonable assumption made from the proportional *CF* provides us the information about the subcarrier allocation in the outer region and can be formulated as $N_\phi = [\phi_k N]$ where ϕ is the normalized evidential weight of kth cell edge user. This leads to $N^* = N_{outer} - N_\phi$ unallocated subcarrier. The priority of the allotment of the subcarriers to the edge UT is decided based on the highest channel gain and is scheduled accordingly. Once the allotment of all N_ϕ is completed, the unallocated N^* are scheduled and managed so as to maximize the overall capacity of the system. The optimization problem is formulated as in Eq. (8) to maximize the capacity.

$$max_{c_{k,n},Ptr_{k,n}} \frac{B}{N} \sum_{k=1}^{K} \sum_{n=1}^{N} c_{k,n} log_2\left(1 + Ptr_{k,n}\frac{H_{k,n}}{\Gamma}\right) \quad (8)$$

The subcarrier partitioning and allocation are done for CCU with N_{inner} subcarriers, from total available subcarriers. From the remaining subcarriers, at least one subcarrier is allotted to the UT with best channel gain. This avoids assignment of all the remaining subcarriers to the UT with highest channel gain.

2.4 Power Allocation Based on Confidence Factor

After allocation of the subcarriers, the power is allotted to the UT based on the computed *CF*, for each UT. The amount of power to be allotted per UT is obtained from the Lagrangian Equation. Replacing the value of ϕ_k by N_k in the Lagrangian Equation, we get a set of simultaneous linear equations. These can be solved easily because of its spacial structure. The set of reordered linear equations are LU factorized on the coefficient matrix. The forward–backward substitution is done on the matrix to obtain individual powers given by Eqs. (9) and (10).

$$Ptr_{k,n} = (P_{tot} - \sum_{k=2}^{K} \frac{b_k}{a_{k,k}})/(1 - \sum_{k=2}^{K} \frac{1}{a_{k,k}}), \text{ for } k = 1 \tag{9}$$

and

$$Ptr_{k,n} = (b_k - P_1/a_{k,k}), \text{ for } k = 2, \ldots, K \tag{10}$$

where,

$$a_{k,k} = -\frac{N_1 H_{k,1} W_k}{N_k H_{1,1} W_1} \tag{11}$$

$$b_k = \frac{N_1}{H_{1,1}}(W_k - W_1 + \frac{H_{1,1} V_1 W_1}{N_1} - \frac{H_{k,1} V_K W_k}{N_k}) \tag{12}$$

The available power per UT is used to further allocate the power which is in proportion with the normalized evidential weights obtained by DSCR. The power allotted per user is given by,

$$P_k = \varphi_{CF} * Ptr_k \tag{13}$$

Equation (13) provides the power to be allocated for the UT requirements to maintain the required data rate. The normalized weighted power is then allotted across subcarriers per UT by using leveling technique. Above discussed sections provide the details to allocate the resources dynamically depending on the evidences obtained by DSCR.

3 Results and Discussion

The system specifications for simulation are provided in the Table 2.

Consider a hexagonal cell layout with radius of 10 km where 10 UTs are spread across the cell. The base station is situated at the center of the cell, and the UTs are randomly distributed within the cell. We bifurcate the UTs as CCU and CEU based on the CI as shown in Fig. 3. The CI is set by considering the either decision ratio or $CF_{threshold}$. The CI is considered depending on either of these conditions to be true and adapts dynamically for every iteration. The $CF_{threshold}$ is set as 75% and decision ratio as 0.4. This criteria provides the evidences about the users' location in the cell and their respective signal strengths. Figure 4 shows the amount of normalized power allotted as per the evidential weights computed by DSCR for each of the UT as compared to without using DSCR. The U_3 is identified as cell edge user, and the amount of power allotted is changed based on the evidences obtained using DSCR. This in turn enhances the capacity and ensures the ICI avoidance. The

Table 2 System Simulation Parameters

Parameters	Values
Channel bandwidth	10 MHz
Carrier frequency	2 GHz
Number of subcarriers	600
Subcarrier spacing	15 KHz
Traffic density	10 UT per cell
Noise density	−174 dBm/Hz
Cell radius	10 km
MIMO size	2×2
Maximum user occupancy	200 UTs
Path loss model	Cost 231 Hata model
Allowed BER	10^{-6}
Modulation technique	QAM

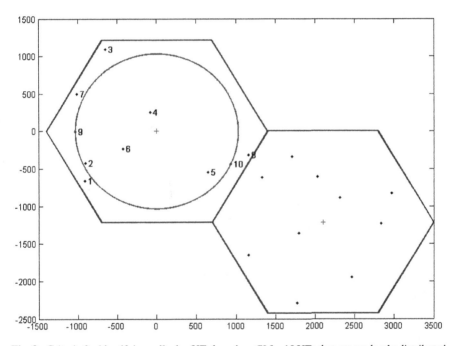

Fig. 3 Criteria for identifying cell edge UTs based on CI for 10 UTs that are randomly distributed in a hexagonal cell layout. The cell radius is set as 10 km

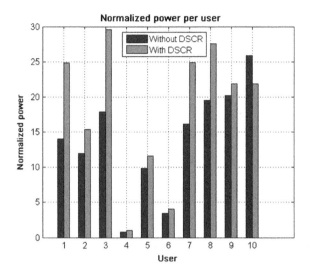

Fig. 4 The amount of normalized power allotted to each UT after computing the *CF* for 10 UTs

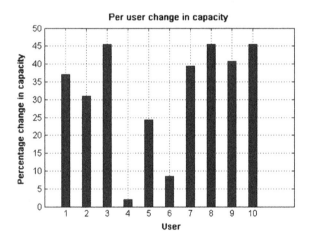

Fig. 5 The percentage increase in the capacity of 10 UTs after applying DSCR for resource partitioning and management

allotted power is decided such that it does not lead to interference with the neighboring cells. Figure 5 is a plot for the achieved enhancement in the capacity after dynamic allocation of the required power to subcarrier for the UTs to maintain the required data rate. This in turn improves the overall cell capacity. The simulations are carried out for multiple UTs, and the corresponding average cell capacity is computed. Fig. 6 provides the improved capacity as UTs increase with DSCR compared to without DSCR technique.

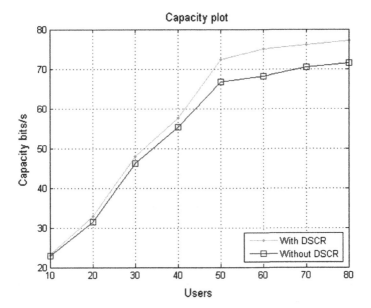

Fig. 6 Total capacity and UTs in an OFDMA system with subcarriers $N = 600$. The capacity achieved with DSCR and without DSCR for a system up to 80 UTs

4 Conclusion

The partitioning and managing of the resources is carried out to avoid ICI in the MIMO-OFDMA system. In this paper, a framework for evidence-based dynamic radio resource allocation technique is proposed. The considered evidence parameters were combined with DSCR to calculate CF. We computed the CI for the estimation of cell edge users based on either decision ratio or threshold CF. The CI was used to decide the number of subcarriers to be allotted to the inner and outer regions of the cell. The obtained CF was used to fix the amount of extra power to be given to the UTs. The obtained evidences were the decisive factors to dynamically allocate the resources to avoid ICI. The simulation results demonstrate the improved cell edge performance and in turn enhance the overall cell capacity.

References

1. A. Daeinabi, K. Sandrasegaran, X. Zhu, *"Survey of Intercell Interference Mitigation Techniques in LTE Downlink Networks,"* 978-1-4673-4410-4/12/31.00 2012 Crown ed., 2012.
2. X. Zhang, Ch. He, L. Jiang, and J. Xu, *"Inter-cell interference coordination based on softer frequency reuse in OFDMA cellular systems,"* International Conference on Neural Networks and Signal Processing: June 2008, pp. 270–275.

3. D. Kimura, Y. Harada, and H. Seki, *"De-Centralized Dynamic ICIC Using X2 Interfaces for Downlink LTE Systems,"* IEEE 73rd VTC, 2011, pp. 1–5.

4. Huawei, R1-050507, *"Soft Frequency Reuse Scheme for UTRANLTE,"* Interference mitigation by Partial Frequency Reuse," 3GPP RAN WG141, Athens, Greece, May 2005.

5. S-E. Elayoubi, O. BenHaddada, and B. Fourestie, *"Performance Evaluation of Frequency Planning Schemes in OFDMA-based Networks,"* IEEE Transaction on Wireless Communications, vol. 7, no. 5, May 2008, pp. 1623–1633.

6. Ch. You, Ch. Seo, Sh. Portugal, G. Park, T. Jung, H. Liu and I. Hwang, *"Intercell Interference Coordination Using Threshold-Based Region Decisions,"* Wireless Personal Communications, vol. 59, no. 4, 2011, pp. 789–806.

7. L. Liu, G. Zhu, D. Wu, *"A Novel Fractional Frequency Reuse Structure Based on Interference Avoidance Scheme in Multi-cell LTE Networks,"* 6th CHINACOM, August 2011, pp. 551–555.

8. Q. Duy La, Y. Huat Chew, and B. Soong, *"An interference minimization game theoretic sub-carrier allocation for OFDMA-based distributed systems,"* IEEE GTC, 2009, pp. 1–6.

9. A. L. Stolyar and H. Viswanathan, *"Self-organizing dynamic fractional frequency reuse for best-effort traffic through distributed inter-cell coordination,"* in Proc. IEEE INFOCOM 2009., 2009, pp. 1287–1295.

10. M.C. Necker, *"A novel algorithm for distributed dynamic interference coordination in cellular networks,"* in Proc. KiVS, 2011, pp. 233–238.

11. S. Ko, H. Seo, H. Kwon, and B. Gi Lee, *"Distributed power allocation for efficient inter-cell interference management in multi-cell OFDMA systems,"* in 16th Asia-Pacific Conference on Communications, 2010, pp. 243–248.

12. Tabib Ramesh et al, *"Decision fusion for robust horizon estimation using Dempster Shafer Combination Rule,"* NCVPRIPG, 2013 Fourth National Conference on IEEE, 2013, pp. 1–4.

13. Rakowsky Uwe Kay, *"Fundamentals of the Dempster-Shafer theory and its applications to system safety and reliability modeling,"* RTA 3-4 Special Issue Dec. 2007.

Query Prioritization for View Selection

Anjana Gosain and Heena Madaan

Abstract Selection of best set of views that can minimize answering cost of queries under space or maintenance cost bounds is a problem of view selection in data warehouse. Various solutions have been provided by minimizing/maximizing cost functions using various frameworks such as lattice, MVPP. Parameters that have been considered in the cost functions for view selection include view size, query frequency, view update cost, view sharing cost, etc. However, queries also have a priority value indicating the level of importance in generating its results. Some queries require immediate response time, while some can wait. Thus, if views needed by highly prioritized queries are pre-materialized, their response time can be faster. Query priority can help in selection of better set of views by which higher priority views can be selected before lower priority views. Thus, we introduce query priority and cube priority for view selection in data warehouse.

Keywords View selection · Query priority · Lattice framework · Cube priority

1 Introduction

Data warehouse is a huge storehouse of data which is integrated, subject oriented, time variant and non-volatile by nature [1]. It aims at providing decision support to top management users. They fire 'who' and 'what' kind of queries on data warehouse to gain an insight and explore improvement scope for the enterprise growth. OLAP acts as a platform on which users communicate with the data warehouse to search answers for their decision support queries. These queries demand data aggregated along various dimensions and when run directly on data warehouse consume lot of time. This delay in response time is unacceptable for decision

A. Gosain · Heena Madaan (✉)
USICT, GGSIPU, New Delhi, India
e-mail: heenamadaan100@gmail.com

A. Gosain
e-mail: anjana_gosain@yahoo.com

© Springer Nature Singapore Pte Ltd. 2018
P.K. Sa et al. (eds.), *Progress in Intelligent Computing Techniques: Theory,
Practice, and Applications*, Advances in Intelligent Systems and Computing 518,
DOI 10.1007/978-981-10-3373-5_40

403

makers. Answers to such queries can be provided in the minimum possible time on cost of storage space if the required aggregates are already stored as materialized views. Selection of optimal set of views that maintains a trade-off between space acquired by them and time required to produce results for queries has been described as view selection problem [2]. Various solutions have been proposed for view selection on different frameworks such as MVPP, lattice, AND-OR graph. Harinarayan et al. [3] proposed lattice framework to represent all possible aggregates along various dimensions and dependencies among them. Then, they selected a set of aggregates on the basis of view size by maximizing the benefit provided by them. They considered query processing cost for selection of views. But the selected views need to be maintained timely with the updating of data in data warehouse to prevent outdated state of views. Gupta et al. [2] incorporated the view maintenance cost for view selection in addition to query processing cost. They used AND-OR graph framework that merges AO-DAG expression of each query. Each node formed in this graph represents a candidate view for materialization and is associated with parameters such as query frequency, view size and update frequency. Their approach aimed at minimizing costs of processing queries and maintaining selected views using these parameters under space constraints. They proposed algorithms for view selection but did not provide any experimental proofs. Yang et al. [4] proposed a new framework named as MVPP (Multiple View Processing Plan) that dealt with the local plans of queries and constructed a global query execution plan by identifying shared operations on data where each node formed a candidate view to be materialized. They implemented the same cost function defined by [2] on MVPP framework. They also proposed a new cost function including shared view cost which denotes the benefit of views being shared by several queries. Since then lot of work has been done to solve this problem where authors have applied different algorithms such as greedy algorithm [2, 3, 5], genetic algorithm [6–9], simulated annealing [10, 11] using the above cost functions and frameworks to improve the set of chosen materialized views. Parameters that have been considered in the cost functions so far include view size, query frequency, view update cost and frequency, shared view cost, etc. However, a query can possess a priority value with itself to indicate its importance level in terms of immediate response time to be provided to users. This query priority can be dependent upon the domain and user of the query [12]. Queries having high importance value being fired by an important user of a domain have to be responded first than less prioritized queries. Such important users can be analysts or statisticians or executives in the organization who need quick answers to take decisions or formulate goals. Delay to such queries can doubt the quality of data warehouse. Cost functions formulated so far did not incorporate query priority as a parameter for view selection. Thus, when such a situation arises where queries have low frequency value but high priority value, these cost functions might fail to select a right set of views delaying the response time of highly prioritized queries. So in our paper, we introduce a new parameter that is the priority value of a query that should be considered along with other parameters for the selection of materialized views.

We choose lattice framework for our work as it can easily model queries posed by users incorporating dimension hierarchies of the warehouse to materialize aggregated views [3]. We also defined cube priority by using the priorities of queries belonging to the cubes. In Sect. 2, we brief about lattice framework. In Sect. 3, we summarize various parameters and cost functions formulated for view selection. Formulation of query priority and cube priority is done in Sect. 4. Lastly, we provide conclusion for our work.

2 Lattice Framework

Decision support queries demand data aggregated along various dimensions. All the possible aggregations that can be formed along the given number of dimensions can be easily represented in the form of lattice. Lattice framework [3] captures all the possible 2^N cubes for N dimensions where each cube represents multidimensional aggregated data. Each level in a lattice framework has cuboids at some level of aggregation starting from base cuboid (no aggregation) to apex cuboid (full aggregation) [13]. These cuboids are constructed by grouping of attributes achieved using group-by clauses in SQL [14]. As the level of aggregation increases, size of cube decreases. Consider an example of a 3-D cube for sales measure with three dimensions as part (p), supplier (s) and customer (c), a lattice framework for it is represented in Fig. 1. Cube {psc} is the base cuboid with no aggregation, and cube {} is the apex cuboid with full aggregation. Each edge between the two cubes corresponds to a dependency relationship between them represented by dependency ($\leq$) operator. Dependency operator forces partial ordering on queries. As shown in Fig. 1, cube {s} $\leq$ cube {ps} and cube {s} $\leq$ cube {sc} denotes that cube {s} can be derived from cube {ps} or cube {sc}. Thus, a query demanding cube {s} can get its answer from cube {s} or {ps} or {sc} or {psc}. Appropriate cubes resulting in minimum query cost must be chosen to be stored as materialized views in data warehouse.

Fig. 1 Lattice framework for a 3-D cube with dimensions part (p), supplier (s) and customer (c)

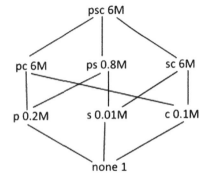

3 Parameters and Cost Functions for View Selection

A query requiring aggregation can derive its result from several views, and thus, the most appropriate views need to be chosen to get results in minimum possible time. In the literature, various parameters that have been considered for view selection are query frequency [2], view size [2, 3], cube size and cube frequency [6, 7], shared view cost [4], view update frequency [2], view update size [2]. These parameters help to define cost functions that are minimized or maximized to select best set of views that will reduce the cost of answering queries.

Harinarayan et al. in [3] justified the linear relationship between size and time of executing a query. So, answering time of a query is equal to the size of the view needed by the query. View size is defined by the number of rows present in the view [3]. Thus, they used view size as a cost function parameter for selecting materialized views by maximizing the benefit of selecting a view. This benefit of a view is calculated by the difference in the number of rows reduced for it and its child views if materialized. Mathematically, this cost function is given as

$$
\begin{aligned}
B_w &= C(v) - C(u) \\
B(v, S) &= \sum_{w \leq v} B_w
\end{aligned}
\tag{1}
$$

where B_w stands for the benefit of materializing view w and $C(v)$ represents number of rows in view v.

A new function was formulated by Gupta et al. in [2] which considered parameters such as query frequency (f_Q), view update cost ($U(V_i, M)$) and view update frequency (g_v) along with the view size for view selection. Here query frequency refers to how many times and how frequently query is being fired. View update cost refers to how many rows are being updated, and view update frequency refers to how frequently views are being updated. The cost function takes into account both query processing and view maintenance cost under space constraint. The authors applied this cost function on AND-OR graph and is mathematically defined as-

For an AND-OR graph G with set of views/nodes M = {V_1, V_2, ..., V_m} and space S, minimize cost function

$$
\tau(G, M) = \sum_{i=1}^{k} f_{Q_i} Q(Q_i, M) + \sum_{i=1}^{m} g_{V_i} U(V_i, M) \text{ under space constraint } \sum_{v \in M} S_v \leq S
\tag{2}
$$

This function was then adapted on other frameworks such as MVPP [4] and lattice [6, 7]. For adaption to lattice framework L, a little modification was made in [6, 7] where query frequencies (f_q) were combined to get cube frequency (f_c). Here cube size (|c|) refers to view size. Cube frequency is calculated by considering all the queries that are using a particular cube to find their answers. According to it,

say a query q from query set Q demands cube c^q from cube set C. Then invoking frequency f_c for cube c is defined as

$$f_c = \sum_{q \in Q \wedge c = c^q} f_q. \tag{3}$$

and overall cost function becomes

$$\tau(L, M) = \sum_{i=1}^{k} f_{C_i} Q(c_i, M) + g_u \sum_{c \in M} U(c, M) \text{ under constraint } \sum_{c \in M} |c| \leq S. \tag{4}$$

Queries may share views or part of views among them. This sharing of views can also reduce the cost of answering queries. Cost of shared views deals with the cost of accessing a view along with how many queries are referring to that view. Thus for a MVPP framework in [4], shared view cost was also included in formulating cost function which is mathematically defined as

$$E \cos t(v) = \left\{ \sum_{q \in R} f_q C_a^q(v) \right\} / n_v. \tag{5}$$

where f_q = query frequency, $C_a(v)$ = cost of accessing view v and n_v = number of queries using view v.

4 Query Priority and Cube Priority

Different queries have different priorities based on the urgency of responding them. Some queries require a very fast response time, while some can wait due to the decisions which have to be undertaken based on their results. These queries may be demanded from a very important domain/user of the organization. Thus, priority of a query depends on the domain from which it is called and level of user querying it. In our work, we designed a hierarchy between domain and user level to formulate query priority where domain is kept at higher level followed by the level of user inside that domain.

Domain denotes the departments within an organization/business for which data warehouse has been constructed. Domains may be already known through ETL process along with the priority according to business use [15]. Authors in [15] defined departments in an enterprise with priority as human resources, finance, material management, operations, marketing and so on. Authors in [12] defined departments as sales, acquisition, purchasing, etc., and assigned priority according to the wholesale chain business.

Similarly, users of the data warehouse have been categorized. Some class of users need immediate summarized data to analyze long trends; some need detailed data to set goals, while some want to explore data. Authors in [16] classified users as tactical decision makers, knowledge workers, strategic decision makers, operational applications, EDI partners and external partners according to their needs and expectations from the data warehouse. Authors in [17] categorized users as statisticians, knowledge workers, executives and information users. Thus, categories of users can be assigned priority according to the level of importance in the organization.

Thus, priority to a query is assigned on the basis of department and users posing that query. Complete query/cube priority is formulated in three steps

(a) Local priority of a query (local_priority(Q))—It refers to how fast a user posing the query needs its results. So a priority is given to a query by its user in the range of 1–5 indicating its importance level which is given as follows: 1—very low importance (general query), 2—low, 3—moderate, 4—high and 5—very high importance (urgently needed)

(b) Global priority of a query (global_priority(Q))—It refers to how fast a query should be answered depending on the department and user posing the query and the local priority of the query. So we also prioritize department and user on the scale of 1–5 leading from the least important department/user to the most important department/user. Following the hierarchy of domain and user, a weight matrix is formed (Table 1) which assigns weight (weight$_Q$) to each query Q_i using the priority of domain and priority of user within that domain according to the formula 6

$$weight_{Q_i} = domain\_priority(Q_i)^2 + user\_priority(Q_i). \qquad (6)$$

Global priority of a query is calculated taking into weight value decided from the domain and the user who have fired it along with local priority of the query. It is defined as

$$global\_priority(Q_i) = weight_{Q_i} * local\_priority(Q_i) \quad \forall i = 1, \ldots, n \qquad (7)$$

Say query Q_1 is of high importance thus local_priority(Q_1) = 4. If it is fired from domain_priority(Q_1) = 4 and user_priority(Q_1) = 3, then global_priority (Q_1) = [19 * 4] = 76.

(c) Cube priority (cube_priority(c))—Query requiring summarized data can derive its answer from the aggregated cube in the lattice framework. So, different cubes can be accessed by different number of queries. Thus, cube priority depends upon

Table 1 Weight matrix example	Domain	User				
		U1	U2	U3	U4	U5
	D1	2	3	4	5	6
	D2	5	6	7	8	9
	D3	10	11	12	13	14
	D4	17	18	19	20	21
	D5	26	27	28	29	30

the number of queries using that cube along with the global priorities of those queries.

This cube priority parameter can then be used as a parameter to select a right set of aggregated views.

5   Conclusion

View selection in data warehouse selects a set of views by optimizing the cost of processing queries and maintaining views following space/maintenance cost constraints. Query frequency, view size, shared view cost, view update cost are the parameters that have been considered for the view selection cost functions. None of these functions considered query priority as the parameter. Query priority is one of the important parameter which deals with how immediate query has to be answered. It depends upon the importance level of the strategic decisions answered by queries being demanded from an important domain and user. A global priority of a cube is designed which further can be used to define cube priority for cubes in the lattice framework. Thus, cube priority can be considered as one of the selection parameter to select highly prioritized views before low prioritized views.

References

1. Inmon, W.: Building the data warehouse. Wiley Publications (1991) 23.
2. Gupta, H.: Selection of views to materialize in a data warehouse. In: Proceedings of the Intl. Conf. on Database Theory. Delphi Greece (1997).
3. Harinarayan, V., Rajaraman, A., Ullman, J.D.: Implementing data cubes efficiently. In: Proceedings of the 1996 ACM SIGMOD International Conference on Management of Data, Montreal, Que., Canada (1996) 205–216.
4. Yang, J., Karlapalem, K., Li, Q.: Algorithm for materialized view design in data warehousing environment. In: Jarke M, Carey MJ, Dittrich KR, et al (eds). Proceedings of the 23rd international conference on very large data bases, Athens, Greece (1997) 136–145.
5. Kumar, TV Vijay., Ghoshal, A.: A reduced lattice greedy algorithm for selecting materialized views. Information Systems, Technology and Management. Springer Berlin Heidelberg (2009) 6–18.
6. Lin, WY., Kuo, IC.: OLAP data cubes configuration with genetic algorithms. In: IEEE International Conference on Systems, Man, and Cybernetics. Vol. 3 (2000).
7. Lin, WY., Kuo IC.: A genetic selection algorithm for OLAP data cubes. Knowledge and information systems 6.1 (2004) 83–102.
8. Zhang, C., Yao, X., Yang, J.: An evolutionary approach to materialized views selection in a data warehouse environment. IEEE Transactions on Systems, Man, and Cybernetics, Part C: Applications and Reviews, 31.3 (2001) 282–294.
9. Horng, J-T., Chang, Y-J., Liu, B-J.: Applying evolutionary algorithms to materialized view selection in a data warehouse. Soft Computing 7.8 (2003) 574–581.
10. Derakhshan, R., et al.: Simulated Annealing for Materialized View Selection in Data Warehousing Environment. Databases and applications. (2006).

11. Derakhshan, R., et al.: Parallel simulated annealing for materialized view selection in data warehousing environments. Algorithms and architectures for parallel processing. Springer Berlin Heidelberg (2008) 121–132.
12. Vaisman, A.: Data quality-based requirements elicitation for decision support systems. Data warehouses and OLAP: concepts, architectures, and solutions. IGI Global (2007) 58–86.
13. Han, J., Kamber, M., Pei, J.: Data mining: concepts and techniques. Elsevier (2011) 113.
14. Gray J, Chaudhuri S, Bosworth A, et al.: Data cube: A relational aggregation operator generalizing group-by, cross-tabs and subtotals. Data Mining and Knowledge Discovery 1(1) (1997) 29–53.
15. Kimball, R., Caserta, J.: The data warehouse ETL toolkit. John Wiley & Sons (2004) 63.
16. Silvers, F.: Building and maintaining a data warehouse. CRC Press, (2008) 277–287.
17. Browning, D., Mundy, J.: Data Warehouse Design Considerations. https://technet.microsoft.com/en-us/library/aa902672(v=sql.80).aspx#sql_dwdesign_dwusers (2001).

Device Fragmentation: A Case Study using "NeSen"

Rakesh Kumar Mishra, Rashmikiran Pandey, Sankhayan Choudhury and Nabendu Chaki

Abstract Remote and eHealthcare Systems are designed to provide healthcare solutions catering to wide variety of requirements ranging from highly personalized to domain-specific systems. Often, a smartphone is used as an aid to port data from embedded or external sensors to remote repository. A majority of smartphones are equipped with multiple network interfaces including provisions for dual subscriber identity modules (SIMs) and a variant of Android as the operating system. Android being an open source system allows customization by the vendor or chipset manufacturer. This raises a serious concern in terms of fragmentation—a form of portability issue with application deployment. For example, App developed on API 16 from MediaTek behaves or crashes over a phone of API 16 from QualComm. We have developed a mobile App called "NeSen" to assess the parameters of all prevalent networks in an area. NeSen uses only the standardized telephony framework and is tried over various smartphones from vendors including Samsung, HTC, LG, iBall, Lava, Micromax, Karbonn, Xiaomi, and Gionee having chipset from MediaTek, QualComm, SpreadTrum, and BroadComm. In this paper, using NeSen, we have conducted first ever evaluation of fragmentation in Android's basic framework. During experimental trails, several issues concerning device fragmentation are noted.

Keywords Fragmentation · NeSen · Android · Network parameters · RHM systems

R.K. Mishra (✉) · R. Pandey
Feroze Gandhi Institute of Engineering & Technology, Raebareli, Uttar Pradesh, India
e-mail: rakesh.mishra.rbl@gmail.com

R. Pandey
e-mail: rashmikiran@hotmail.com

S. Choudhury · N. Chaki
Deparment of Computer Science & Engineering, University of Calcutta, Kolkata, West Bengal, India
e-mail: sankhyan@gmail.com

N. Chaki
e-mail: nabendu@ieee.org

© Springer Nature Singapore Pte Ltd. 2018
P.K. Sa et al. (eds.), *Progress in Intelligent Computing Techniques: Theory, Practice, and Applications*, Advances in Intelligent Systems and Computing 518, DOI 10.1007/978-981-10-3373-5_41

411

1 Introduction

Android is an open source and customizable operating system allowing the manufactures to modify the core libraries of Android. These modifications in Android OS raise a serious concern in the form of fragmentation, i.e., inability of a code to exhibit the homogeneous behavior in different Android platforms. Android compatibility definition document (CDD) [1] provides certain standards and policies to avoid the fragmentation issues. CDD is able to control fragmentation to some extent but failed to evade completely. Manufactures are also twisting basic essence of CDD in several forms. This has been exposed while deploying NeSen among various smartphones from different vendors. Fragmentation is continually reported as a serious concern for the App development community [2].

1.1 TelephonyManager Framework

Android provides an informative manager class that supplies information about telephony-related operations and details on the device. An application interacts with the TelephonyManager framework of the Android. The TelephonyManager framework is direct reflection of native telephony manager in radio interface layer (RIL). There is a mapping between the application framework and native TelephonyManager. Native TelephonyManager opens connection with RIL daemon and extending the connection down to kernel drivers.

TelephonyManager Application Framework is supposed to make the platform-specific variations transparent to the overlying application. The native TelephonyManager is a platform-dependent component and parts of it will have to be adjusted to work with the potentially proprietary vendor radio interface layer (RIL). Figure 1 contains a graphical representation of the various blocks that compose the telephony component. The RIL interactions start right above the baseband, which is the firmware-specific platform to perform various cellular communication-related actions such as dialing numbers, hanging up calls, and accepting calls, and perform the callbacks thereupon. On the other side, the Android package com.android.internal.telephony contains various classes dealing with controlling the phone.

The android.telephony package contains the TelephonyManager class which provides details about the phone status and telephony-related data. TelephonyManager can be used to access phone properties and obtain phone network state information. PhoneStateListener—an event listener—may be attached to the manager enabling an application to get aware of when the phone gain and lose service, and when calls start, continue, or end [13].

Fig. 1 Radio interface layer and TelephonyManager [3]

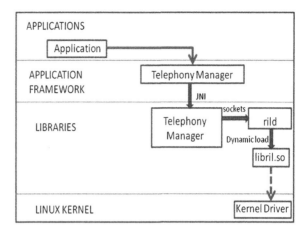

1.2 NeSen-The App

"NeSen" App [4] is developed to assess and record network parameters such as bit error rate (BER), signal strength, cell stability, and service connectivity. All above attributes are usually referred as performance quality parameters. NeSen is a service augmenting RHM (Remote Healthcare Monitoring) system with the capabilities of assured quality-based connection. The targeted service is supposed to harness the availability of best cellular network with dual SIM smartphones.

Vendors refer to the product marketing companies of the devices such as Samsung, Micromax, Sony, and Lava, while the manufactures refer to the companies which manufacture mobile chipsets such as MediaTek and QualComm. To further investigate the issues, another App is designed and deployed for introspecting telephony framework of each phone. The App is designed to reveal the information such as manufactures, vendor, Build version, model no, device id, and radio version. This information is extracted through Build class within the Android basic framework. It has been seen that different manufactures and vendor's ported their arbitrary information-defined formats. It is directly written in Android compatibility document [1] that such information has to be presented in homogeneous manner and is presently violated by the vendors/manufactures of mobile phone.

In one of the recent works [5], possible heterogeneities with the TelephonyManager framework of Android are discussed. Here, a first of a kind case study is presented to expose the fragmentation within the basic framework. Fragmentation is predominantly identified as API fragmentation under device fragmentation category. Lack of Google specification for dual SIM telephony framework is often considered as the major cause behind fragmentation. This has been handled by various techniques including introspection as illustrated in Table 1.

The paper is organized as follows: Related work is described in Sect. 2 and API Fragmentation manifested with NeSen is detailed in Sect. 3, while discussion with conclusions is presented in Sect. 4.

Table 1 Configuration of equipment

Manufacturer	API level	Approach for adaptation	Vendor of product	Remarks
MediaTek	16–20	SDK based	iBall, Lava, Gionee	BER is captured as −1 till API 17 and thereafter as 99
QualComm	15–16	Service call for instantiation and introspectation for invocation	Samsung, Karbonn, LG, Xiaomi	–
SpreadTrum	19	Introspection for both instantiation and invocation	Karbonn, Intex	–
BroadComm, QualComm	19	Introspection of fields for instantiation and invocation	Micromax, Xiaomi	Unusual approach but worked
QualComm	18	Service call for instantiation and introspection for invocation with different names	HTC	–

2 Related Work

A learning-based energy-efficient approach is implemented using Android App in [6] for network selection. The algorithm primarily focuses over lower power consumption and high quality of services using parameters such as network availability, signal strength of available networks, data size, residual battery life, velocity, location of users, and type of application. Battery life, location, and application type are used to determine optimal performing network based on certain predefined rules. App-based monitoring of the network is implemented in [7]. This reports the network parameters perceived by the user equipment to the network-side entity for QoE assessment.

Authors in [8] have detailed the causes of fragmentation. Two vendors, i.e., HTC and Motorola, have been chosen here to analyze the bug reports. Bug reports submitted by Android users help to identify fragmentation. Topic analysis methods have been used to extract the set of topics from the bug reports. Two topic analysis methods opted for fragmentation analysis are Labeled Latent Drichlet Allocation (Labeled LDA) and LDA. Topics extracted from the Labeled LDA and LDA are compared to find out the unique bug report topics, and these topics manifest the fragmentation.

Park et al. [9] proposed two methods to detect the fragmentation problem in Android. The methods used are code-level check and device-level test. Code-level checking method analyzes the code and finds out the part of code where the fragmentation occurs by converting code into itemized values. The itemized values are then mapped on predefined set of rules to correct the code accordingly. Code-level check methods are generally dealt with the hardware fragmentation. Other method is used to analyze the fragmentation at API level. The method collects the test results of APIs and store in the database along with the functions of APIs. Then, these two methods compare the API's functions used by developer with the corresponding functions stored in the database and find the fragmentation.

In [10], a behavior-based portability analysis methodology is proposed to overcome the problems of portability issues in Android. The methodology lets the developer to extract the ideal behavior of application to compare it with similarity in application flows. The entire analysis includes behavior extraction, test execution, and log analysis to identify ideal pattern of App operation flow.

In another work [11], focus is on the change and fault proneness of underlying API. Two case studies are planned: First case study orients toward the change of user ratings in Google Play Store, while the second study emphasizes on experience and problems faced by Android developers and their feedbacks affecting the user ratings. Among the techniques discussed so far, work in [9] is very close and seems to be appropriate for identifying and locating the fragmentation issues of NeSen, but the technique relies on the list-based approach. In case of NeSen, this is not possible because the TelephonyManager is exhibiting valid results as expected from the single SIM phone telephony framework. There is no specification at Google repository for the dual SIM telephony framework. Thus, a list for mapping cannot be prepared. Further, customizations from manufacturers resulted in non-standardized nomenclature of the public interfaces. Hence, a generic list cannot also be prepared even after introspection. A case study-based approach using NeSen for fragmentation with telephony framework is conducted below.

3 Device Fragmentations: API Fragmentation with NeSen

Android is an open source readily available Linux-based operating system. On the one hand, openness of Android allows customizing the framework, keeping the basic library intact; on the other hand, a vendor opts Android to cut down the cost of the products.

NeSen [4] is an Android-based tool incorporated at the Mobile Terminal (MT) which is capable of assessing both static and dynamic network parameters of the smartphone. The real-time values of these parameters are logged at the file system of the MT. The customization from the manufactures of the different vendors posed NeSen a serious challenge of portability among the smartphones from different vendors as well as to different APIs of the same vendors. Fragmentation is the one of the major reason why Android Apps misbehaves and shows inconsistency in the functioning [8].

This flexibility to customize the APIs results into differences within existent framework from different manufacturer as well as platforms from same manufacturer with different APIs. The fragmentation is categorized in two types, viz. operating system (OS) fragmentation and device fragmentation [9]. Device-level fragmentation happens because of the difference in design of underlying hardware of phones as well as customization of the APIs by the manufacturers. Device-level fragmentation is further classified into hardware-level fragmentation and API-level fragmentation. NeSen exhibited device-level fragmentation, particularly API-level fragmentation.

Initially, NeSen was tried on Samsung and Karbonn smartphones with Qual-Comm chipset and Android API 15/16. The objects were instantiated through service instance invocation method with service name as parameter. The telephony objects thus created were given independent state listeners, and the data is logged in different files storing values of signal strength received, bit error rate, cell identity, location area code, etc. The list of phones tested with the first version is in Table 2.

NeSen is tested over different vendors and manufacturer of smartphone and has the manifestation of fragmentation. Fragmentation is observed during testing of App over several manufactures as well as API levels. Figure 2 illustrates the success and failure experience with NeSen's basic version over different phones. NeSen's initial version is tested with API 15–21, phones of 11 vendors and 4 manufacturers. NeSen is successfully installed over some phones, while on others, it is either completely failed or only GUI appeared. NeSen's reasons for the unsuccessful run or failed installation are identified as object instantiation failure, method invocation

Table 2 List of vendors and manufactures used for NeSen

Chipset manufacturer	Example
MediaTek	iBall, Lava, Gionee
QualComm	Samsung, Karbonn, LG, Xiaomi
SpreadTrum	Karbonn, Intex
BroadComm	Micromax, Xiaomi

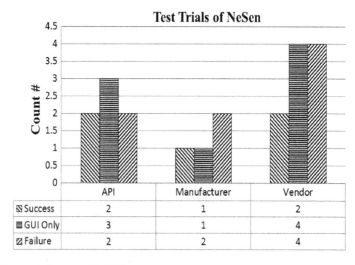

Test Trials of NeSen

	API	Manufacturer	Vendor
Success	2	1	2
GUI Only	3	1	4
Failure	2	2	4

Fig. 2 Fragmentation posed to NeSen

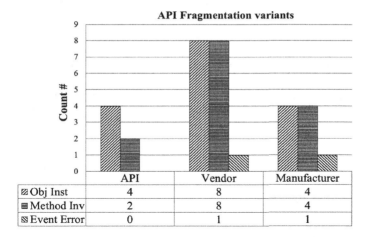

Fig. 3 Different types of fragmentation seen with NeSen

failure and event error. Certain worth mentioning failures of NeSen with different manufacturers are on account of change in the API level which elucidates in Fig. 3.

Methodology opted for the purpose was to perform introspection for each smartphone's library framework. Introspection of phone telephony framework reveals several challenges for NeSen with respect to object instantiation, method invocation, and listeners assignment.

3.1 Library Manifestation

Resources claimed that there is a gap in the Android library specifications and underlined hardware of the smartphones, hence raising the serious compatibility issues [12]. Hardware enhancements usually do not have direct support from standard library, thereby exposing the scope for the customization of Android platform by the manufacturer.

Similar instance has been observed with NeSen over several smartphones platform. Generally, as per the standard specification of Android, TelephonyManager class can be instantiated by system service invocation method with service kind literal as a parameter. This does not work for NeSen trails beyond basic platform. Meanwhile, the crucial points observed are like even a class of Android. That is, TelephonyManager has its four variants. These four variants of the class implied to their corresponding smartphone manufactures. These are being used by the different manufactures as listed in Table 3. Library has been incorporated into the framework in four different variants by the manufactures.

Table 3 TelephonyManager
class variants

Chipset manufacturer	TelephonyManager class
MediaTek	TelephonyManagerEx (through SDK)
QualComm	TelephonyManager, MSimTelephonyManager, MultiSimTelephonyManager
SpreadTrum	TelephonyManager
BroadComm	TelephonyManager

3.2   Object Instantiation

Instantiation of object for NeSen require several specific ways as per their library and underlined hardware. Each manufactures force to dig out its way of object instantiation. This process makes the App development more chaotic as well as time taken. In case of NeSen, there are five different ways of instantiation were identified where each is unique to the specific API and manufacture. These are listed as in Table 4. A few of the manufacturers such as MediaTek provided their own SDK (Software Development Kit) framework which is freely available to be integrated with IDE, while the rest of the manufactures as well as vendors does not provide any such SDK. Standard Android framework library documentation is not sufficient to explore the dedicated framework libraries from different manufactures. The identification of a mechanism for instantiation of objects, for different manufactures except MediaTek, was really a thought-provoking, time-taking, and investigative task. Each smartphones' library needs to be introspected for the purpose, and each alternative is to be explored for method invocations and listener binding.

Table 4 Object instantiation
mechanism

Chipset manufacturer	Instantiation mechanism
MediaTek	getSystemServices()
QualComm	TelephonyManager, static method from(), static fields instance instantiation through introspection
SpreadTrum	Instantiation through introspection
BroadComm	TelephonyManager

3.3   Method Invocation

All the smartphones enlisted here are explored deeply with the help of introspection to investigate the methods of particular class. Generally, all the required methods are present in the library, but they are also overridden or renamed in some forms. NeSen study reveals that though all the methods are present still, we need to dig out them as per the requirement. Methods are renamed by extending the standard method name of with word such as "Gemini." For example, getNetworkOperator() is renamed as getNetworkOperatorGemini(), etc. It is also observed that in some cases, few of the methods generate ambiguous or erroneous results which cannot be directly applied for NeSen. Another way to invoke such methods is through introspection of the library for the corresponding method variants and identifying the suitable candidates for the purpose.

3.4   Listener Assignment

Listener object is bound to the class object to listen the events and reciprocate with action. The required listener binding method for the NeSen was found in the library in three different variants with no particular trends of APIs and manufactures. An introspection of the library is required to choose the corresponding method for the listener accordingly. The listener binding method is found to have changed name and/or parameters; e.g., the listen method in standard framework has PhoneStateListener and an integer as parameter. The original version is sometime complimented with listenGemini method with one additional integer parameter in some libraries. In other cases, the name remains the same, but signature becomes like that of listenGemini.

3.5   Data Logs and Event Triggers

During data recording by NeSen, it was observed that BER reported by event listener in smartphone manufactured by MediaTek was incorrect as that was not similar to values in other smartphones. Parameters taken into account for comparison are operator name, time and LAC were same. MediaTek SDK is providing suspicious value for BER, while all others are as per expectation during comparison. Valid values for the BER are [0–7, 99] as in [13]. BER of −1 is reported when phone is in "out-of-service" state. One of the Sony smartphones deployed with NeSen is failed to provide CID/LAC for second interface.

Its exclusive problem observed with MediaTek smartphones, wherein the null object is being thrown by the event monitor for the listener. The problem is removed when the empty SIM slot of the smartphone is provided with the SIM. This problem is not manifested with the smartphones of the other manufactures such as Qual-Comm, SpreadTrum, and BroadComm. In these smartphones, events for individual interfaces are generated mutually independently. This affects cost in terms of efforts to understand and debug the problem.

4 Conclusions

Generally, most of the RHM systems require some network access and exploit all possible communication interfaces for extending communication capability with assured reliability of the data communication. NeSen is developed for this very purpose. Majority, i.e., 86%, of Android developer are considering fragmentation for rise in the cost of development [14]. Google initiatives such as Android Source Code, Compatibility Definition Code, and Compatibility Test Suite are existing to tackle the issue; however, such initiatives are far away from resolving the fragmentation problems of the domain [9]. Android compatibility document classifies the customization restrictions into MUST, REQUIRED, OPTIONAL, SHALL, SHOULD, and REC-OMMENDED for ensuring the compatibility [1]. During the course of tackling the deployment issues with NeSen, we have encountered issues such as erratic exhibition of the BER value by different phone of same vendors, cell_id undetected in Sony phone C2004, and different hardware and device information by two phones with SpreadTrum chips.

In our endeavor to resolve the reported fragmentation, another tool called "Intros" reveals the platform information using the Build class. The App itself faces the challenges of fragmentation. This is an opinion that fragmentation is appearing on the account of the lack of specification for API by Google itself for multi-SIM phones and strict binding over the semantic meaning of information from vendors and manufactures of the phones.

References

1. Android Compatibility Definition Document: http://static.googleusercontent.com/media/source.android.com/en//compatibility/android-cdd.pdf. (Accessed on Oct, 2015).
2. Malavolta, I., Ruberto, S., Soru T., Teragani, V.: Hybrid Mobile Apps in Google play Store: An Exploratory Investigation. In: 2nd ACM International conference on Mobile Software Engineering (MOBILESoft). pp. 56–59 (2015).
3. http://www.nextinnovation.org/doku.php?id=android_ril. (Accessed on Jan, 2016).
4. Mishra, R. K., Pandey, R., Chaki, N., Choudhury, S.: "NeSen" -a tool for measuring link quality and stability of heterogenous cellular network. In: IEEE International Conference on Advanced Networks and Telecommuncations Systems (ANTS). pp. 1–6. IEEE (2015).
5. http://www.ltfe.org/objave/mobile-network-measurements-using-android/. (Accessed on Jan, 2016).

6. Abbas, N., Taleb, S., Hajj, H., Dawy, Z.: A learning-based approach for network selection in WLAN/3G heterogeneous network. In: Third International Conference on Communications and Information Technology (ICCIT). pp. 309–313. IEEE (2013).

7. Poncela, J., Gomez, G., Hierrezuelo, A., Lopez-Martinez, F. J., Aamir, M.: Quality assessment in 3G/4G wireless networks. In: Wireless Personal Communications, 76(3), pp. 363–377 (2014).

8. Han, D., Zhang, C., Fan, X., Hindle, A., Wong, K., Stroulia, E.: Understanding android fragmentation with topic analysis of vendor-specific bugs. In: 19th Working Conference on Reverse Engineering (WCRE). pp. 83–92. IEEE (2012).

9. Park, J. H., Park, Y. B., Ham, H. K.: Fragmentation Problem in Android. In: International Conference on Information Science and Applications (ICISA). pp. 1–2 (2013).

10. Shin, W., Park, D. H., Kim, T. W., Chang, C. H.: Behavior-based portability analysis methodology for Android applications. In: 5th IEEE International Conference on Software Engineering and Service Science (ICSESS). pp. 714–717. IEEE (2014).

11. Bavota G, Linares Vasquez M: The Impact of API Change- and Fault-Proneness on the User Ratings of Android Apps. In: IEEE Transaction on Software Engineering. vol. 41(4). pp. 384–407 (2015).

12. SushrutPadhye: https://dzone.com/articles/major-drawbacks-android. (Accessed on Sep, 2015).

13. Signal Strength in Android Developer: http://developer.android.com/reference/android/telephony/SignalStrength.html. (Accessed on Sep, 2015).

14. W. Powers: Q1'11 - Do you view Android Fragmentation as a Problem? Baird Research (2011).

15. http://marek.piasecki.staff.iiar.pwr.wroc.pl/dydaktyka/mc_2014/readings/Chapter_7_Telephony_API.pdf. (Accessed on Jan, 2016).

Automated Classification of Issue Reports from a Software Issue Tracker

Nitish Pandey, Abir Hudait, Debarshi Kumar Sanyal and Amitava Sen

Abstract Software issue trackers are used by software users and developers to submit bug reports and various other change requests and track them till they are finally closed. However, it is common for submitters to misclassify an improvement request as a bug and vice versa. Hence, it is extremely useful to have an automated classification mechanism for the submitted reports. In this paper we explore how different classifiers might perform this task. We use datasets from the open-source projects HttpClient and Lucene. We apply naïve Bayes (NB), support vector machine (SVM), logistic regression (LR) and linear discriminant analysis (LDA) separately for classification and evaluate their relative performance in terms of precision, recall, *F*-measure and accuracy.

Keywords Bug classification · Naïve Bayes · Support vector machine · Precision · Recall · F-measure

1 Introduction

Software evolves continuously over its lifetime. As it is developed and maintained, bugs are filed, assigned to developers and fixed. Bugs can be filed by developers themselves, testers or customers, or in other words by any user of the software.

N. Pandey (✉) · A. Hudait · D.K. Sanyal
School of Computer Engineering, KIIT University, Bhubaneswar 751024, Odisha, India
e-mail: nitish5808@gmail.com

A. Hudait
e-mail: abirhudait@gmail.com

D.K. Sanyal
e-mail: debarshisanyal@gmail.com

A. Sen
Dr. Sudhir Chandra Sur Degree Engineering College, Kolkata 700074, West Bengal, India
e-mail: amitavasen@yahoo.com

© Springer Nature Singapore Pte Ltd. 2018
P.K. Sa et al. (eds.), *Progress in Intelligent Computing Techniques: Theory,
Practice, and Applications*, Advances in Intelligent Systems and Computing 518,
DOI 10.1007/978-981-10-3373-5_42

For open-source projects, defect tracking tools like GNATS [1], JIRA [2] or Bugzilla [3] are commonly used for storing bug reports and tracking them till closure. Proprietary software also uses similar tools. However, along with bugs (for corrective maintenance), it is common to file change requests that ask for adaptation of the software to new platforms (adaptive maintenance) or to incorporate new features (perfective maintenance). Similarly, there may be requests to update the documentation, which may not be clubbed as a bug due to its usually less serious impact. Requests for code refactoring, discussions and request for help are other categories for which users may file an issue. However, the person filing the reports may not always make a fine-grained distinction between these different kinds of reports and instead record them as bugs only. The consequence could be costly: developers must spend their precious time to look into the reports and reclassify them correctly. Hence, it is worthwhile to explore whether this classification could be performed automatically.

Machine learning, especially techniques in text classification and data mining, provides invaluable tools to classify bug reports correctly. We call a report in a software defect tracking system as an *issue report* irrespective of whether it refers to a valid bug or it is related to some other issue (as discussed above). In this paper, we study how machine learning techniques can be used to classify an issue report as a *bug* or a *non-bug* automatically. One simple way to distinguish between these two kinds of requests is to look for the text patterns in them. Certain words that describe errors or failures in the software are more common in descriptions that truly report a bug. This suggests that supervised learning techniques like classification can be used. A classifier is initially trained using the data of an issue tracker and subsequently used to label a new issue report.

Contribution: We use a collection of issue reports from the open-source projects HttpClient [4] and Lucene [5] as present in the issue tracker JIRA [2]. Only summary part of each report is parsed and used. We study the performance of various classifiers on these issue summaries. More specifically, we apply naïve Bayes (NB) classifier, support vector machine (SVM), logistic regression (LR) and linear discriminant analysis (LDA) separately for classification and evaluate their relative performance in terms of precision, recall, F-measure and accuracy. In an attempt to find the best classifier, we observe that, in terms of F-measure, SVM followed by NB performs significantly better than other classifiers for both HttpClient and Lucene projects. The classification accuracies obtained by NB and SVM are also better than those of other classifiers for each project. For each of F-measure and accuracy, the values for NB and SVM are close to each other. Hence, NB or SVM appears to be a better choice compared to other classifiers for automatic issue report classification.

Roadmap: A brief background of the current research is provided in Sect. 2. Related work is reported in Sect. 3. Our proposed approach is outlined in Sect. 4, while Sect. 5 describes the experiments, results and threats to validity of the results. The conclusion appears in Sect. 6.

2 Background

Software issue reports capture crucial information about the problem faced by the user who filed the report. A host of issue tracking tools is available, each with varying degrees of sophistication in recording the issue filed. The variations occur in the number of fields that the user needs to fill into the number of stages that the issue goes through before it is declared closed. Note that closing could refer to either fixing the issue or declaring it as void (i.e., invalid). We used issue reports from the issue tracking tool JIRA [2]. An issue in JIRA could report a host of different things like bug, maintenance, improvement, document update, code refactoring. In our discussion we will categorize reports into two classes: *bug* and *non-bug* (note: we use *non-bugs* to refer to all reports that are not categorized as *bug* in JIRA). We use supervised learning tools [6] to automatically segregate the reports into these two categories. Supervised learning involves two steps: (1) training a classifier using labeled samples and (2) classifying an unknown test case after it is trained. We use four kinds of supervised learning algorithms: (1) naïve Bayes (NB) classifier, (2) support vector machine (SVM), (3) logistic regression (LR) and (4) linear discriminant analysis (LDA).

3 Related Work

Analysis of software issue reports submitted to issue tracking tools is a common research area due to its applications in triaging issue reports [7], grouping bugs into different types, estimating issue resolution time and providing feedback on the quality of reports. The extent and cost of misclassification are studied in [8]. Researchers have suggested various methods to automatically classify the reports so that even if the original issue type reported by the user is incorrect, the right type can be inferred and used for further analysis by application engineers. Antoniol et al. [9] manually classified 1800 issues collected from issue trackers of Mozilla, Eclipse and JBoss projects into two classes: bug and non-bug. They investigated the use of various information contained in the issue reports for the classification. They also performed automatic classification of (a smaller subset of) issue reports using naïve Bayes, ADTree and linear logistic regression classifiers. Recently, Ohira et al. [10] manually reviewed 4000 issue reports in JIRA from four open-source projects and manually classified them based on their impact (e.g., security bug, performance bug) on the project. Pingclasai, Hata and Matsumoto [11] reported results of automated bug report classification done with topic modeling (using latent Dirichlet allocation) followed by application of one of the three techniques—ADTree, naïve Bayes classifier and logistic regression—on the issue repositories of three open-source projects. Chawla and Singh [12] proposed a fuzzy logic-based technique to classify issue reports automatically. They have reported higher values of *F*-measure compared to [11] for each of the same three projects. However, [12] used a smaller dataset; so one might wonder whether the results would hold when the repository is much larger. Wu et al. [13] developed the BugMiner tool that uses data mining on historical bug databases to derive valuable information that may be used to improve the quality of the reports as well as detect duplicate reports. Zhou et al. [14] employed text mining followed by data mining techniques to classify bug reports. Like the preceding works, we too study automatic classification of issue reports but use a partially different collection of classifiers (e.g., SVM and LDA are added). However, we do not use topic modeling but a simple term–frequency matrix as an input to the classifiers. We use the highly reliable R [15] environment for experiments. We attempt to identify the classifiers that can be used with satisfactory performance.

4 Our Approach

We now describe the approach to classify the unseen reports into their belongingness. The approach is shown schematically in Fig. 1. The issue reports are first parsed, and only the summary from each report is taken. The body of the report as well as other details like heading, ID, category, description are ignored since it is time-consuming to process them and is usually not found to be of added value to classification [9]. The summary is then preprocessed: common words and stop words are removed from each summary, each word is stemmed and tokenized into terms, and finally the frequency of each term is computed to create the term–frequency matrix (*tfm*) for each report. As argued in [9], *tfm* is probably better suited compared to *tf-idf* (term frequency–inverse document frequency) indexing for software issue classification. Note that we remove terms with frequency lower than a threshold. The preprocessed reports (or alternatively, the *tfm*) are divided into training and testing sets. The training set is used to train the classifier, while the testing set is used to study how well—with respect to chosen metrics—the classifier performs the task of classification.

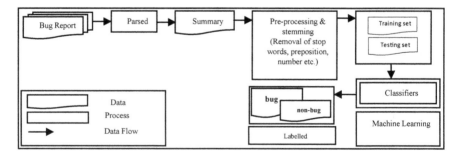

Fig. 1 Proposed approach for issue classification

5 Experiments, Results and Discussion

We use a subset of manually classified issue reports from HttpClient [4] and Lucene [5] projects as provided by [8]. The issue reports were, in turn, extracted from JIRA by researchers [8]. Only the summary part of the reports is taken into consideration for classification. The count of reports used is shown in Table 1.

For each project, we randomly partition the dataset into two subsets for training and testing, respectively, in the ratio 80:20. Only report summary is used for processing. The classifiers are trained first. In the testing phase, each issue report is first preprocessed just like we do in the training phase so that we have only unique terms in the datasets now. These preprocessed datasets are then given to the classifiers to assign labels. The classifier uses past knowledge (i.e., of training phase) to find the belongingness of the reports. The experiments for each project are conducted ten times, each time with a random 80:20 partition to compute the parameters of interest (as explained in the next subsection), and their average values are reported.

We use the R [15] language and environment to perform the experiments. We used R version 3.2.1 which contains implementations of the classifiers NB, SVM, LR and LDA.

5.1 Performance Measures

To measure the performance of the classifiers, we use the metrics: precision, recall, *F*-measure and accuracy. Before we define them, we look at four important quantities that measure how the classifier classified the test inputs as belonging to or not belonging to the (positive) class *bug*.

1. True positive (TP): number of reports correctly labeled as belonging to the class.
2. True negative (TN): number of reports correctly rejected from the class.
3. False positive (FP): number of reports incorrectly labeled as belonging to the class.
4. False negative (FN): number of reports incorrectly rejected from the class.

The entries of the confusion matrix, in terms of the above vocabulary, are indicated in Table 2.

Using the measurements and the following formulae, we calculate precision, recall, *F*-measure and accuracy.

Table 1 Projects and their issue reports

Project name	#Total reports	#Bug	#Non-bug
HttpClient	500	311	189
Lucene	253	110	143

Table 2 Confusion matrix of classifier

		Results of classifier	
		Bug	Non-bug
True classification	Bug	TP	FN
	Non-bug	FP	TN

(a) **Precision**: It is the ratio of the number of true positives to the total number of reports labeled by the classifier as belonging to the positive class.

$$precision = \frac{TP}{TP + FP} \qquad (1)$$

(b) **Recall**: It is the ratio of the number of true positives to the total number of reports that actually belong to the positive class.

$$recall = \frac{TP}{TP + FN} \qquad (2)$$

(c) **F-measure**: It is the harmonic mean of precision and recall.

$$F = 2 \times \frac{precision \times recall}{precision + recall} \qquad (3)$$

(d) **Accuracy**: It measures how correctly the classifier labeled the records.

$$accuracy = \frac{TP + TN}{TP + FP + FN + TN} \qquad (4)$$

5.2 Results and Discussion

We classify the issue reports on HttpClient and Lucene in JIRA as bug or non-bug, i.e., we perform binary classification. The performance of each classifier in terms of the above metrics for each project is given in Tables 3 and 4. The highest and

Table 3 Precision, recall, F-measure and accuracy for NB, SVM, LR and LDA on HttpClient project

	NB	SVM	LR	LDA
Precision	0.752658	0.736468	0.676794	0.709587
Recall	0.802054	0.843185	0.599136	0.692016
F-measure	0.775167	0.784175	0.633574	0.699153
Accuracy	0.714	0.711788	0.564	0.621396

Table 4 Precision, recall, *F*-measure and accuracy for NB, SVM, LR and LDA on Lucene project

	NB	SVM	LR	LDA
Precision	0.771069	**0.811171**	0.476134	0.504317
Recall	0.51246	0.538948	**0.541953**	0.510138
***F*-measure**	0.612841	**0.640172**	0.500015	0.503319
Accuracy	**0.721569**	0.719608	0.560785	0.568627

second highest values of each metric are highlighted for each project. In case of HttpClient, SVM performs best while NB is the second best in terms of precision, recall and *F*-measure. The performance of the other classifiers is far worse. Highest accuracy is provided by NB in classification of HttpClient reports with SVM at the second position. For Lucene, in terms of precision and *F*-measure, SVM again performs best while NB is second best. LR, however, gives the highest recall value in case of Lucene. The highest accuracy is again achieved by the NB classifier, while SVM is behind by a small margin. Overall, both SVM and NB perform very well for each project.

5.3 Threats to Validity

The results obtained in this paper are sensitive to the choice of the datasets. In particular, if the datasets are changed, the outcomes, i.e., precision, recall, *F*-measure and accuracy, may change. The datasets we used are small, and hence, the results might not be reflective of the outcome for a larger sample. We used data from only two open-source projects. The results may be different for other projects.

6 Conclusion

We used four classifiers, namely NB, SVM, LR and LDA, to classify issue reports into bug and non-bug categories. The experiments were conducted using issue reports of the projects HttpClient and Lucene from the JIRA issue tracker. In terms of *F*-measure, SVM followed by NB performs significantly better than other classifiers for HttpClient and Lucene. The classification accuracies obtained using NB and SVM are comparable. They are far better than those of other classifiers for each project. Hence, NB or SVM appears to be a good choice for automatic issue report classification. Since the datasets are not quite large and belong to only two projects, we refrain from making general comments on the exact numerical results. Implementations of NB and SVM are widely available. Our results suggest it might be profitable to classify the reports using these classifiers before further analysis by developers or managers. In future, we plan to use other classifiers including ensemble classifiers and expand the dataset to larger number of reports and projects.

References

1. https://www.gnu.org/software/gnats/.
2. https://www.atlassian.com/software/jira.
3. https://www.bugzilla.org/.
4. https://hc.apache.org/httpclient-3.x/.
5. https://lucene.apache.org/.
6. F. Sebastiani. Machine learning in automated text categorization. *ACM Computing Surveys (CSUR)* 34.1:1–47, 2002.
7. D. Čubranić. Automatic bug triage using text categorization. In *Proceedings of the 16th International Conference on Software Engineering & Knowledge Engineering (SEKE'2004)*, 2004.
8. K. Herzig, S. Just, and A. Zeller. It's not a bug, it's a feature: How Misclassification Impacts Bug Prediction. In *Proceedings of the 35th IEEE/ACM International Conference on Software Engineering*, 2013.
9. G. Antoniol, et al. Is it a bug or an enhancement? A text-based approach to classify change requests. In *Proceedings of the 2008 Conference of the Center for Advanced Studies on Collaborative Research: Meeting of Minds (CASCON'2008)*, ACM, 2008.
10. M. Ohira, et al. A dataset of high impact bugs: manually-classified issue reports. In *Proceedings of the IEEE/ACM 12th Working Conference on Mining Software Repositories (MSR'2015)*, 2015.
11. N. Pingclasai, H. Hata, K. Matsumoto. Classifying bug reports to bugs and other requests using topic modeling. In *Proceedings of 20th Asia-Pacific Software Engineering Conference (APSEC'2013)*, IEEE, 2013.
12. I. Chawla, S. K. Singh. An automated approach for bug classification using fuzzy logic. In *Proceedings of the 8th ACM India Software Engineering Conference (ISEC'2015)*, 2015.
13. L. L. Wu, B. Xie, G. E. Kaiser, R. Passonneau. BugMiner: software reliability analysis via Data Mining of Bug Reports. In *Proceedings of the 23rd International Conference on Software Engineering & Knowledge Engineering (SEKE'2011)*, 2011.
14. Y. Zhou, Y. Tong, R. Gu, H. Gall. Combining text mining and data Mining for bug report classification. In *Proceedings of 30th IEEE International Conference on Software Maintenance and Evolution (ICSME'2014)*, 2014.
15. https://www.r-project.org/about.html.

Memory-Based Load Balancing Algorithm in Structured Peer-to-Peer System

G. Raghu, Neeraj K. Sharma, Shridhar G. Domanal
and G. Ram Mohana Reddy

Abstract There are several load balancing techniques which are popular used in Structured Peer-to-Peer (SPTP) systems to distribute the load among the systems. Most of the protocols are concentrating on load sharing in SPTP Systems that lead to the performance degeneration in terms of processing delay and processing time due to the lack of resources utilization. The proposed work is related to the sender-initiated load balancing algorithms which are based on the memory. Further to check the performance of the proposed load balancing algorithm, the experimental results carried out in the real-time environment with different type of network topologies in distributed environment. The proposed work performed better over existing load balancing algorithm such as Earliest Completion Load Balancing (ECLB) and First Come First Serve (FCFS) in terms of processing delay and execution time.

Keywords Structured Peer-to-Peer (SPTP) systems · Load balancing · Sender-initiated algorithm · Network topologies · Distributed environment

G. Raghu (✉) · N.K. Sharma · S.G. Domanal · G. Ram Mohana Reddy
National Institute of Technology Karnataka, Surathkal, Mangalore 575025,
Karnataka, India
e-mail: raghugolla.22@gmail.com
URL: http://www.nitk.ac.in/

N.K. Sharma
e-mail: neeraj16ks@gmail.com

S.G. Domanal
e-mail: shridhar.domanal@gmail.com

G. Ram Mohana Reddy
e-mail: profgrmreddy@gmail.com

© Springer Nature Singapore Pte Ltd. 2018   431
P.K. Sa et al. (eds.), *Progress in Intelligent Computing Techniques: Theory,*
Practice, and Applications, Advances in Intelligent Systems and Computing 518,
DOI 10.1007/978-981-10-3373-5_43

1 Introduction

The importance of the SPTP system is increasing day by day for sharing the files on the Goggle Drive, Skype, etc. To sharing the files in the SPTP system, we are facing various challenging problems such as reduction of bandwidth of Internet, results in the degradation of throughput of overall systems. To overcome these problems, there are many solutions already existing in the SPTP system such as Hash Table method in Distributed environment, virtual servers, ECLB, and FCFS. In the case of DHT, it works on the name space methodology. The DHT technique gives the unique id to each and every node in the system, and it stores the related information about all other nodes in the identifiers space. Further, the routing table maintained by all the nodes in the system using previous routing logs. To store the routing table, the space of the Hash Table method is O(log n). Where n represents the nodes in the SPTP system. Hence, its difficult to maintain the information regarding all the nodes in the system. The virtual server technique is a centralized technique, and the working process of the virtual server technique is similar to the Hash T. The ECLB technique is static in nature. Therefore, its difficult to delete or add the nodes in the system [1]. In the case of FCFS load balancing technique, it is also static in nature, and the late coming process suffered by starvation. Hence, to overcome all these problems such as processing delay, minimization of process execution time, a load balancing algorithm is required in the SPTP system.

In the proposed work for load distribution in the SPTP system for solving the processing delay and process execution problems. We used normal distribution for sharing load among different nodes in the system. The advantage for using normal distribution in our proposed load balancing algorithm is, it creates equal size of chunks of incoming process data. Hence by dividing the incoming process data into equal size of chunks, we can reduce the processing delay as well as process execution time in the SPTP system. Further, it leads the overall throughput of the SPTP system.

The remaining paper is structured as follows: Sect. 2 deals with the Related Work; Sect. 3 explains the Proposed Methodology; Sect. 4 deals with Experimental Results and Analysis; Finally Concluding Remarks with future directions in Sect. 5.

2 Related Work

To address the problem of load balancing in SPTP system w.r.t. minimization of processing delay and improving the process execution time, there are several algorithms are already exist. The load balancing algorithm is classified into two types such as SIA (initiated by sender) and RIA (initiated by receiver). Two major polices play a major role while doing load balancing in the SPTP system such as transfer policy and location policy. The existing methods have some advantages and disadvantages. The key load balancing algorithms with their advantages and disadvantages are described as follows:

W. Haque et al. [2] suggested load balancing approach in SPTP system. In the suggested approach, they considered all the database is fully replicated. The main limitation of the suggested work is to maintain the replicated data on all nodes in the system.

Lin Xia et al. [3] suggested a typical virtual server-based load balancing technique. In this method, every node has its own id and it has information of all the nodes. In this method, load is transferred from heavily loaded machine to low loaded machine. Here, drawback of this method is node must maintain all the node information and that information will be stored statically.

B. Sahoo et al. [4] proposed different methods. One of them is greedy-based method. This method maintains a matrix for arrival time of each and every request. All the requests are stored in priority queue. And based on that matrix order, we will assign the request to appropriate node to execute that task. The main drawback of this approach is its centralized approach. In this method, arrival time of the request is used to allocate the resources. This method is based on the Non-Preemptive Scheduling Mechanism. Another is the Randomly Transfer Method, where the load will transfer randomly to the computing Machine. This approach also suffers from the issue of being a centralized approach.

Mohammad Haroon et al. [5] proposed the Dynamic load balancing algorithm. This method is not suitable for homogeneous systems and grid computing. And that too, it lacks the spirit as it is capable to exchange the information only among the adjacent or neighboring nodes.

Amit Chhabra et al. [6] proposed a typical method for load balancing technique in Distributed Computing Environment. Their method requires the prior information of the processing time and communication time. So it is not suitable for real-time distributed environment.

Elias D. Nino et al. [7] proposed an optimized load balancing algorithm which produces better results in static environment but not in the dynamic situations. Table 1 summarizes the merits and demerits of the existing load balancing algorithms.

Table 1 Existing load balancing algorithms

Existing methodology			
Authors	Approaches	Merits	Limitations
J.C Patni et al. [8]	Locally initiated approach	Reduced cost of communication	Static load balancing approach
R.A. Khannak et al. [9]	decentralized load balancing approach	Heterogeneous method	Processing time is more
L.T. Lee et al. [10]	Threshold function	Less processing time	Static method
J. Gehweiler et al. [11]	distributed heterogeneous hash tables	Minimizing the processing time	Centralized approach

3 Proposed Methodology

The load balancing approach in the SPTP network is shown in Fig. 1. In this figure, the user-requested process distributed among three different nodes in the SPTP system. In our proposed load balancing algorithm for the SPTP system, we consider memory-based load balancing approach. The flow chart of our proposed algorithm is shown in Fig. 2. The proposed algorithm is classified as a sender-initiated load

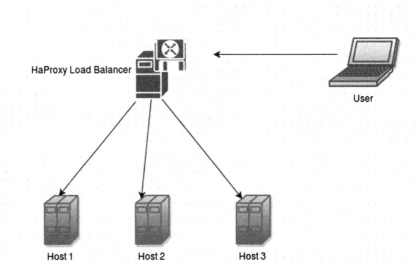

Fig. 1 Load balancing

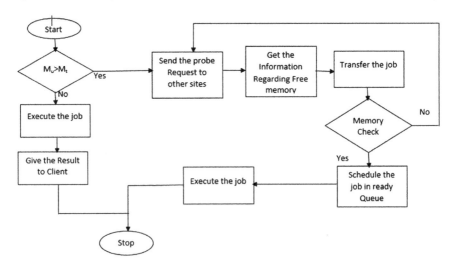

Fig. 2 MLASPS

balancing algorithm in the SPTP network. In our proposed algorithm, a node with higher load initiates the load balancing process. Also, before executing the process, it will check whether to keep the process at the same node or it transfer to the other node in the SPTP system. In the proposed load balancing algorithm, we set the threshold memory value for each node in the system. The threshold memory value of all the nodes in the SPTP system is decided based on the RAM capacity of the node. If the current used memory is more then threshold memory value of the node, in this condition our proposed algorithm transfer the process to the other node in the system. Otherwise, it will execute the process on the same node. In the case of used memory, value is more compared to threshold memory value, the concerning node send the probe message to all the nodes in SPTP system. In response to probe message, all the nodes give the reply message to the probe message generator node. The reply message contains the node current status information. After getting all the reply messages from the nodes in the system, the probe message generator node decide the location of the node where process to be transfer.

For example, whenever a new process arrives to the system, it performs Memory Check $M_u > M_t$, if this condition fails, then we will send the probe message to other sites/nodes i.e., $REQ_{message}(Node_{id},Information)$, all the nodes gives the replay to the requested site that message contains the $Node_{id}$ and states of the system (used memory, free memory) i.e., $REP_{message}(Node_{id},free\ memory, Used\ memory)$. After receiving the replay message, it will extract the free memory details of all the other sites. Among those, it will select the site which having huge amount of free memory. Then, the job will transfer to that node (say $node_x$). Again memory Check operation is performed in the site ($node_x$). That operation is successfully performed, and then, we will schedule that process in Ready Queue. After that, site will produce the result and it will give it to the original Site.

Algorithm 1 describes the overall procedure of Memory-based Load balancing Algorithm in Structured peer-to-peer system (MLASPS).

4   Experimental Results and Analysis

To calculate the performance, processing delay, and processing time of the Memory-based load balancing in structured peer-to-peer system, we used 5 nodes; the overall description and required parameters are specified in the Table 2. Our proposed Memory-based Load balancing Algorithm in Structured peer-to-peer system (MLASPS) periodically checks whether our system is balanced or not. To show results, we are comparing with some of the existing algorithms and we are using Normal distribution to compare the workload rate. We are comparing based on the Workload, processing time, and number of jobs successfully transfer from the client. Our system gives better results compare to the existing method, reason is that in our proposed methodology to distribute the workload among the nodes, we are using Normal Distribution with $\sigma = 4$ and $\mu = 5$ *to* 12, job is divided into chunks depends upon the line rate, and initially, the low capacity server sends the first chunk to one

---

Algorithm 1: Memory-based Load Balancing Algorithm

Data: Set of tasks, Set of nodes, free memory details
M_T: Threshold value of Memory
M_F: Free Memory
M_U: Used Memory
Result: Load balancing in peer-to-peer system with low latency delay, low processing time
while *Processes are not empty* **do**
    1. Set of processes arrive to the system.
    2. perform the Memory check operation.
    3. **for** *every $\langle from, to \rangle \in loadTransfer$* **do**
        if $\langle to, from \rangle \in loadTransfer$ **then**
            $M_F \leftarrow$ loadToFreeMemory.get($\langle from, to \rangle$)
            $M_T \leftarrow$ loadToThresholdMemory.get($\langle to, from \rangle$)
            $M_U \leftarrow$ loadToUsedMemory.get()

        if $M_U > M_T$ **then**
            loadToTransfer.put($\langle from, to \rangle$, job)
            loadToTransfer.remove(job)

        else
            loadToExecute.put(readyQueue)

return *Load balancing in peer-to-peer system with low latency delay, low processing time*

Table 2 System configuration and algorithm parameters

Parameter	value
Processor	Intel(R) i5CPU@2.40 GHz
Memory, threshold memory	200 GB, 2000 MB
RAM	4 GB
OS	Ubuntu 64 bit
Total job load	1000 KB
Type of distribution	Normal distribution with $\mu = 5$ to 12, $\sigma = 4$
Nodes	Fixed 10 and we can extent upto 100
Connection type	Sockets

of the Server peer, then in a parallel next chunk of job is forwarded to the another server, and this will continue until whole job transfer, and total number of messages for communication between server and client will be very less.

Figure 3 shows that comparison of overall execution time of our proposed algorithm with TEST [3], here we are transferring the process periodically in slow rate. Our proposed methodology gives better results compared to TEST [3] Methodology because total no. of messages required for communication is comparatively low.

Figure 4 shows the comparison of overall execution time of our proposed algorithm with ECLB, FCFS, Random Algorithms in the case of slow arrival rate of requests. Here, we are transferring the process periodically in slow rate. Our proposed methodology gives the better results, because we used Normal distribution

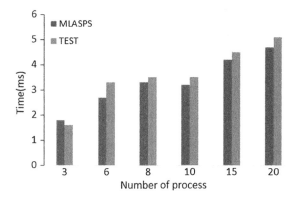

Fig. 3 Processing delay

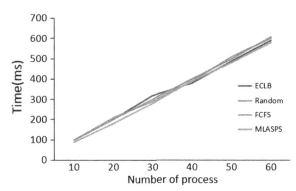

Fig. 4 Execution time for slow arrival rate

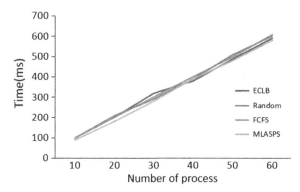

Fig. 5 Execution time for medium arrival rate

for transferring the load among the nodes, so that by seeing Fig. 4, our proposed method process execution time is comparatively low.

Figure 5 shows that comparison of overall execution time of our proposed algorithm with ECLB, FCFS, Random Algorithms in the case of medium arrival rate of requests. Here, we are transferring the process periodically in medium rate. Our proposed methodology gives the better results because we used Normal distribution for

Fig. 6 Execution time for
fast arrival rate

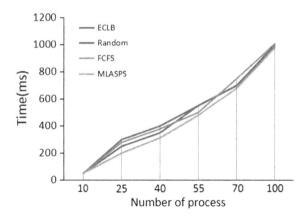

transferring the load among the nodes, so that by seeing Fig. 5, our proposed method process execution time is comparatively low.

Figure 6 shows the comparison of overall execution time of our proposed algorithm with ECLB, FCFS, Random Algorithms in the case of fast arrival rate of requests. Here, we are transferring the process periodically in fast rate. Our proposed methodology gives the better results because we used Normal distribution for transferring the load among the nodes, so that by seeing Fig. 6, our proposed method process execution time is comparatively low.

5 Conclusion

In this paper, we proposed Memory-based Load balancing in structured peer-to-peer system. Experimental results demonstrated our proposed algorithm is giving better results when compared to existing algorithms based on ECLB, Random method, and FCFS. In terms of processing delay and execution time, our proposed methodology takes less time which exhibits its efficiency. As future work, we will try to reduce the number of messages by making use of Hamiltonian cycles in which we will connect the servers as of Hamiltonian cycle and the clusters of clients connected to it.

References

1. Nino, Elias D., Carolina Tamara, and Karen Gomez. "Load Balancing Optimization Using the Algorithm DEPRO in a Distributed Environment." P2P, Parallel, Grid, Cloud and Internet Computing (3PGCIC), 2012 Seventh International Conference on. IEEE, 2012.
2. Haque, Waqar, Andrew Toms, and Aaron Germuth. "Dynamic Load Balancing in Real-Time Distributed Transaction Processing." Computational Science and Engineering (CSE), 2013 IEEE 16th International Conference on. IEEE, 2013.

3. Xia, Lin, et al. "Heterogeneity and load balance in structured P2P system." Communications, Circuits and Systems (ICCCAS), 2010 International Conference on. IEEE, 2010.
4. Sahoo, Bibhudatta, Dilip Kumar, and Sanjay Kumar Jena. "Performance analysis of greedy Load balancing algorithms in Heterogeneous Distributed Computing System." High Performance Computing and Applications (ICHPCA), 2014 International Conference on. IEEE, 2014.
5. Haroon, Mohammad, and Mohd Husain. "Interest Attentive Dynamic Load Balancing in distributed systems." Computing for Sustainable Global Development (INDIACom), 2015 2nd International Conference on. IEEE, 2015.
6. Chhabra, Amit, and Gurvinder Singh. "Qualitative Parametric Comparison of Load Balancing Algorithms in Distributed Computing Environment." Advanced Computing and Communications, 2006. ADCOM 2006. International Conference on. IEEE, 2006.
7. Han, Jiawei, Micheline Kamber, and Jian Pei. Data mining: concepts and techniques. Elsevier, 2011.
8. Patni, Jagdish Chandra, and Mahendra Singh Aswal. "Distributed load balancing model for grid computing environment." Next Generation Computing Technologies (NGCT), 2015 1st International Conference on. IEEE, 2015.
9. Al-Khannak, R., and B. Bitzer. "Load balancing for distributed and integrated power systems using grid computing." Clean Electrical Power, 2007. ICCEP'07. International Conference on. IEEE, 2007.
10. Lee, Liang-Teh, et al. "An extenics-based load balancing mechanism for distributed computing systems." TENCON'02. Proceedings. 2002 IEEE Region 10 Conference on Computers, Communications, Control and Power Engineering. Vol. 1. IEEE, 2002.
11. Gehweiler, Joachim, and Gunnar Schomaker. "Distributed load balancing in heterogeneous peer-to-peer networks for web computing libraries." Distributed Simulation and Real-Time Applications, 2006. DS-RT'06. Tenth IEEE International Symposium on. IEEE, 2006.

Performance Analysis and Implementation of Highly Reconfigurable Modified SDM-Based NoC for MPSoC Platform on Spartan6 FPGA

Y. Amar Babu and G.M.V. Prasad

Abstract To meet today's demanding requirements such as low power consumption and high performance while maintaining flexibility and scalability, system-on-chip will integrate several number of processor cores and other IPs with network-on-chip. To implement NoC-based MPSoC on an FPGA, NoCs should provide guaranteed services and be run-time-reconfigurable. Current TDM- and SDM-based NoCs take more area and would not support run-time reconfiguration. This paper presents modified spatial division multiplexing-based NoC on FPGA; in this we have modified complex network interface and proposed flexible network interface and efficient SDM-based NoC. This proposed architecture explored feasibility of connection requirements dynamically from soft cores during run-time.

Keywords NoC · SDM · VHDL code · Microblazes

1 Introduction

According to Moore's law, chip density is increasing exponentially, allowing multiple processor system-on-chip to be realized on today's FPGA. The main challenge in today's MPSoC is communication architectures among processors. The conventional way of utilizing bus architectures for inter-IP core communication has many limitations. Mainly, it is not scalable well with increasing soft cores on single FPGAs. The design flow of computationally complex and high-bandwidth

Y.A. Babu (✉)
LBR College of Engineering, Mylavaram, Andhra Pradesh, India
e-mail: amarbabuy77@gmail.com

G.M.V. Prasad
B.V.C Institute of Technology & Science, Batlapalem, Andhra Pradesh, India
e-mail: drgmvprasad@gmail.com

© Springer Nature Singapore Pte Ltd. 2018
P.K. Sa et al. (eds.), *Progress in Intelligent Computing Techniques: Theory, Practice, and Applications*, Advances in Intelligent Systems and Computing 518, DOI 10.1007/978-981-10-3373-5_44

on-chip communication of the MPSoC platform takes long design time and is very expensive. Network-on-chip has become the only alternative to solve these problems [1].

Time Division Multiplexing based Network-on-chip uses packet switching techiques to transfer data from source node to destination node. MANGO and Xpipes NoCs are good packet-based NoCs which provide best effort service [2]. In TDM based NoC, no need to establish path from source node to destination node. But in SDM based NoC, we need to fix links between source to destination through different routers. SPIN and PNoC [3] are based on circuit switching method. Today's multimedia-based system demands predictable performance as node link between soft cores is tightly time-constrained. For such multi-core systems, it is compulsory to provide guaranteed throughput service before run-time. To meet these constraints, link allocation should be done in advance during design flow time.

TDM-based NoC provides guaranteed throughput where different time slots are used on the same link. One disadvantage of TDM-based NoC is that configuration of router switching needs to be updated for given time slot. This unique feature needs time slot table memory that requires huge area leading to power consumption in every router. Nostrum and AEthereal are based on TDM NoCs; those architectures have to maintain time slot tables. Spatial division multiplexing-based NoC is a best method where node physical connections, which interlink the routers, are granted to different connections. Number of wires for every link has been allocated to them. The serialized data are sent from sender on the assigned wires, and those are deserialized by the receiver for making the data format of the destination IP core. The main advantage of SDM-based NoC compared with other techniques is that SDM-based NoC removes the need of memory required for time slot tables that leads to power optimization, but area complexity is moved to the serializer and deserializer of network interface.

In this paper, we provide best methods to the above problems. We have proposed a novel design methodology and modified logic structure for network interface to handle the complexity of serializer and deserializer which is common in SDM-based NoC [4]. A simple router with less area complexity is proposed which optimizes area at the higher cost of routing flexibility. This unique feature mainly reduces reordering of the data problem when data reach the destination network interface. Number of channels required between routers and width of each channel depends on application. How tasks in application transfer data from another. We have modeled VHDL code for SDM-based NoC. We have connected microblazes in an MPSoC using modified SDM-based NoC with Xilinx EDK, and an emulation prototype has been realized on a Xilinx Spartan6 FPGA SP605 development board (SP605). The Xilinx soft processor 32-bit RSIC microblazes have been utilized to evaluate the run-time dynamic configurability of the NoC as well as for on-chip data communication between each other nodes.

2　Modified SDM-Based NoC Architecture

The proposed architecture has been modeled as a dual-layer structure where first layer is utilized for data transfer and the second layer is used for configuring router links. The network topology used for the architecture is mesh which is best for multimedia applications. Figure 1 shows basic architecture of modified SDM-based NoC.

2.1　Dual-Layer Structure

The second layer is a simple network mainly used to program the NoC as application demanded bandwidth. To program the NoC, links between routers should be fixed as soft cores which require data from other IP or soft cores. Number of programming byte required to program the NoC depends on size of mesh network (i.e., 2 × 2 to maximum 7 × 7) that can mapped onto target FPGA. For each router, there will be one soft IP which was internally connected through network interface. Network interface in each IP serializes soft IP data from sender and deserializes at receiver side in order to receive data from source IP. All routers and soft IP with network will be placed in the first layer which is responsible to transfer data from any source NoC node to any destination NoC node through router links. Designer fixes the number of wires demanded to transfer data from source NoC node to destination NoC node as per bandwidth requirements which is programmable at design time. Figure 2 shows dual layers in detail.

2.2　Modified Network Interface

The modified network interface logic architecture for the spatial division multiplexing-based NoC has a special control block that will be used to control incoming 32-bit data from different channels of soft IP cores. This intelligent

Fig. 1 Modified SDM-based NoC

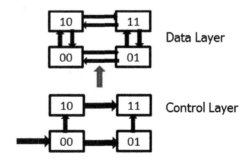

(a)   **(b)**

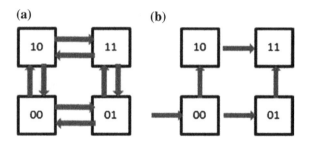

Fig. 2 **a** Data layer, **b** control layer

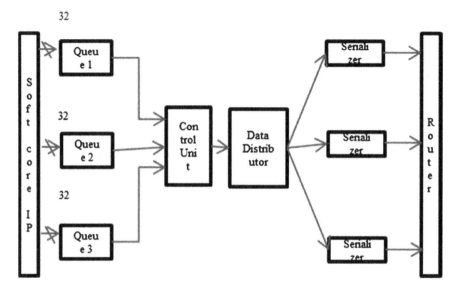

Fig. 3 Network interface

control unit replaces multiple data distributors at transmitter side and multiple data collectors at receiver side with only one data distributor and one data collector. Proposed control unit can be used for fault tolerance to minimize faults at transmitter and receiver blocks of network interface. The modified network interface has many features; one of the features is huge area saving which is a main problem in any network-on-chip architectures and fault tolerance that is very demand for multi-core system-on-chip in embedded applications. We have modified 32-bit to 1-bit serializer with intelligent control unit in network interface. Figure 3 shows network interface.

2.3 Router for Modified SDM-Based NoC

We have targeted Xilinx FPGA to implement SDM-based NoC for MPSoC platform, so router architecture was modified just like architecture of Xilinx switch by avoiding unnecessary complex logic. Modified router for proposed network-on-chip has five ports which include north, south, east, west and local. Soft IP cores are connected through local port. From local port, designer can send data to adjacent router through other four ports. This feature is very unique when compared to any other network-on-chip architectures, which provides more flexibility, scalability and huge area saving. Port size is function of number of sending channels and receiving channels and width of each channel. Figure 4 shows router architecture in detail. Each side has one input port of size 8 bits, one output port of size 8 bits, one out allocated input port of size 3 bits and one out allocated input index port of size 3 bits. The size of out allocated input port size 3 bit because in this, only 5 possible direction data can be sent from any side. The size of out allocated input index port depends on size of input port and output port on each side.

Fig. 4 Router

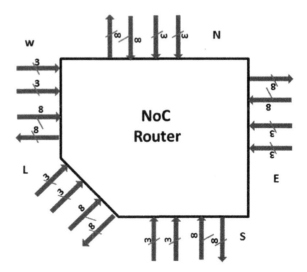

3 Results

3.1 Simulation Results

We have set up 2 × 2 NoC architecture with proposed blocks which are modeled using VHDL and simulated using Xilinx ISE simulator ISIM. Figure shows network interface results and data sent from transmitter of network interface and data received by soft IP core into receiver of network interface. Figure also shows top-level 2 × 2 NoC architecture results with all four routers (Fig. 5).

3.2 Synthesis Results

For our experimental test setup, 2 × 2 modified SDM-based NoC for MPSoC platform synthesis reports is generated using Xilinx synthesis tool XST. Figure shows synthesis report which has available resource on Spartan6 FPGA used for our test setup and percentage utilization of available resources. Our report concludes that area is optimized at network interface level and router side which can be compared with any other network-on-chip architectures for area optimization.

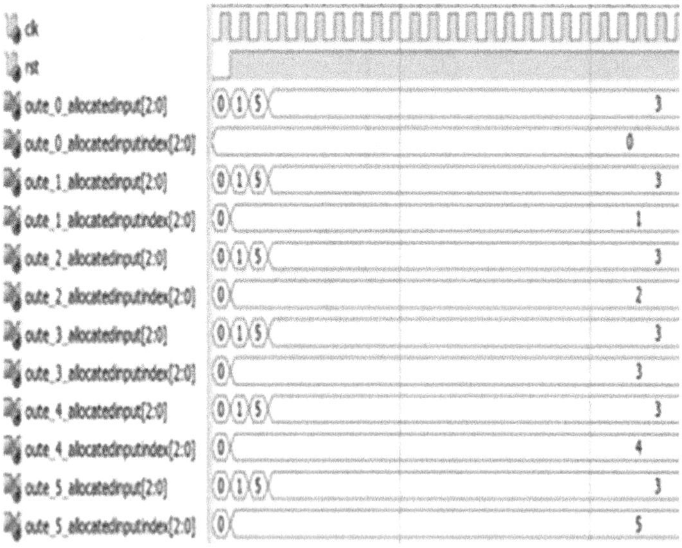

Fig. 5 Simulation results

Table 1 Synthesis report

Spartan6 FPGA utilization summary			
Logic utilization	Used	Available	Utilization (%)
Number of slice registers	6062	54576	11
Number of slice LUTs	5257	27288	19
Number of fully used LUT-FF pairs	2457	8862	27
Number of bonded IOBs	288	296	97
Number of BUFG/BUFGCTRLs	2	16	12

Table 2 Device utilization summary

Device utilization summary			
Logic utilization	Used	Available	Utilization (%)
Number of slice registers	1321	54576	2
Number of slice LUTs	1645	27288	6
Number of fully used LUT-FF pairs	999	1967	50
Number of bonded IOBs	103	296	34
Number of BUFG/BUFGCTRLs	1	16	6

3.3 Implementation Results

We have integrated proposed modified SDM-based NoC for MpSoC platform with four microblaze soft Xilinx IP RSIC cores and 9 fast simplex links (FSL) using Xilinx embedded development kit (EDK) 13.3 ISE design suite. Table 1 shows Spartan6 FPGA utilization summary after implementation. Table 2 shows device utilization summary of proposed modified SDM-based NoC.

4 Performance Analysis

To analyze the performance of proposed NoC architectures on Spartan6 FPGA [5–9], we have selected some case studies which include advanced encryption standard (AES) algorithm, JPEG compression, JPEG2000 compression and H263 video compression standards. We have evaluated the application programs on proposed NoC architecture and other well-popular TDM-based NoCs, SDM-based NoCs, shared bus architectures and advanced extensible interconnect (AXI) architectures and compared them. Our NoC architectures show better results than other popular architectures in terms of area, power, execution time and reconfiguration time as shown in Tables 3, 4 and 5.

Table 3 Device utilization in slices

Application	Proposed NoC architecture	SDM-based NoC	TDM-based NoC	PLB shared bus	AXI architecture
AES	6000	6500	7250	6250	6450
JPEG	6300	7000	7780	6700	6900
JPEG2000	7000	7500	8300	7400	7600
H263	7900	8200	9000	8200	8500

Table 4 Power consumption

Application	Proposed NoC architecture (mW)	SDM-based NoC (mW)	TDM-based NoC (mW)	PLB shared bus (mW)	AXI architecture (mW)
AES	200	250	300	400	390
JPEG	236	290	320	410	400
JPEG2000	300	360	390	490	480
H263	435	450	490	560	550

Table 5 Execution time of application on test architectures

Application	Proposed NoC architecture (s)	SDM-based NoC (s)	TDM-based NoC (s)	PLB shared bus (s)	AXI architecture (s)
AES	2	4.5	6	59	50
JPEG	3	6.3	8	90	80
JPEG2000	4	7	9	120	90
H263	5	8	12	200	170

5 Conclusion

In this paper, we have analyzed performance of various computationally complex applications and proposed a novel design and flexible network interface architecture for existing SDM-based NoC to improve performance and provide guaranteed service for multimedia applications. This architecture saves huge area and required only 5% of existing architectures. In future, multiple applications can be evaluated concurrently on modified SDM-based NoC to explore high scalability and performance.

References

1. International Technology Roadmap for Semiconductors: Semiconductor Industry Association, Dec 2015.
2. K. Goossens, J. Dielissen, and A. Radulescu, "Æthereal network on chip: concepts, architectures, and implementations," IEEE Design & Test of Computers, vol. 22, pp. 414–21, 2005.
3. E. Bolotin, I. Cidon, R. Ginosar, and A. Kolodny, "QNoC: QoS architecture and design process for network on chip," Journal of Systems Architecture, vol. 50, pp. 105–128, 2004.
4. C. Hilton and B. Nelson, "PNoC: A flexible circuit-switched NoC for FPGA-based systems," IEE Proceedings: Computers and Digital Techniques, vol. 153, pp. 181–188, 2006.
5. A. Leroy, D. Milojevic, D. Verkest, F. Robert, and F. Catthoor, "Concepts and implementation of spatial division multiplexing for guaranteed throughput in networks-on-chip," IEEE Transactions on Computers, vol. 57, pp. 1182–1195, 2008.
6. J. Rose and S. Brown, "Flexibility of interconnection structures for field programmable gate arrays," IEEE Journal of Solid-State Circuits, vol. 26, pp. 277–282, 1991.
7. A. Kumar, S. Fernando, Y. Ha, B. Mesman, and H. Corporaal, "Multiprocessor system-level synthesis for multiple applications on platform FPGA," in Proceedings–2007 International Conference on Field Programmable Logic and Applications, FPL, 2007, pp. 92–97.
8. A. Javey, J. Guo, M. Paulsson, Q. Wang, D. Mann, M. Lundstrom, and H. Dai. High-field quasiballistic transport in short carbon nanotubes. Physical Review Letters, 92(10), 2004.
9. V. Agarwal, M. S. Hrishikesh, S.W. Keckler, and D. Burger. Clock rate versus ipc: the end of the road for conventional microarchitectures. In ISCA'00: Proceedings of the 27th Annual International Symposium on Computer Architecture, pages 248.259. ACM.
10. R. H. Havemann, and J. A. Hutchby, "High Performance Interconnects: An Integration Overview", Proceedings of the IEEE, vol. 89, No. 5, May 2001.
11. D. Bertozzi, A. Jalabert, S. Murali, R. Tamhankar, S. Stergiou, L. Benini, and G. De Micheli, "NoC synthesis flow for customized domain specific multiprocessor systems-on-chip," IEEE Transactions on Parallel and Distributed Systems, vol. 16, pp. 113–129, 2005.
12. T. Bjerregaard and J. Sparso, "A router architecture for connection-oriented service guarantees in the MANGO clockless network-on-chip," in Proceedings -Design, Automation and Test in Europe, DATE'05, 2005, pp. 1226–1231.
13. D. Castells-Rufas, J. Joven, and J. Carrabina, "A validation and performance evaluation tool for ProtoNoC," in 2006 International Symposium on System-on-Chip, SOC, 2006.
14. A. Lines, "Asynchronous interconnect for synchronous SoC design," IEEE Micro, vol. 24, pp. 32–41, 2004.
15. M. Millberg, E. Nilsson, R. Thid, and A. Jantsch, "Guaranteed bandwidth using looped containers in temporally disjoint networks within the Nostrum network on chip," in Proceedings–Design, Automation and Test in Europe Conference and Exhibition, 2004, pp. 890–895.

A New Evolutionary Parsing Algorithm for LTAG

Vijay Krishna Menon and K.P. Soman

Abstract Tree adjoining grammars (Tags) are mildly context-sensitive psy-cholinguistic formalisms that are hard to parse. All standard TAG parsers have a worst-case complexity of $O(n^6)$, despite being one of the most linguistically relevant grammars. For comprehensive syntax analysis, especially of ambiguous natural language constructs, most TAG parsers will have to run exhaustively, bringing them close to worst-case runtimes, in order to derive all possible parse trees. In this paper, we present a new and intuitive genetic algorithm, a few fitness functions and an implementation strategy for lexicalised-TAG parsing, so that we might get multiple ambiguous derivations efficiently.

Keywords Tree adjoining grammar · Evolutionary parsing · Genetic algorithm · Genetic operators · NLP · Syntax analysis · Derivation · Parse tree · Lexicalisation · Crossover · Mutation

1 Introduction

Tree adjoining grammars (Tags) were proposed by Joshi et al. [1], to be used for natural language representation and processing. The grammar is a non-Chomskian formalism which is mildly context sensitive in nature. Unlike string generation grammars, Tags use trees to be their elementary constructs. The benefit of using Tags over generally popular grammars, such as context-free grammars (CFGs), is

V.K. Menon (✉)
Amrita School of Engineering, Center for Computational Engineering & Networking (CEN), Coimbatore, Tamil Nadu, India
e-mail: m_vijaykrishna@cb.amrita.edu

K.P. Soman
Amrita Vishwa Vidyapeetham, Amrita University, Coimbatore, Tamil Nadu, India
e-mail: kp_soman@amrita.edu

© Springer Nature Singapore Pte Ltd. 2018   451
P.K. Sa et al. (eds.), *Progress in Intelligent Computing Techniques: Theory, Practice, and Applications*, Advances in Intelligent Systems and Computing 518,
DOI 10.1007/978-981-10-3373-5_45

that TAGs are able to capture a lot more linguistically and lexically relevant features which are normally lost in plain CFG models. TAGs have an *extended domain of locality* which captures furthest dependencies in a single rule. Furthermore, they are able to factor domain dependencies into sub-rules without affecting the parent's template. This gives TAGs an edge over other formalisms; one can model a language using fewer rules and capture its semantics (at least in a limited way) without a separate dependency parsing [2]. The TAG derivation is in fact a good dependency structure that we can use instead. In cases where a probabilistic parse is done, TAGs can almost compete with CFGs and the additional dependency parse required complementing the syntax trees.

The problem we face with TAGs is its exponentially worse parsing complexity for longer sentences; with multiple parse trees (ambiguous), this would be an exhaustive problem. We want all ambiguous parses of a given phrase or sentence. This will push the parser to worst-case scenarios. This was the main motivation to consider alternates that fish out multiple solutions (or optimums in certain cases); genetic algorithm seemed a good candidate.

TAG G is defined as a quintuple in [3] as follows:

$$G_{TAG} = (N, L, T_I, T_A, S) \tag{1}$$

where N is the set of all non-terminals, L is the set of all terminals, T_I is the set of all *initial trees,* T_A is the set of *auxiliary trees* and S is a sentential start symbol. Trees that have a root node named S are called *sentential trees.* TAGs generate trees and not strings. The trees conjoin, using two operations, namely *adjunction* and *substitution*. Substitution is a nominal operation where two initial trees merge at a node that is marked for substitution. This is the same substitution that results in the middle-out growth of a CFG sentential form; essentially, it is a CFG operation. One tree is a parent and the other is the child. The parent's node which is marked for substitution is replaced by the child's root node, *attaching* the entire child tree with the parent. This is essentially possible *iff* the substitution node is a leaf (external) node.

Adjunctions on the other hand are *inserting* auxiliary trees to initial trees. An auxiliary tree has got a root node and an identical *foot node.* The concept of adjunction is splitting a node of the parent tree horizontally into a root and foot nodes of an auxiliary tree. The criterion is the same as substitution except that it can be done on any node (mostly internal nodes). While the root of the auxiliary tree replaces the adjunction node in the parent, the foot node of the auxiliary tree will adopt the sub-tree of the same node being replaced, so essentially inserting the auxiliary tree into the initial tree.

Adjunctions make TAG mildly context sensitive. Figure 1 illustrates how both operations on the tree forms eventually affect the string yield from the final derived tree. Figure 2 pictures the physical process of these operations, a simple attachment for substitutions and a partly complex insertion for adjunctions. For more details on TAGs, refer works of Aravind Joshi, Vijay-Shanker, Yves Schabes and M Shieber [3–5].

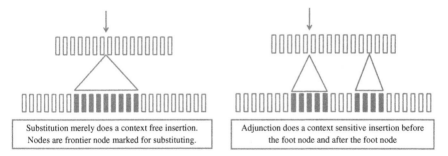

Fig. 1 Yield of substitution operations between two initial trees (*left*) and yield of adjunction operation between an initial and an auxiliary tree (*right*). Clearly, adjunction does a context-sensitive insertion preserving the foot node and its yield intact while insert before and after it

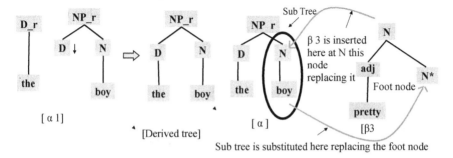

Fig. 2 Substitution is a single attachment, while adjunction does two separate attachments. The classical parsing treats adjunction as partial jump and completion process requiring a *left* completion and a *right* prediction before the insertion is completed

2 Prior Works on Evolutionary Parsing

There are many classical and conventional algorithms to parse TAGs. The popular algorithms are detailed in [4], a CYK-type parser, and [6], an Earley-type parser. Our comparing implementation is the later with an obvious difference that it is a multithreaded parser rather than the backtracking version detailed in [3, 6]. For more details on this please refer our prior published works [7, 8].

The main work of relevance focused on combining EA and TAG parsing is $EATAG_P$ [9], where an evolutionary algorithm has been proposed to parse TAGs which they demonstrate on a simple copy language. The implementations are done on different platforms so they have given a comparison of the number of computations required in each case. Their results show $EATAG_P$ to be much more efficient not just for one parse but also asymptotically too as the EA is able to fit a linear order while the classical TAG parsing will have an exponential order.

The EA version is a randomised search so evidently it gives different values as for the population size and for the selection in each generation; computations vary for different executions of the parser with the same string. To finalise the number of computations, it is averaged across minimum 10 runs, for comparison. The main focus of the paper is to find the right derived (parse) trees using *gene pools* created from (tree, node) pairs that will be progressively added to a chromosome eventually tracing out the *derived tree* with the desired yield. The fitness function is rather vaguely defined, but the strategy is clear. They use multiple fitness scores with decreasing priority like a three-tuple score vector *(matches, coverage and yield)*. The matches record the continuous matches of words in the input string order, the coverage is the total word matches, and the yield tracks the length overshoot of the chromosome over the real string with a negative value. They claim to have used cubic order fitness first and successively reduced the order eventually using a linear function which gave best results. It also merits mentioning some other basic works on evolutionary parsing such as the genetic algorithm-based parser for stochastic CFGs detailed in [10]. This work describes a lot of vague areas when it comes to GA-based parsing. The grammar is considered to be probabilistic giving an easy way to evaluate the sentential forms, in order to rank and compare different individual parses.

The concept of *coherent genes* is introduced in [10] and is a concept that is absent in the $EATAG_P$ where they have eliminated the non-viable gene issue by carefully biasing the initial gene pool and using it in a circular manner. In CFG, however, the sentential forms are string and this will not be a problem. In fact, this gives a better fitness criterion to validate and evaluate individuals in a population based on relative coherence. However, this approach works mostly on non-lexicalised grammars by grouping the words based on POS categories.

3   Genetic Algorithm (GA) Model and Operators

Our approach to evolutionary parsing is to generate random derivations with pre-assembled genes. In TAG parsing, derived trees or parse trees give only syntax information, while other relevant linguistic attributes, such as dependency, semantics and lexical features, are all lost. Undoubtedly, the more useful parse output is not the parse tree but the know-how of creating one. In TAG terminology, we call this the derivation structure, which is normally represented as a tree. It gives obvious advantages to fetch derivations rather than just parse trees.

We rank derivations of an input string initially generated randomly and pruned using the GA process to some threshold fitness value. The random derivations are indicative of individual parses in our algorithm. To represent various genes is the challenge as the derivation nodes contain a lot of information. Thus, we defined TAG parsing as an eight-tuple GA process.

$$\text{GATAG}_P = \{W_D, \Gamma_I, \Gamma_C, f_{cover}, f_{cohere}, \Omega_{LR}, P_{derivation}, C_{stop}\} \qquad (2)$$

where

- W_D is a complete lexical dictionary of words. The lexicalisation can be treated as an indexing of W_D. These aspects are discussed in detail in a later section.
- Γ_I is set of genes that start a parse (sentential genes).
- Γ_C is the set of all other genes that relates to each lexical item.
- f_{cohere} is a fitness score that measure total coherence between genes in a chromosome.
- f_{cover} is the coverage of lexicons by various chromosomes.
- Ω_{LR} is a genetic operator that only yields one child (the left-right child).
- $P_{derivation}$ is initially a random population of chromosomal individuals.
- C_{stop} is a termination criterion for the GA, also called the stopping condition. This is usually a preposition that needs to be realised for the GA process to stop.

The *stop condition* for the GA process is that average fitness be greater than a threshold value. This value, however, needs to be estimated based on empirical observation of multiple runs on the process itself. There are some other such parameters that too require similar estimations. To understand more of these parameters and their statistical properties, we can define a convex problem for it that gives better mathematical grounds for analysing them. This, however, can be tabled for another publication as it is not in focus here.

3.1 Genes, Gene Pools and Coherency

For a complete representation of a derivation tree in TAG, we need information as to the main tree that will be lexicalised with the matrix verb of the sentence. These are sentential constructs which are also initial type trees; we call them *sentence-initial* trees. In order for us to initiate a parse on a sentence, we need such a tree. In an ambiguous grammar, there can be multiple trees which can initiate a parse on the sentence. So this has to be forced into the algorithm that the first tree it selects will be a sentence-initial tree. Hence, genes which shall code for these trees are unique and needs to be handled separately.

For representing a gene in any derivation string, we need the following two-tuple and four-tuple structures.

$$\gamma_I = \{t_I, l_{W_D}\}, \gamma_I \in \Gamma_I, \gamma_I \in \Gamma_I, T_I^S \subset T_I, l_{W_D} \in W_D \qquad (3)$$

$$\gamma_C = \{t_N, l_{W_D}, t_p, n_{t_p}\}$$
$$\gamma_C \in \Gamma_C, t_N \in (T_I \cup T_A - T_I^S), T_I^S \subset T_I, l_{W_D} \in W_D, t_p \in (T_I \cup T_A) \qquad (4)$$

The genes are of two types the sentence-initial (SI) genes in (3) and the common genes in (4). The sets are generated into separate gene pools. The SI-type gene γ_I is simple enough; they encode the root node of derivation with a sentence-initial tree and a possible lexicalisation with any occurring word l_w. The common gene is a bit more complicated. This encodes all the other nodes of the derivation and has two extra items: the parent tree t_p and the node of the parent tree n_p, where this given tree t_N has attached itself and lexicalised with l_w.

The genes in true nature represent all possible nodes in all derivation trees. This is the reason we must try and fit the nodes in the right tree. The creation of these is a batch process that is $O(w.k^2)$ in time where there are k trees and w words. This can also be done in a single initialisation process and not repeat it for every parse again and again. We have evaluated our algorithm purely based on the fitness scoring and selection process and have omitted gene creation from it, in order to get an idea of the efficiency as a function of the length of the sentence, and compare it to classical TAG parsing.

Coherency in a gene is defined as the viability of the gene to exist in a real derivation. This is a binary property, and the gene can be non-coherent in many ways: if the tree is not lexicalised by the paired lexicon, if the parent tree does not have a parent node specified, or if the main tree can never conjoin with the specified parent site due to adjunction constraints or lack of substitution marker. These can be verified while the gene is being created as a set of check heuristics, thus never creating a non-viable gene in the gene pools. The benefit of this process is we can bias it with more heuristics if required. This justifies our call to make this a batch process for the bulk of trees and lexicons, as once the coherent genes are created, then it is just a matter of indexing them to create local gene pool for each sentence.

3.2 Chromosomes and Fitness

A chromosome is a complete derivation tree. It contains genes that code for every word of the given sentence. While the coherence of genes can be ensured, chromosomes are randomised and may be non-coherent. This can be refined only through genetic evolution (GA) process. Since the gene that starts the parse needs to be added separately, the chromosomes are built by this bias. A chromosome is defined as an ordered set of genes as follows

$$\kappa_i = \{\gamma_I \prod_{j=1}^{w-1} \gamma_{C_j} : \gamma_I \in \Gamma_I \text{ and } \forall j \ \gamma_{C_j} \in \Gamma_C\}, w \text{ is length of the sentence} \qquad (5)$$

Note that it is concatenation of genes and the order of genes is the same as the order of the lexicons (lexicons) in the sentence. We are not doing any other kind of biasing for continuous genes in the above formation. We have empirically found that a population of over twenty thousand per instance is a good pool to comprehensively collect most syntactic ambiguities.

Coherence over chromosomes is defined as a fitness score unlike in genes. The max score is $|w\text{-}1|$, and min score is 0. It is a count of the number of cohering genes; if a gene with parent t_p exists, then there should be a gene with tree t_p in the same chromosome. If a single non-coherent gene exists, the derivation becomes non-viable. However, coherence can be improved by genetic evolution, by applying genetic operators.

The fitness calculation is done using a search and a matching heuristic that iterates over the entire chromosome to find gene pairs that match. It is possible to do it exhaustively using an $O(n.log_2(n))$ pair comparison like a sort process. This can be replaced by a shallow parse that will take an $O(n^3)$ complexity. However, it can also be achieved with a linear heuristic function. Out of all these, the shallow parse version is the most accurate and gives almost all the ambiguous derivations. The other techniques yield good results faster but also some false positives that need to be additionally filtered.

Coherence is secondary here, and there is a more important fitness concern that we need to solve, the *word coverage* problem. To make sure that the chromosome has complete word coverage, that is, to check that there is a gene in the chromosome for every word in the given sentence, we employ a simple linear search logic that marks words as each gene is iterated on. The coverage problem is given due importance in the earlier generations, and later on, the coherence will take over while coverage will still be active at a minimal level.

3.3 The Left-Right Genetic Operations

The main biasing of our GA model is that there are fewer random operators than most classic GAs. The operator we have used is a crossover and biased mutation for the main selection process. Our crossover genetic operator works by finding exactly one child per pair of parents. This is the leading child or as to call it, the LR child. Since our parse and coherency model works from left to right, we get better and faster convergence from it. The operator can be defined as follows

$$\Omega_{C_{LR}}(\kappa_i) = \kappa_i^{LR} \text{ such that } \kappa_i^{LR} = \{\gamma_l \prod_{j=1}^{h} \gamma_{C_j} \prod_{k=h+1}^{w} \gamma_{C_k} : \forall j \; \gamma_{C_j} \in$$
$$\kappa_i \text{ and } \forall k \, \gamma_{C_k} \in \kappa_{Random}\}, \text{ h is the coherence index of } \kappa_i \tag{6}$$

As it follows, the crossover needs a good husband and randomly selects a wife to crossover, preserving the husband's coherence and fitness. The idea is never to create a less fit offspring from any husband. So naturally the evolution will go forward. Similarly, biased mutation is also defined. The operator never lowers the fitness of a chromosome it is operating on. The mutation can be defined as follows.

$$\Omega_{M_{LR}}(\kappa_i) = \kappa_i^M \text{ such that}$$

$$\kappa_i^M = \{\gamma_I \prod_{j=1}^{h} \gamma_{C_j} \prod_{k=h+1}^{m-1} \gamma_{C_k} \gamma_{C_{Random}} \prod_{k=m+1}^{w} \gamma_{C_k} : \forall j \; \gamma_{C_j} \in \kappa_i \text{ and}$$

$$\forall k \; \gamma_{C_k} \in \kappa_i, \; \gamma_{C_{Random}} \in \Gamma_C\}, \text{h is the coherence index of } \kappa_i$$

and m is the mutation index such that h < m < w

(7)

4 Implementation

The primary implementation is done using a pruned XTAG subset for English [8]. The main assumption as we have discussed earlier, maintains that the TAG should be single anchored and is lexicalised with only one word. This is why we associate a tree with just one lexicon in our gene model. We have used the $O(n.\log(n))$ fitness strategy which we introduce earlier.

The *howCover* method for estimating word coverage works by directly counting the genes that code for each word. The total count must be equal to the count of the words in the input sentence. This algorithm works linearly by using a word-hashed search of the required gene pool.

```
method howCover(k_i, w)
    cover <- 0
    for j from 0 until |w|
        if k_i contains w_j then increment cover
    rof
    return cover
end
```

The *howCohere* method reads continuous coherence and returns when any gene in the sequence is non-cohesive. Since it employs a *binary search* on genes, the chromosome is required to be sorted on parent-node ordering. This is an additional overhead for the sake of lessoning this computation. In real practice, we can avoid this sorting too. This will be demanding as the length of the string increases; we would need a real big population size to make sure enough viable chromosomes exist. The above method is of *O(n.log(n))*.

```
method howCohere ( kᵢ )
    cohere <- 0
    for gⱼ ∈ kᵢ, j = 1, 2 … |kᵢ|
        if bSearch (kᵢ, gⱼ.tₚ, gⱼ.nₚ)
            increment cohere
        else return cohere
        fi
    rof
end
```

The selection is yet again a straight process. Once the genetic operations are defined, then the selector simply calls them to incrementally evolve and transform the population. Typically within 10–15 generation, we observed convergence. The challenge the selector faces is how to eliminate duplicate chromosomes and duplicate solutions. The first way is to make the population into a set so duplicates will never be stored. The second way is to eliminate duplicates when they are created during the process, by making the solution holder a binary tree or a set. One observation we have made is that the duplication of initial chromosomes is less as it is created in a sequentially random process (no re-seeding of the random generator). So we have used a binary tree to check for duplicates in the solution holder. We can extend our model for a wide coverage grammar like XTAG [11], but requires some overhaul of our base GA definition and will incur considerable complexity for coherency computations.

```
method selector ( P_D, sols_set, good_set )
    for kᵢ ∈ P_D, i = 0, 1, 2 … |P_D|
        if howCover( kᵢ ) and howCohere( kᵢ ) are maximum
            add kᵢ to sols_set
        else if only howCover(kᵢ) is maximum
            add kᵢ to good_set
        else add mutateLR(kᵢ) to newP_D
        fi
    rof
    if good_set is not empty
        newP_D <- crossoverLR( good , P_D )
    else newP_D <- crossoverLR( Random(P_D) , P_D)
    fi
    return newP_D
end
```

5 Results and Conclusion

The proposed algorithm theoretically fares best when we have a word size of seven or more words. The overall complexity of $GATAG_P$ is $O(n.log(n).|P_D|)$. The implementation front we incur some extra overheads and the efficient size of the input sentence is observed to be eight. Our main obstacle to this comparison is the false positives that we incur which needs to be filtered using shallow parsing. We have presented in Table 1, the theoretical speed up between classical and evolutionary TAG parsing. For this analysis, we assume the population size to be 10000 with ten generations. We have also empirically observed that the number of trees selected per sentence is on an average 41. The worst-case complexity for

Table 1 A theoretical computations chart for $GATAG_P$ and Earley-type parser for different input lengths. The calculations assume a population of 10000 chromosomes and a tree pool of 41 trees on an average. The complexity of former is $O(n.log_2(n).|P_D|)$ and latter is $O(|G|.n^6)$

|w|	GATAG$_P$	Earley type	|w|	GATAG$_P$	Earley type
7	2665148	4823609	12	5501955	122425344
8	3200000	10747904	13	6110572	197899169
9	3752933	21789081	14	6730297	308710976
11	4905375	72634001	15	7360336	467015625

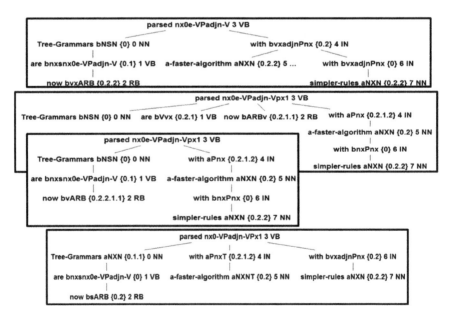

Fig. 3 Some derivations created by parsing the example sentence 'Tree Grammars are now parsed with a faster algorithm with simpler rules'. The sentence yielded 65 derivations and parsed in less than 10 s. The same sentence gave 40 derivations in the classical Earley-Type parser and took 25 s

Earley-type TAG parser is $O(|G|^2.n^9)$ [6], but ours being a multithreaded parser we can assume the minimum worse case (time) for TAG parsing as $O(|G|.n^6)$ [7]. Figure 3 gives screen shots of a few syntactically ambiguous derivations given by the GATAG$_P$ for the example sentence *'Tree Grammars are now parsed with a faster algorithm with simpler rules'*. The trees are generated from solution set chromosomes, using our *TAG Genie* [7] tree viewer.

References

1. Joshi, A. K., Levy, L. S., Takahashi, M.: Tree adjunct grammars. Journal of Computer and System Sciences. 10, 1, 136–163 (1975).
2. Schuler, W.: Preserving semantic dependencies in synchronous Tree Adjoining Grammar. In: Proceedings of the 37th annual meeting of the Association for Computational Linguistics on Computational Linguistics (ACL'99). Association for Computational Linguistics, Stroudsburg, PA, USA, pp. 88–95 (1999).
3. Joshi, A. K., Schabes, Y.: Tree-adjoining grammars. Handbook of formal languages, vol. 3, Rozenberg, G. and Salomaa, A., (eds.). Springer-Verlag New York, Inc., USA, pp. 69–123. (1997).
4. Vijay-Shankar, K., Joshi, A. K.: Some computational properties of Tree Adjoining Grammars. In: Proceedings of the 23rd annual meeting on Association for Computational Linguistics (ACL'85). Association for Computational Linguistics, Stroudsburg, PA, USA, pp. 82–93. (1985).
5. Shieber, S. M., Schabes, Y.: Synchronous tree-adjoining grammars. In: Proceedings of the 13th conference on Computational linguistics - Volume 3 (COLING'90), Karlgren, H. (Ed.). Association for Computational Linguistics, Stroudsburg, PA, USA, pp. 253–258. (1990).
6. Schabes, Y., Joshi, A. K.: An Earley-type parsing algorithm for Tree Adjoining Grammars. In: Proceedings of the 26th annual meeting on Association for Computational Linguistics (ACL'88). Association for Computational Linguistics, Stroudsburg, PA, USA, pp. 258–269 (1988).
7. Menon, V. K.: English to Indian Languages Machine Translation using LTAG. CEN, Amrita Vishwa Vidyapeetham University. Master's Thesis. Coimbatore, India, doi:10.13140/RG.2.1. 5078.5048 (2008).
8. Menon, V. K., Rajendran, S., Soman, K. P.: A Synchronised Tree Adjoining Grammar for English to Tamil Machine Translation. In: International Conference on Advances in Computing, Communications and Informatics (ICACCI), Kochi, India, pp. 1497–1501, doi:10.1109/ICACCI.2015.7275824. (2015).
9. Dediu, A. H., Tîrnauca, C. I.: Parsing Tree Adjoining Grammars using Evolutionary Algorithms. In: ICAART. (2009).
10. Araujo, L.: Evolutionary Parsing for a Probabilistic Context Free Grammar. In: Revised Papers from the Second International Conference on Rough Sets and Current Trends in Computing (RSCTC '00). Ziarko, W. and Yao, Y. Y., (eds.). Springer-Verlag, London, UK, pp. 590–597. (2000).
11. The XTAG Research Group, University of Pennsylvania, http://www.cis.upenn.edu/~xtag/tech-report/.

Classification of SQL Injection Attacks Using Fuzzy Tainting

Surya Khanna and A.K. Verma

Abstract The embellishment of the Internet has escalated the need to resolve cyber security issues. SQL injection attacks (SQLIAs) being one of the oldest yet tenacious attacks captivating the Web applications pose a serious threat. Various techniques are introduced over the years to tackle this problem, but there are times when it becomes difficult to meticulously define the fine line separating a valid input from a malicious one. This work proposes a SQL injection (SQLI) threat level indicator based on fuzzy logic for handling SQL injection attacks. The fuzzy tainting approach helped in ruling out the possibilities of false positives.

Keywords SQL injection · Fuzzy logic · Tainting · Web security · Threat level indicator

1 Introduction

The past decade has seen a rapid shift of products and services to the world of Internet. Whether it is the commercial sector or financial, even the government is going online. All this hype of being a part of the online world has led these services to become the focal point of attacks. Hence, the need of the moment is protect the user's data from maleficent attacker(s).

SQL injection is among the top rankers in the threat rating lists as analyzed by various organizations like OWASP [1], MITRE [2]. An average number of 6,800 SQL injection attacks (SQLIAs) per hour were observed in Imperva's Web application attack report edition #6 [3]. Even alarming observation was an increase of

S. Khanna (✉) · A.K. Verma
Computer Science & Engineering, Thapar University, Patiala, Punjab, India
e-mail: surya.khanna88@gmail.com

A.K. Verma
e-mail: akverma@thapar.edu

© Springer Nature Singapore Pte Ltd. 2018 463
P.K. Sa et al. (eds.), *Progress in Intelligent Computing Techniques: Theory,*
Practice, and Applications, Advances in Intelligent Systems and Computing 518,
DOI 10.1007/978-981-10-3373-5_46

29.63% in SQLIA than the previous year. Further, Verizon's 2015 Data Breach Investigations Report (DBIR) enlists 2,122 confirmed data breaches out of 70 organizations [4]. Information is the most valuable asset to the organizations. So, the databases should be hardened against injection attacks.

2 Motivation

The first point of initiating this work was to legalize valid inputs containing keywords usually considered insecure. Hence, the keywords alone cannot be relied upon. They give us results that are partially true. Now, fuzzy logic is a technology that can handle partial truth values. If that be combined with the known attacks, we can resolve our problem. Work has been done using fuzzy logic to assess the security risks [5] using various density functions but validation of such inputs has not been pondered upon.

3 SQL Injection Attacks

This section provides a brief background on SQL injection. SQLIAs can be broadly categorized as:

- First Order/Direct Attacks
  - Through user input
  - Through cookies
  - Through server variables

- Second Order/Indirect Attacks, i.e., the malicious input is inserted at a place different from where the attack is intended to be performed.

4 Related Work

This section acknowledges the work of various researchers who contributed to provide solution for SQLIA.

In 2003, Huang et al. [6] proposed a black box approach to assess the security of Web application using fault injection and behavior monitoring. It uses Web crawler to determine SQLIA target points of a Web application. A knowledge expansion model is used to learn the behavior of malicious pattern and determine high confidence terms. Lastly, the negative response extraction (NRE) algorithm is used to determine the impact of input on the results. Later on, Stephen W. Boyd and Angelos D. Keromytis 2004 [7] used a secret key to randomize the SQL queries with the help of a proxy. The security of this technique depends on the strength of the secret key.

In 2005, William G.J. Halfond and Alessandro Orso [8] built SQL query models combining static as well as dynamic monitoring techniques to detect injection. The SQL query model is a non-deterministic finite-state automaton which checks the dynamically generated queries to figure out any violation. Further, X. Fu et al. 2007 [9] proposed a white box model for static analysis of byte code to determine vulnerabilities at compile time. A hybrid constraint resolver following string analysis approach is used to decide security breaches. Further, in 2010, Bisht et al. brought forward CANDID [10], a query structure mining approach to determine the deviation of queries formed by candidate input from programmer-intended queries.

In 2013, A.S. Gadgikar [11] came forward with a negative tainting approach to detect SQLIA. In this approach, all entry points are checked for known attack keywords. Recently, in 2015, B. Hanmanthu and colleagues [12] proposed data mining technique using decision trees to classify attack signatures, thereby preventing SQL injection. Decision making is based on associative classification rules.

5 Methodology

We propose a fuzzy logic-based tainting solution to determine SQLI vulnerabilities. In this technique, we initially use negative tainting [11] to determine any malicious string in input parameters and attempt to determine their attack types according to which we can associate a severity level with them as shown in Table 1. The risk level is based on the hindrance; it can cause in the smooth working of a system.

Table 1 Severity level of different types of SQLIAs

Attack type	Example	Risk level
Tautology	' or a = a –	Medium
Logically incorrect queries	Pass character in integer data type, e.g., pin = '#123\' or pin = convert (int, (select top 1 name from sysobjects where xtype = 'u'))	Low
Union query	' UNION SELECT accNo from users where uid = 1349 –	Medium
Piggybacked queries	(I)'; insert into users values (666, 'attacker', 'admin', 0xffff) — (II)'; Drop table accounts;–	High
Stored procedures	Pass following as parameters (I)'; SHUTDOWN; – (II)'; Drop table users;–	Medium, high
Alternate encoding	'; exec (char(0x73687574646f776e)) –	Medium, high
Inference (I) Blind injection (II) Timing attacks	(I) john' and 1 = 0 – john' and 1 = 1– (II) id = 1') or sleep(25) = 0 limit 1–	Low

Scores are calculated from the taints found against the known attacks for all input parameters of a query. These scores are used to indicate the risk level of the malicious input strings. Attackers use logically incorrect queries or inference techniques to determine information about the database though error messages. Not much information can be retrieved through these methods and is usually considered as a part of database fingerprinting. So, no sensitive data breach happens; therefore, they are enlisted to pose a low-level threat, whereas attacks like tautology, union, stored procedure may lead to breach of confidentiality and hence given a relatively higher-level threat indication.

Then, we look over the programmer-intended query corresponding to the input parameter and determine the probability with which our system will be affected if any malicious input is given to the query. Malicious content in a DDL poses highest level of threat. Any modification at the schema level can lead to huge loss/leakage of sensitive data as they can affect the entire table, e.g., the attacker can create another table in which he/she can dump sensitive information about the system and make it available for access by unauthorized users or a table containing monetary records can be deleted affecting large no. of users.

Fuzzy logic basically reduces to paradoxes or multivalued logic to half-truths (or half-falsities). A quote (') is considered as a dangerous parameter for SQL inputs, although it is partially true. We encountered places where the quote is a part of a perfectly valid input. Due to the various guidelines laid down in the defense strategies of SQLIAs, these legal inputs are also considered dangerous. To resolve such problems, we have designed a set of fuzzy rules in their deductive form. A fuzzy rule set used to determine the overall threat level is shown in Table 2. An example rule (as shown in 4th row of Table 3) is "IF input_risk_level IS high AND statement_type IS DDL THEN SQLI_risk IS severe."

For each query, the set of rules is applied to obtain the degree of membership value which is then defuzzified to produce the output. To predict the exposure of query to SQL injection attacks, we use center of gravity (COG) method of defuzzification [13] given by the algebraic expression:

$$x^* = \frac{\int \mu(x)x\,dx}{\int \mu(x)\,dx} \tag{1}$$

Table 2 Examples of fuzzy rules

Input risk level	Statement type	SQLI risk
Valid	Any	None
Medium	DML	Average
High	DML	Severe
Medium	DDL	Severe

Fig. 1 Various sorts of input to a query

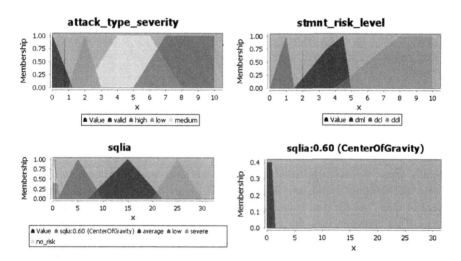

Fig. 2 Input: O'Neal

Consider the different kinds of input values in a SQL query as shown in Fig. 1. First one is a valid query despite containing quotation mark. The second form of input leads to breach in confidentiality of the system, while the last one results in the loss of sensitive data.

Initially, the negative taint values of all inputs are calculated followed by determining the type of query, in this case, a DML. Lastly, these values are fuzzified to indicate SQLIA. Our system shows appropriate indications for threat levels produced by the different inputs as shown in Figs. 2 and 3.

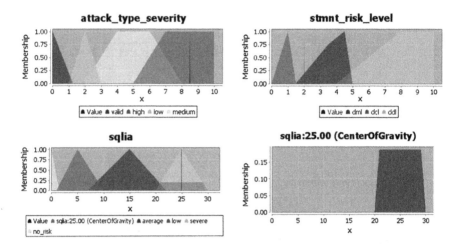

Fig. 3 Input: '; drop table users—

6 Conclusion and Future Work

In this paper, we have classified SQL injection threat level with the help of fuzzy logic and tainting techniques. The proposed model detects the impact of a full fledged query on the data source and works on successfully removing false positives. This approach uses the commonly known COG method, while implementation of other method can be explored in the near future. Security of the application can be enhanced by including more encoding patterns. This method can further be extended to solve the issue of no SQL injection.

References

1. Top 10 2013-Top 10. In: - OWASP. https://www.owasp.org/index.php/top_10_2013-top_10. Accessed 4 Mar 2016.
2. Common Weakness Enumeration. In: CWE - 2011 CWE/SANS Top 25 Most Dangerous Software Errors. http://cwe.mitre.org/top25/. Accessed 4 Mar 2016.
3. 2015 Web Application Attack Report (WAAR) - Imperva. http://www.imperva.com/docs/hii_web_application_attack_report_ed6.pdf. Accessed 5 Mar 2016.
4. 2015 Data Breach Investigations Report. In: Verizon Enterprise Solutions. http://www.verizonenterprise.com/dbir/2015/. Accessed 5 Mar 2016.
5. Shahriar H, Haddad H (2014) Risk assessment of code injection vulnerabilities using fuzzy logic-based system. Proceedings of the 29th Annual ACM Symposium on Applied Computing - SAC '14.
6. Huang Y-W, Huang S-K, Lin T-P, Tsai C-H (2003) Web application security assessment by fault injection and behavior monitoring. Proceedings of the twelfth international conference on World Wide Web - WWW '03 148–159.

7. Boyd SW, Keromytis AD (2004) Sqlrand: Preventing SQL Injection Attacks. Applied Cryptography and Network Security Lecture Notes in Computer Science 292–302.
8. Halfond WGJ, Orso A (2005) Combining static analysis and runtime monitoring to counter SQL-injection attacks. SIGSOFT Softw Eng Notes ACM SIGSOFT Software Engineering Notes 30:1–7.
9. Fu X, Lu X, Peltsverger B, Chen S, Qian K, Tao L (2007) A Static Analysis Framework For Detecting SQL Injection Vulnerabilities. 31st Annual International Computer Software and Applications Conference - Vol 1- (COMPSAC 2007) 1:87–94.
10. Bisht P, Madhusudan P, Venkatakrishnan VN (2010) Candid. ACM Transactions on Information and System Security TISSEC ACM Trans Inf Syst Secur 13:1–39.
11. Gadgikar AS (2013) Preventing SQL injection attacks using negative tainting approach. 2013 IEEE International Conference on Computational Intelligence and Computing Research 1–5.
12. Hanmanthu B, Ram BR, Niranjan P (2015) SQL Injection Attack prevention based on decision tree classification. 2015 IEEE 9th International Conference on Intelligent Systems and Control (ISCO) 1–5.
13. Ross, T. J.: Fuzzy Logic with Engineering Applications. 2nd edn. Wiley (2004).

CMS: Checkpoint-Based Multi-versioning System for Software Transactional Memory

Ammlan Ghosh, Rituparna Chaki and Nabendu Chaki

Abstract In this paper, we propose a novel checkpoint-based multi-versioning concurrency control system (CMS) for software transactional memory (STM). A checkpoint is defined to refer a set of versions. CMS maintains a list of live read-only transactions for every checkpoint. An appropriate object versioning technique decides on how many older versions are to be kept or to be removed. Moreover, the proposed method does not require any global synchronization point, such as centralized logical clock for versioning. Due to its efficient garbage-collecting mechanism, CMS is suitable for both read and write-dominated workload. We formally prove that CMS complies useless-prefix garbage collection (UP-GC). On the basis of number of memory accesses for searching consistent version, CMS performs better in comparison with conventional multi-version permissive STM.

Keywords Software transactional memory (STM) · Concurrency control · Checkpoint · Versioning · Multi-version concurrency control

1 Introduction

Software transactional memory (STM) [1] is an emerging technique for concurrent computing in multi-core architecture. STM speculatively allows multiple transactions to execute concurrently. However, in order to ensure data consistency, aborting transactions may occur. Reducing transactional abort for STM is one of the key challenges. The permissiveness property of STM requires that transactions should not be aborted unless it is necessary to ensure consistency. Multi-version permissive-

A. Ghosh (✉) · R. Chaki · N. Chaki
University of Calcutta, Kolkata, West Bengal, India
e-mail: ammlan.ghosh@gmail.com

R. Chaki
e-mail: rchaki@ieee.org

N. Chaki
e-mail: nabendu@ieee.org

© Springer Nature Singapore Pte Ltd. 2018   471
P.K. Sa et al. (eds.), *Progress in Intelligent Computing Techniques: Theory,*
Practice, and Applications, Advances in Intelligent Systems and Computing 518,
DOI 10.1007/978-981-10-3373-5_47

ness uses multi-version concurrency control technique [MVCC], one of the classical approaches of database management system, to isolate concurrent updates by maintaining old versions. This allows read-only transactions to secure consistent snapshots corresponding to an arbitrary point of time. Thus, by keeping multiple versions, it is possible to ensure that every read-only transaction can successfully commit without any abort. Multi-versioning technique has been adopted by various STM algorithms, e.g., [2–6].

The most important aspect for maintaining multiple versions is to delete older versions that will no longer be required. The mechanism for removing unwanted versions is known as garbage collection (GC). However, it is quite difficult to predict which versions are to be accessed by the live transactions in future. Consequently, keeping unnecessary versions will lead to inefficient storage utilization. On the other hand, deletion of older versions may lead to abort of the read-only transactions due to unavailability of consistent data version. Thus, garbage-collecting mechanism is a crucial issue, and there are numerous garbage-collecting algorithms proposed by different multi-version STM techniques to overcome existing challenges. Some of these STM techniques maintain a pre-configured constant number of versions for every object [7]. Although such technique requires simple implementation mechanism, it suffers from inherent memory consumption problem with increasing number of transactional objects. Moreover, pre-configured constant number of versions may not be sufficient, and read-only transactions are required to be aborted due to the unavailability of consistent data version. In [4], a selective multi-version model (SMV) is proposed that only keeps the useful versions and removes obsolete one. The current versions (useful) of each object are referred by strong references, and other versions are referred by weak references. In [5], the mechanism maintains at least same number of versions of maximum live transactions in the system. Perelman et al. have elaborately discussed the principles of garbage collection for multi-versioning STM in [8], where useless-prefix GC (UP-GC) is presented for garbage collection of old versions as many as a system can do.

The state-of-the-art study leads to the conclusion that multi-versioning STM faces the challenges of space utilization, garbage collection, and maintenance of number of versions. In this paper, we propose CMS—a new checkpoint-based multi-versioning mechanism—to alleviate the memory management challenges for multi-versioning STM.

This paper is organized as follows. Challenges in multi-versioning STM are discussed in Sect. 2. We propose the new checkpoint-based multi-version STM (CMS) in Sect. 3. In Sect. 4, we evaluate the performance of proposed CMS by formal analysis as well as using controlled execution of the algorithm. The conclusions appear in Sect. 5.

2　Challenges in Maintaining Multi-versions

In this section, we present brief notes on the basic challenges of multi-versioning STM. In Sect. 3, we would revisit each of these three aspects to compare how CMS handles these challenges.

2.1　Memory Constraint Due to K-Versioning

In [7], it is shown that maintaining a constant number of versions, say K, for n transactional objects has a worst space complexity $\Omega(K^n)$ versions as all older versions continue to refer updated objects. Although this technique has simple implementation mechanism, it requires huge memory space.

2.2　Space Utilization

It has been established in [8] that multi-versioning STM does not provide online space optimization. Thus, better space utilization remains a basic criterion even if the online space optimization cannot be achieved. An STM is said to be space optimal if it is designed to store only the minimum set of object versions at all the given instances.

2.3　Garbage Collection

Although no MV-Permissiveness is online space optimal, however, it is possible to manage older versions better than constant K-versioning. In [8], authors define an achievable GC property, useless-prefix garbage collection (UP-GC), which keeps only those object versions that might be needed by existing read-only transactions and removes rest of the versions. Satisfying UP-GC is a challenging task and imposes cost to an STM, e.g., disjoint access parallelism and invisible read [5]. Another challenge for satisfying UP-GC is that system keeps all the versions that have been added since the inception of the oldest active read-only transaction. The situation deteriorates further, from memory management perspective, when a read-only transaction is stuck to commit for long time.

3   Checkpoint-Based Multi-versioning System (CMS)

We present a checkpoint-based multi-versioning system for STM to alleviate various memory management challenges, which are faced by multi-versioning STM.

3.1   Design Principles

Checkpoints for Read-only Transactions. Read-only transactions always commit. Every read-only transaction is associated with a checkpoint, which refers to a set of data object versions, to observe a consistent snapshot corresponding to its start time.

Versioning Techniques. CMS stores multiple versions of data objects and allows read-only transactions to observe consistent data version. Each data object version number is set as the update transaction id, which has committed write on that data object. CMS uses unique global counter to generate transaction id. The transaction id for an update transaction is set as per their commit order, and for a read-only transaction it is set at the time of its initiation. The global counter value is incremented inside the critical region. This mechanism helps read-only transaction T_x to observe a consistent snapshot by retrieving the most recent data object version, say O_i^j, where $j < x$.

3.2   Data Structure for Proposed CMS

Figure 1a shows the data structure for version controlling in CMS. The checkpoint version box, i.e., cp-version contains a list of checkpoints to refer the older versions of data objects for respective check points. Each checkpoint contains a checkpoint version number, i.e., cp-version; indexed list of data objects and their versions, i.e., object-index and a reference to the previous state of the checkpoint version. Each object in the object list has object identifier, Obj_i, version, value, and a reference to the previous version of that object. The object identifier Obj_i denotes the ith object. Each data object version is set as per the update transaction id by which this version is written.

A read-only transaction reads values from the most recent data object versions in the latest checkpoint that existed when the transaction is initiated. Thus, read-only transactions always have consistent object versions and never conflicts with other transactions.

Figure 1b illustrates a snapshot of object version using checkpoint-based multi-versioning technique. In this figure, the object-index filed of checkpoint version (cp-version: 2) refers to three objects Obj_1, Obj_2, and Obj_3, which have stored multiple data versions updated by different transactions. The Obj_1 has been updated by transactions 2, 5, 6, and 11; Obj_2 is updated by transactions 3, 5, and 9; and Obj_3 is updated

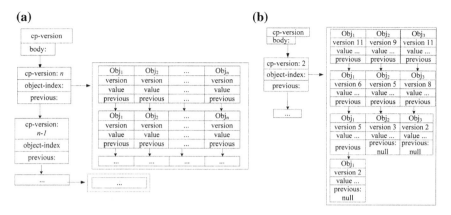

Fig. 1 **a** Data structure of CMS **b** Snapshot of data object versions using checkpoint-based multi-versioning

by transactions 2, 8, and 11. It is important to mention here that versions are ordered by decreasing version number, as it required less effort for read-only transaction to search consistent version if the latest versions are nearer to the head of the list.

Let us assume that a read-only transaction, say T_7, is initiated after successful commit of update transaction T_6. Transaction T_7 has successfully read data from Obj_1^6 (version 6 of data object 1). Now T_7 wants to read from Obj_2 and Obj_3, which are already written in between by transactions T_9 and T_8, T_{11}, respectively. To avoid spurious abort, transaction T_7 will search the most recent version from the older version list where object versions are less than its transaction id (i.e., <7, in this case). T_7 reads object version Obj_2^5 and Obj_3^2 and commits successfully.

3.3 Garbage Collection Mechanism

In CMS, garbage collection mechanism employs two independent operations: On the one hand, unwanted object versions are removed based on real-time order (horizontal truncation), and on the other hand, object versions of a checkpoint are reconciled to create a new checkpoint (vertical truncation).

Removal of Older Version by Real-Time Order. CMS uses an active read-only transactional record, motivated from [4] to maintain list active read-only transactions, for every checkpoint (Fig. 2). Whenever a read-only transaction commits successfully, CMS checks for the oldest transaction in the list, i.e., the transaction with smallest time stamp (id). Now, for every data object, all versions less than transaction id, except the latest, are deleted.

Fig. 2 Checkpoint-based
active read-only transaction
List

Let us take an example. Suppose active read-only transaction list contains trans-actions T_7, T_{10}, T_{12}, and T_{13} for the checkpoint snapshot depicted in Fig. 1b. Trans-action T_7 commits successfully. CMS searches the oldest transaction from the list and finds T_{10}, which may require the most recent object versions less than 10 for all the available objects. Thus, CMS removes unwanted versions, viz. versions 2 and 5 from Obj_1; versions 2 and 5 from Obj_2; and version 3 from Obj_3 (Fig. 1b). This process continues until all the read-only transaction commits, and the list contains a NULL set.

Creation of New checkpoint and Reconciliation. In CMS, an indexed list of all the data objects accessed by transactions is maintained for each checkpoint. This list is going on increasing over time and may have huge memory overhead. In order to overcome the problem, every checkpoint is reconciled after some preconfigured update operation. In the new checkpoint, recent object versions, those are accessed by the live update transactions, are being copied. Once a new checkpoint is created, all the transactions started thereafter are referred by this checkpoint.

3.4 Algorithms

In this section, we present description of CMS algorithms. The pseudocode for ver-sion mechanism is presented in Algorithm 1, and pseudocode for garbage collection mechanism is presented in Algorithm 2.

setTransactionVersion method

- **Check at initiation time**: If the transaction type is read only, then set transaction version by incrementing the global counter by one. Include the read-only trans-action in the active read-only transaction list for the current checkpoint version. (Line 2–7; Procedure 1)
- **Check at commit time**: If transaction type is write, then set transaction version by incrementing the global counter by one. (Line 8–12; Procedure 1)

Garbage Collection Algorithm In the proposed method, object versions are recon-ciled after every successful commit of every read-only transaction [Algorithm 2].

Algorithm 1 Methods for setTransactionVersion

> ▶ to be checked at the time of transaction initiation
> 1: **Procedure** setTransactionVersion(T_x, T_x.Type){
> 2:   **if** (T_x.Type = read-only) **then**
> 3:   newTxVersion ← globalCouner + 1 ;
> 4: T_x.Version ← newTxVersion;
> 5:   globalCounter ← newTxVersion;
> 6:   Include T_x in Active read-only transaction record
> 7: **end if**
> ▶to be checked at the time of transaction commit
> 8: **if** (T_x.Type = write) **then**
> 9:   newTxVersion ← globalcouner + 1 ;
> 10: T_x.Version ← newTxVersion;
> 11:   globalCounter ← newTxVersion;
> 12: **end if**
> 13: **return** T_x.Version;
> 14: }

Algorithm 2 Methods for Garbage Collection in CMS

> 1: **Procedure** removeOlderVersion(T_x, cp-version){
> 2: **for** object in cp-version.object-index **do**
> 3:   **for** versions in object.version **do**
>   recent-version ← search for most recent version where object.version < T_x.ID;
>   **create** newCheckpoint;
> 4:   **end for**
> 5: **end for**
> 6: **Procedure** creationNewCheckpoint(cp-version, Active-Read-Set){
> 7: **if** (update_count = preconfigured_update_count) **OR** Active-ReadSet is Empty **then**
> 8:   create newCheckpoint;
> 9: **end if**
> 10: }

4 Performance Evaluation

4.1 Improvement Analysis

STM satisfies multi-version permissiveness (MV-Permissiveness) when no read-only transactions are ever aborted unless it is required to ensure data consistency [8]. Besides, an STM system satisfies useless prefix garbage collection (UP-GC) if at any point in a transaction history H, an object version O_x^j is stored, provided there exists an extension of H with a live transaction T_i, such that T_i can read O_x^j and cannot read any versions thereafter [8].

Property 1 *Every transaction T_i is assigned a unique time stamp* i. *For a read-only transaction, time stamp is assigned at its start time and time stamp for update transaction is assigned at its commit. Every object version is set as the update transaction's time stamp.*

Property 2 *If a read-only transaction T_j begins after another read-only transaction T_i, then* i < j. *If an update transaction T_i commits after another update transaction T_j commits, then* j < i.

Property 3 *If a transaction T_i reads from a jth version of data object O_x, i.e., O_x^j, then (1) O_x^j is written by committed transaction T_j in history* H *and (2)* j *is the largest time stamp, which is smaller than* i.

Property 4 *In CMS, checkpoint refers to $O_i^{j_i}$ versions, where* i *varies from* 1, ... , n *and the number of versions for each object vary in a set of positive integers.*

Lemma 1 *CMS algorithm is MV-Permissive.*

Proof In CMS, older versions of data objects are always kept until there exist no live read-only data, which may access these versions. Thus, every read-only transaction will always be able to find consistent data version, and hence read-only transaction never aborts. Hence, CMS algorithm is MV-Permissive.

Lemma 2 *CMS satisfies useless-prefix (UP) garbage collection.*

Proof We will prove this by contradiction. Assume that there exists data object version O_x^j in a point of history H and there exists a live read-only transaction T_i in an extension of history H, is unable read O_x^j. In CMS, a check point maintains a list of versions ranging from j to j + n, where n > 0 for the object O_x. Thus, all the versions written by T_j to T_{j+n} are stored in this list. As per our assumption, T_i is initiated after T_j and j < i. Thus, T_i can always find the object version that is largest among the versions less than j and greater than i. This contradicts our assumption. Thus, CMS satisfies useless-prefix.

Lemma 3 *CMS maps the versions to a reduced set $O_i^{j_i}$.*

Proof A checkpoint CP_{t_0}—checkpoint at time instance t_0—in CMS can point to n number objects and j_i number of versions for each object. Thus, CP refers to $O_1^{j_1}, O_2^{j_2}$, ..., $O_n^{j_n}$ => CP refers to $O_i^{j_i}$, where i = 1, ..., n and $j_i \in \mathbb{Z}$.

Suppose a read-only transaction T_x wants to access a consistent version from a data object O_i, which is referred by a checkpoint. Thus, T_x searches for the object version $O_i^{j_{k'}}$, where $j_{k'} < x$ and $j_{k'} = \max(k')$ where k < k'.

Now, for the availability of all such versions for all the objects, O_i must keep versions j_k where k < x $\forall i$, i = 1, ..., n.

After a preconfigured number of updates P (where P>1) for CP_{t_0}, a new checkpoint CP_{t_1} is created based on (T_{p+1})th update.

Initially, CP_{t_1} refers to a set of $O_{i'}^{j'}$ objects. Here, $i' < i$ in comparison with CP_{t_0} and CP_{t_1} such that $|CP_{t_1}| < |CP_{t_0}|$, and $|X|$ denotes the cardinality of set X. Thus, every newer checkpoint has always a set with reduced cardinality.

4.2 Experimental Results

We consider the performance issue on the basis of number of memory access for searching consistent version from a set of older versions. The execution of the proposed algorithm is done in two different contexts. In the first situation, the total number of objects is increased with every experiment. As the number of sharable objects increases, more versions are supposed to be stored with every update. Higher number of versions results in increased effort for read-only transaction to find the consistent version. Three different scenarios with respect to total number of sharable objects are considered here; the total number of sharable objects is comparatively lower, medium, and higher. For each scenario, the number of concurrent access of sharable objects is increased. When sharable objects are accessed concurrently, older versions are required to be stored for live read-only transactions, and thus the volume of older versions increases. On the basis of these two conditions, i.e., total number of sharable objects and number of objects accessed by transactions, the execution length of the transaction is considered to be lower, medium, and higher. When transaction execution length is higher, the older versions of objects are to be kept, which adversely affect the number of memory access ratios.

The performance is evaluated using a micro-benchmark based on the number of sharable objects, which are concurrently accessed by transactions. The characterization of the micro-benchmark is done by randomly generating a set of transactions, by varying their execution length; number of sharable objects; and number of objects transaction will access.

The performance of the proposed CMS, on the basis of number of memory access for searching consistent version, is compared with conventional K-versioning and UP-GC technique. Figure 3 depicts the execution scenario for the short-, medium-, and long-running transactions for lower number of total sharable objects, where the number of sharable objects accessed by transactions is increased gradually from 20% to 100% of total available sharable objects. In Fig. 4, result shows the same experimental scenario, which are done in medium level of total number of sharable object. When total number of sharable objects is high, the situation is depicted in Fig. 5. In all the three cases, CMS performs better than UP-GC and K-versioning.

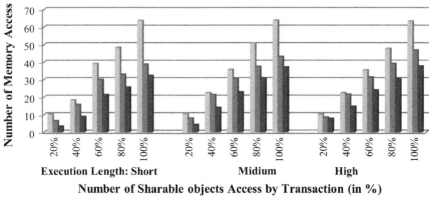

Fig. 3 Performance on memory access when number of sharable objects is low

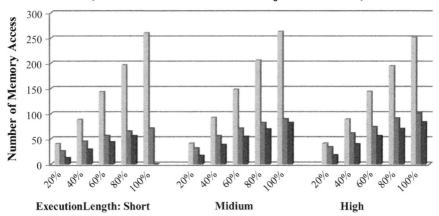

Fig. 4 Performance on memory access when number of sharable objects is moderate

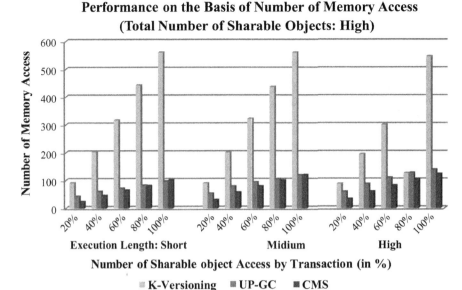

Fig. 5 Performance on memory access when number of sharable objects is high

5 Conclusion

In this paper, we propose a novel checkpoint-based multi-versioning system toward efficient version controlling for STM. The proposed CMS algorithm is simple and does not require complex mechanism for referring older and recent versions as in some other existing works. The proposed algorithm uses checkpoint to refer a set of data object versions. An older checkpoint and its versions are deleted whenever all read-only transactions accessing this checkpoint are committed successfully. The experimental results, on the basis of number of memory access for searching consistent version, establish that the proposed CMS performs better in comparison with conventional multi-version permissive STM.

References

1. Shavit N., Touitou D.: Software transactional memory, in: ACM SIGACT-SIGOPS Symposium on Principles of Distributed Computing, pp. 204–213, (1995)
2. Cachopo, J., Rito-Silva, A.: Versioned boxes as the basis for memory transactions. Science of Computer Programming 63(2), 172–185, (2006)
3. Fernandes, S.M., Cachopo, J.A.: Lock-free and Scalable Multi-Version Software Transactional Memory. In ACM SIGPLAN Notices 46(8), pp. 179–188, (2011)
4. Perelman, D., Byshevsky, A., Litmanovich, O., Keidar, I.: SMV: Selective multi-versioning STM. In: Peleg, D. (ed.) DISC 2011. LNCS, vol. 6950, pp. 125–140. Springer, Heidelberg, (2011).

5. Torvald, R., Felber, P., Fetzer, C.: A lazy snapshot algorithm with eager validation. In: Distributed Computing, pp. 284–298. Springer Berlin Heidelberg, (2006)
6. Lu, Li., Michael L. Scott.: Unmanaged multi-version STM. In: 7th ACM SIGPLAN Workshop on Transactional Computing (TRANSACT) (2012)
7. Guerraoui Rachid, and Paolo Romano, Eds. Transactional Memory. Foundations, Algorithms, Tools, and Applications: COST Action Euro-TM IC1001. Vol. 8913. Springer, (2014)
8. Perelman, D., Fan, R., Keidar, I.: On maintaining multiple versions in STM, In: Proceedings of the 29th ACM SIGACT-SIGOPS symposium on Principles of distributed computing, ACM, pp. 16–25, (2010)

Estimation of Incomplete Data in Mixed Dataset

Suhani Sen, Madhabananda Das and Rajdeep Chatterjee

Abstract This paper puts forward a fresh approach which is a modification of original fuzzy kNN for dealing with categorical missing values in categorical and mixed attribute datasets. We have removed the irrelevant missing samples through list-wise deletion. Then, rest of the missing samples is estimated using kernel-based fuzzy kNN technique and partial distance strategy. We have calculated the errors at different percentage of missing values. Results highlight that mixture kernel gives minimum average of MAE, MAPE and RMSE at different missing percentage when implemented on lenses, SPECT heart and abalone dataset.

Keywords Fuzzy sets · Fuzzy knn · Kernel functions · Partial distance strategy · Hellinger distance

1 Introduction

Missing data estimation has been an important matter of concern in the field of machine learning and pattern recognition as missing records introduce a bias affecting the classification performance, aspect of derived patterns. Many distinct miscellaneous methods have been developed that effectively handle missing numerical values, but these are not applicable to many real-time datasets that contain categorical values also. To comply with this requisite, this paper puts forward a novel setting of incomplete value estimation. It first removes the missing records whose entropy is less than specified threshold. Then, rest of the incomplete samples is imputed using modified kernel-based fuzzy kNN technique.

S. Sen (✉) · M. Das · R. Chatterjee
School of Computer Engineering, KIIT University, Patia, Bhubaneswar, Odisha, India
e-mail: suhanisen1991@gmail.com

M. Das
e-mail: mndas_prof@kiit.ac.in

R. Chatterjee
e-mail: cse.rajdeep@gmail.com

© Springer Nature Singapore Pte Ltd. 2018   483
P.K. Sa et al. (eds.), *Progress in Intelligent Computing Techniques: Theory,*
Practice, and Applications, Advances in Intelligent Systems and Computing 518,
DOI 10.1007/978-981-10-3373-5_48

Partial distance strategy and Hellinger distance are used as distance measures. Three kernels are compared in terms of mean of MAE, MAPE and RMSE when implemented on lenses, SPECT heart and abalone dataset.

1.1 Overview of Missing Data

Based on the essence, missing data can be categorized into three groups [1]: missing completely at random (MCAR), missing at random (MAR) and missing not at random (MNAR). There are various existing techniques for dealing with incomplete data. The most common technique is to delete the entire sample that contains missing data called list-wise deletion. Another way is to supersede the missing data with the estimated data. Most frequently used imputation methods are hot deck [2], mean substitution [3], multiple imputation [4], maximum likelihood [5] and expectation-maximization [6]. Jose and Jerez et al. [7, 8] discusses many existing techniques. Our previous work deals with numerical missing values [9].

1.2 Techniques Used

- **Fuzzy kNN**: K-nearest neighbour gives equal importance to each of the labelled records. By introducing fuzzy set theory to kNN, fuzzy membership grades are assigned to the labelled samples [10, 11]. Keller et al. [10] proposes three techniques of assigning fuzzy membership grades to the samples.
- **Kernel Functions**: Through Kernel function, input data can be mapped to higher dimensional space using dot product. All the functions must satisfy Mercers condition [12, 13]. Kernel has been used in many disciplines [14]. The choice of kernel functions determines interpolation and extrapolation abilities.

1.3 Organization of Paper

Next section designs the new algorithm. Section 3 reports the results and analyses them. Section 4 wraps up the paper with future work.

2 Proposed Approach

Let D be either a categorical or mixed dataset containing 'n' samples and 't' attributes. X_{pb} denotes the value of the variable b at sample p. $X_p = \{x_{p1}, x_{p2}, \ldots x_{pt}\}$ be the pth sample. Out of p sample some samples are missing. In this section, we reveal a new algorithm to estimate the missing data. The proposed approach can impute only categorical missing data.

This technique consists of two steps

1. **List-wise deletion**: In this step, irrelevant missing samples are deleted. Relevant instances are selected based on fuzzy entropy. When the number of samples containing missing values increases, computational complexity of imputation also increases. In order to minimize the complexity, some unimportant missing sample is deleted and classification accuracy is not also hampered. First, all the non-missing instances are put in a set 'L'. For each incomplete instance, k-nearest neighbour in set 'L' is determined, and then membership of the missing sample with respect to each class is computed. If the entropy of the sample is found to be greater than threshold value, it is added to the set 'L', otherwise it is discarded. This set L gives the list of relevant missing samples. Threshold is in between 0.5 and 0.95.

   Entropy is computed as follows

   $$Entr(x) = -\sum_{q=1}^{c} \mu_r(x) \log_2 \mu_r(x) \tag{1}$$

   where x refers to the missing sample and r is the class label.

2. **Imputation**: In this step, all the missing values in the set 'L' are imputed using modified kernel-based fuzzy k-nearest neighbour algorithm. Set L is divided into M and C. Set M consists of missing data, and C consists of complete samples. For each sample in M, kNN is determined in C. All the incomplete samples are categorical. The probability of sample x containing value q is denoted as $\mu_q(x)$ which is given by:

   $$\mu_q(x) = \frac{\sum_{p=1}^{k} \mu_{qp}\left(\frac{1}{1 - K(x, x_q)}\right)}{\sum_{p=1}^{k} \left(\frac{1}{1 - K(x, x_q)}\right)} \tag{2}$$

   Where μ_{qp} is the fuzzy membership grade and $K(x, x_q)$ denotes the kernel function.

Three types of kernel functions are used in this approach:

- Polynomial kernel:

$$K_P(x, x_q) = (x^T x_q + 1)^n \tag{3}$$

- Radial Basis kernel:

$$K_R(x, x_q) = \exp\left(-\frac{||x - x_q||^2}{2\sigma^2}\right) \tag{4}$$

- Mixture kernel:

$$K_M(x, x_q) = \tau . K_P(x, x_q) + (1 - \tau) K_R(x, x_q) \tag{5}$$

$$\mu_{qp} = \sum_{r=1}^{c} \frac{\left\|x_p - x_r'\right\|^{2/m-1}}{\left\|x_p - x_q'\right\|^{2/m-1}} \tag{6}$$

Partial distance strategy is used to compute $\left\|x_p - x_q'\right\|$ which is given by:

$$||x_p - x_q'|| = \frac{1}{\sum\limits_{b=1}^{s} I_y} \sum_{b=1}^{s} (x_{pb} - x_{qb})^2 I_b \tag{7}$$

$$I_b = \begin{cases} 1 & \text{if } x_{pb} \text{ and } x_{qb} \text{ non missing} \\ 0 & \text{otherwise} \end{cases} \tag{8}$$

For mixed dataset with PDS, Hellinger distance is added given by:

$$||x_p - x_q'|| = \frac{1}{\sum\limits_{b=1}^{s} I_y} \sum_{b=1}^{s} (x_{pb} - x_{qb})^2 I_b + [\sum_{q=1}^{q} W_q(\sqrt{x_s} - \sqrt{x_{qs}})^2]^{1/2} \tag{9}$$

where w_q is the inverse of column total converted to row total (Figs. 1 and 2) Then, if Entr $\max(\mu(x_q))) > \emptyset$, i is imputed by q.

Fig. 1 Flow chart of
List-wise deletion

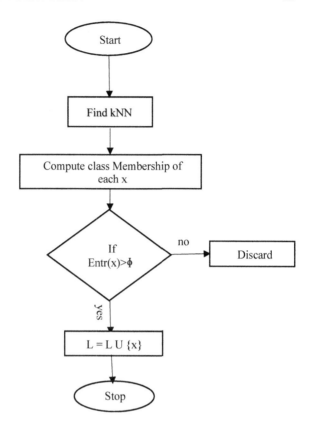

Fig. 1 Flow chart of List-wise deletion

2.1 Steps of Algorithm for List-Wise Deletion

Input: D with missing data
Output: L list of relevant data
Procedure:
Step 1: All non-missing instances are put in a set L
Step 2: For each missing sample 'x' in D
Step 3: K-nearest neighbours are determined in L
Step 4: Class membership grade for x is computed
Step 5: If $Entr(x) > \emptyset$ by (1)
Step 6: $L = L \cup \{x\}$
Step 7: else Discard x.

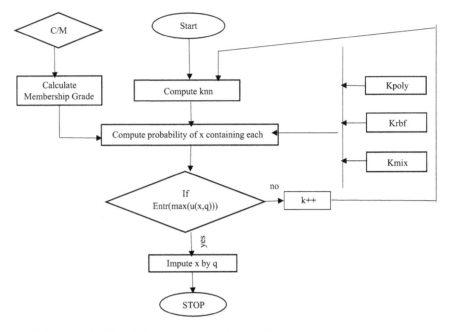

Fig. 2 Flow chart of Imputation using proposed approach

2.2 Steps of Algorithm for Imputation

Input: L with missing data (irrelevant missing data removed)
Output: D' a complete dataset
Procedure:
Step 1: L is divided into M and C. M contains missing samples. C contains complete samples.
Step 2: For each x in M:
Step 3: K-nearest neighbour is determined in C.
Step 4: The probability of x containing value 'q' is computed for each q by (2).
Step 5: Kernel functions and fuzzy membership grade are calculated by (3), (4), (5), (6).
Step 6: If L is categorical dataset, norm is computed by (7) and (8).
Step 7: else if L is mixed dataset, norm is computed by (8) and (9).
Step 8: If $Entr(max(\mu(x_q))) > \emptyset$, i is imputed by q by (1)
Step 9: Else increment 'k' and go to step 4.

3 Results and Discussion

To test the performance of the modified kernel-based fuzzy knn approach, we used three datasets lenses, SPECT heart and abalone, whose details are given in Table 1. Some of the sample values are made missing first 5%, then 10%, then 15% and finally 20%. We imputed these missing data using the proposed method. Firstly, the unimportant missing data samples are list-wise deleted. Rest of the missing data samples are estimated using the proposed approach. After obtaining the k-nearest neighbour, we found the probability of a missing data containing each categorical value by modified kernel-based fuzzy knn approach. The distance measures used are partial distance strategy and Hellinger distance.

Tables 2, 3 and 4 give average of mean absolute error (MAE), mean absolute percentage error (MAPE) and root-mean-square error (RMSE) of lenses, SPECT heart and abalone datasets. All the errors are computed with respect to the four missing percentage. All these tables and Figs. 3, 4 and 5 highlight that mixture kernel outperforms polynomial kernel, rbf kernel and traditional knn approach.

MAE, MAPE and RMSE are given by:

$$MAE = \frac{1}{m} \sum_{i=1}^{m} \|t_i - p_i\| \tag{10}$$

$$MAPE = \frac{1}{m} \sum_{i=1}^{m} \left\| \frac{t_i - p_i}{p_i} \right\| \tag{11}$$

$$RMSE = \frac{\sqrt{\sum_{i=1}^{m} (t_i - p_i)^2}}{m - 1} \tag{12}$$

Table 1 Details of datasets used

Dataset	No. of rows	No. of columns	Type
Lenses	24	4	Categorical
SPECT heart	267	23	Categorical
Abalone	4177	8	Mixed

Table 2 Average error for different percentage of missing values for lenses dataset

% of missing samples	KNN	$K_p(x, y)$	$K_r(x, y)$	$K_m(x, y)$
5	0.0366	0.0192	0.0236	0.0110
10	0.0519	0.0320	0.0383	0.0282
15	0.0618	0.0357	0.0517	0.0317
20	0.0805	0.0472	0.0597	0.0414

490   S. Sen et al.

Table 3 Average error for different percentage of missing values for SPECT heart dataset

% of missing samples	KNN	$K_p(x, y)$	$K_r(x, y)$	$K_m(x, y)$
5	0.0478	0.0268	0.0394	0.0224
10	0.0676	0.0340	0.0492	0.0349
15	0.0917	0.0471	0.0637	0.0428
20	0.1049	0.0545	0.0784	0.0488

Table 4 Average error for different percentage of missing values for abalone dataset

% of missing samples	KNN	$K_p(x, y)$	$K_r(x, y)$	$K_m(x, y)$
5	0.5507	0.2678	0.4382	0.2197
10	0.8354	0.4129	0.6248	0.3600
15	0.9570	0.4654	0.6453	0.4220
20	1.3002	0.6292	0.9807	0.5749

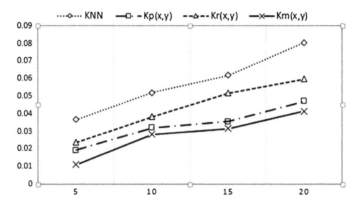

Fig. 3 Average error for different percentage of incomplete data for Lenses dataset

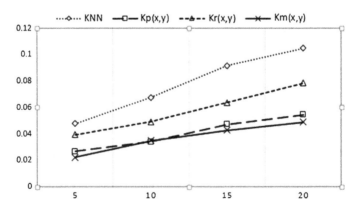

Fig. 4 Average error for different percentage of incomplete data for SPECT heart dataset

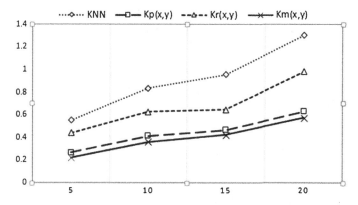

Fig. 5 Average error for different percentage of incomplete data for Abalone dataset

where 't' refers to the target output, 'p' refers to the predicted output and 'm' is the number of samples.

3.1 Discussion

- Comparison of modified fuzzy kNN with kNN: proposed approach gives far better result than kNN.
- Comparison of three different kernels: mixture kernel is slightly better than polynomial and much better than rbf kernel. Polynomial has good extrapolation ability, whereas rbf has better interpolation ability. Mixture has both these abilities better than polynomial and rbf kernel.
- The number of nearest neighbour: for lenses, it is taken as 4, for SPECT heart 6 and for abalone 6.
- Time complexity: PDS is usually used with FCM. But when it is used with FkNN, there is improvement in time complexity

4 Conclusion

In this paper, a novel modified fuzzy k-nearest neighbour approach has been proposed for the imputation of categorical values. After deletion of unimportant missing samples based on threshold, rest of the incomplete samples is estimated using kernel-based fuzzy k-nearest neighbour technique using partial distance strategy and Hellinger distance measure. Three different kernels are compared in terms of average of MAE, MAPE and RMSE. When the proposed approach is implemented on lenses, SPECT heart and abalone datasets, it is found that mixture kernel gives better result than rest two as mixture kernel has better interpolation and extrapolation abilities. In future, we plan to implement this technique in more datasets and for higher percentage of missing data.

References

1. R.J.A. Little, D.B. Rubin.: Statistical Analysis with Missing Data. second ed., Wiley, NJ, USA (2002), pp. 138–149.
2. B.L. Ford, An overview of hot-deck procedures, in: Incomplete Data in Sample Surveys. Academic Press, New York, USA (1983), pp. 185–207.
3. J. Cohen, P. Cohen.: Applied Multiple Regression/Correlation Analysis for the Behavioral Sciences. second ed.(2003), pp. 185–207.
4. F. Scheuren, Multiple imputation: how it began and continues, Am. Statist. 59 (2005) 315–319.
5. J.L. Schafer, Multiple imputation: a primer, Statistical Methods in Medical Research 8 (1) (1999) 3–15.
6. X. Meng, D.B. Rubin, Maximum likelihood estimation via the ecm algorithm: a general framework, Biometrika 80 (2) (1993) 267–278.
7. Jose M. Jerez et.al.: missing data imputation using statistical and machine learning method in a real breast cancer problem. Elsevier (2010), vol. 10, pp-105–115.
8. Esther-Lydia Silva-Ramírez et.al.: Missing value imputation on missing completely at random data using multilayer perceptrons. Elsevier (2011), vol. 24, pp. 121–129.
9. Suhani Sen, Dr. M.N. Das, Rajdeep Chatterjee.: A Weighted kNN Approach to Estimate Missing Data. . 3rd international conference on signal processing and integrated networks, IEEE (2016) [in press].
10. Keller J. M., Gray M. R., Givens J. A.: A fuzzy k-nearest neighbor algorithm. IEEE Transactions on Systems Man and Cybernetics (1985), vol 15, pp. 580–585.
11. T. Warren Liao Damin Li.: Two manufacturing applications of the fuzzy KNN-algorithm. Fuzzy Sets and system (1997), vol. 92, pp. 289–303.
12. M. Aizerman, E. Braverman and L. Rozonoer.: Theoretical foundations of the potential function method in pattern recognition learning. Automation and Remote Control, 25 (1964), pp. 821–837.
13. M. Girolami.: Mercer kernel based clustering in feature space. IEEE Trans. On Neural Nework (2002), Vol. 3, No. 13, pp. 780–784.
14. T. Warren Liao, Damin Li.: Two manufacturing applications of the fuzzy K-NN algorithm. Fuzzy sets and system, (1997) vol. 92, pp. 289–303.

Development of LIN 2.1 Driver with SAE Standards for RL78 Microcontroller

P.G. Vishwas and B.S. Premananda

Abstract Communication Protocols play a vital role in the functioning of various communication systems. LIN is cheaper compared to other communication protocols and is used wherever low costs are essential and speed is not an issue. To enable cold start in case of diesel engines, the cylinders are fitted with glow plugs to preheat the cylinder. The signals for controlling the GLPs are sent using LIN bus from the ECUs. This paper involves design and implementation of LIN 2.1 driver for RL78 microcontroller and configuration of IDs for communication of messages between ECU and the driver. The driver is designed to operate in slave mode with fixed baud rate as per SAE standards and incorporates sleep mode and error handling capability. Embedded C codes are written for various modules and are compiled using IAR Embedded Workbench. The functionality of driver is tested using the CANalyzer tool.

Keywords CANalyzer · LIN · RL78 · SAE J2602 · Sleep mode

1 Introduction

There are innumerable communication devices that are currently present and in order for them to communicate, a common language that can be understood by all the devices must be used. Communication protocols permit these devices to share data and information based on certain rules and regulations. Since these protocols are built on certain standards, each device on the network must follow these rules, so that the communication is possible.

P.G. Vishwas (✉) · B.S. Premananda
Department of Telecommunication Engineering, R.V. College of Engineering,
Bengaluru 560059, Karnataka, India
e-mail: vishwaspg595@gmail.com

B.S. Premananda
e-mail: premanandabs@gmail.com

© Springer Nature Singapore Pte Ltd. 2018
P.K. Sa et al. (eds.), *Progress in Intelligent Computing Techniques: Theory,
Practice, and Applications*, Advances in Intelligent Systems and Computing 518,
DOI 10.1007/978-981-10-3373-5_49

Flex fuel is a technology that permits vehicles to run on ethanol, gasoline, or a varying combination of both. To safeguard the combustion when the engine is cold, Glow Plugs (GLP) are used as an aid [1]. The glow plug increases the temperature in the cylinder prior to heating. The temperature can reach up to 1000 °C, and hence the combustion is possible even at low temperatures [2]. The current levels are controlled by varying the ON and OFF times of the power transistors which monitor the plugs. Further, the fuel injection quantity is also reduced and the lifetime of the component is increased. The temperature of glow plug not only enables cold start but also maintains the quality of the exhaust [3]. The Glow Control Unit (GCU) performs an open/closed loop monitoring and controlling of the GLP. They provide a number of benefits to its customers such as instant cold start, increased the lifetime of GLP, reduced hardware expense, reduced emissions, small volume and light weight.

The microcontroller RL78/F14 is the preferred controller for creating fast, responsive and energy efficient embedded systems. The most important feature of this microcontroller is that it is CAN and LIN incorporated. Hence, it is the pre-ferred controller for embedded system design, analysis, and optimization. Unlike other controllers used for vehicular applications, the RL78 microcontroller has an inbuilt option for checking the break and sync fields. The other important features of this controller are a 16-Bit CPU with CISC architecture, 8 kB flash, 20 kB RAM, 256 kB ROM and a 32 MHz internal clock [4]. LIN, CAN, FlexRay, and MOST are some of the commonly used protocols in automobiles for communication among the components [5, 6].

Excessive wiring increases vehicle weight, weakens the performance, and makes the adherence to reliability standards difficult. For every 100 km travelled, an extra 50 kg of wiring increases the power by 100 W and fuel consumption by 0.2 L [2]. In order to have a standard sub-bus, a new bus, called LIN, was invented to be used in car seats, door locks, sun roofs, rain sensors, mirrors, and so on. LIN is posi-tioned as a Class-A networking solution. SAE defines Class-A communications loosely as general purpose and non-emissions based diagnostic communications at low speeds [7]. Checksum data and parity check guarantee safety and error detection. The LIN standard includes the specification of the transmission protocol, the transmission medium, the system definition language and the interface for software programming [8].

2 Design and Implementation of LIN Driver

The LIN module present in the RL78 microcontroller is shown in Fig. 1. The various sub-modules present include LTXDn and LRXDn which are LIN module I/O pins, LINn Baud rate generator that generates the LIN module communication clock signal, LINn registers which are the LIN/UART module registers, LINn interrupt controller to control the interrupt requests generated by the LIN module.

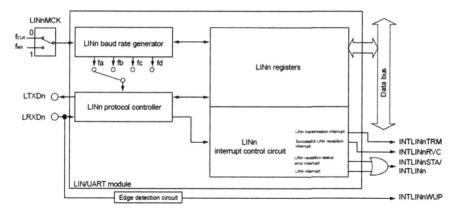

Fig. 1 LIN module [4]

The nominal value for transmission of a frame exactly matches the number of bits. The time taken for transmitting the entire frame is given by Eqs. (1), (2) and (3).

$$T_{Header\_Nominal} = 34 \times T_{Bit}. \tag{1}$$

$$T_{Response\_Nominal} = 10 \times (N_{Data} + 1) \times T_{Bit}. \tag{2}$$

$$T_{Frame\_Nominal} = T_{Header\_Nominal} + T_{Response\_Nominal}. \tag{3}$$

Various registers are configured to enable the communication between the LIN and ECU, interrupt and error handling and also for configuring IDs. The ports configured include P13 and P14 which are the LIN module I/O pins. The registers configured include LTxD0 and LRxD0, LCHSEL, LINCKSEL, LCUC0, LWBR0, LIE0, LEDE0, LSC0, LTRC0 etc. Also, according to SAE standards, the baud rate should be set to 10.417 kBd and asynchronous frames should be used for the header and response fields.

The slave task state machine is shown in Fig. 2. The advantage of RL78 microcontroller is that it performs automatic checking of break and sync field. Hence, the slave task directly starts with PID verification. If the PID is successfully configured, the slave task performs either the transmission or reception of response field and also the checksum verification based on the direction bit in the PID.

The algorithm for implementing LIN driver is as follows:

1. Initialization of registers for LIN module to operate in LIN slave mode with fixed Baud rate and generating interrupts based on the frame.
2. Determine number of bytes to be transmitted/received, type of checksum (classic/enhanced), and the direction of data from the PID.

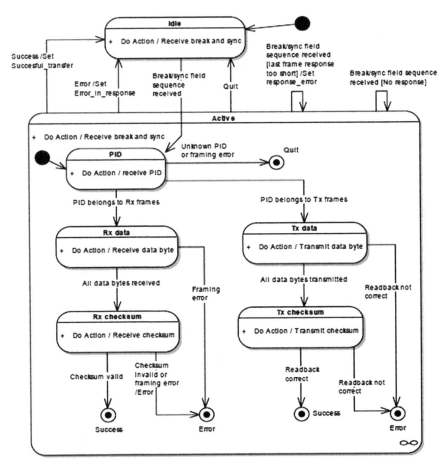

Fig. 2 Slave task state machine [8]

3. On successful transmission, the state is changed to LIN transmit success. After the transmission/reception is complete, the LIN state is changed to wait for interrupt till the next data frame is received.

3 Results

The developed driver is checked for all the functionalities like register configuration, message transmission, and reception, sleep mode configuration, error detection etc. The steps followed are:

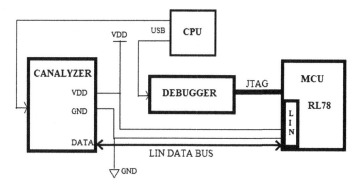

Fig. 3 Hardware setup

1. The hardware setup is done as shown in Fig. 3.
2. The IAR Workbench is placed in Emulator mode and the .exe file is flashed onto the controller.
3. The Baud rate is set to 10.417 kbps in the CANalyzer tool.
4. Data frame is set for the ID and transmitted by the CANalyzer to check the reception functionality of the driver. Another ID is configured for transmission to check the transmission functionality.
5. ID 0x3C is configured to put the driver into sleep mode.
6. The waveforms of the LIN frames transmitted/received are obtained using Picoscope.

The ID 0x3D is configured for data transmission from the driver. The reception of data by the CANalyzer confirms the transmission. On successful transmission, the LIN state is changed to LIN_TX_SUCCESS.

The error detection functionality of the LIN driver is also enabled. The driver throws a checksum error upon receiving a wrong data byte or an extra byte. Similarly, other errors are obtained whenever a fault is detected on the bus.

The ID 0x17 is used for response reception by the LIN driver as shown in Fig. 4. The data bytes are seen in the data buffer register of the driver confirming the reception. The waveform shows the LIN frame with break, sync, PID, data and checksum fields. The PID contains a value 0x97 followed by the 4 data bytes (05, 02, 06 and 11) and a checksum of 0x4A.

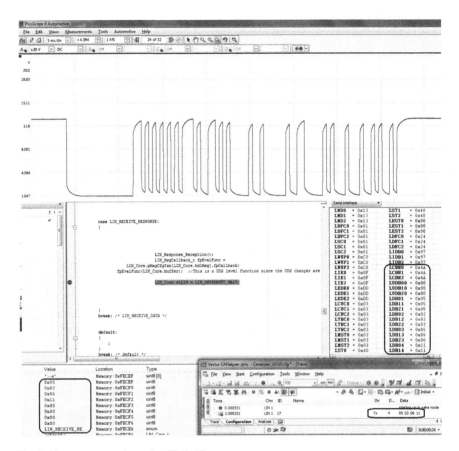

Fig. 4 Response reception using ID 0x17

The ID 0x3C is configured to put the driver into a temporary sleep mode as depicted in Fig. 5. The PID field contains a value 0x3C followed by 8 data bytes containing a value 0. This is followed by a classic checksum field holding a value 0xFF. The slave (driver) goes into sleep mode and halts the data transfer process until a wake-up command is received.

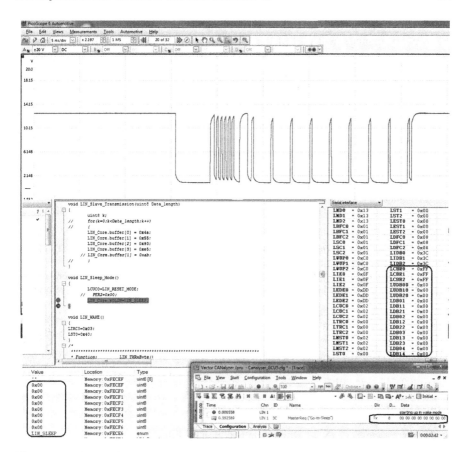

Fig. 5 LIN sleep mode

4 Conclusions and Future Work

The LIN bus employs a single wire implementation for data transfer and hence a master device controls the data transfer process to avoid collisions. The ports P13 and P14 are multifunction pins. The Embedded C functions for transmission and reception are called only upon receiving corresponding interrupts. This eliminates any error that might occur due to improper interpretation of signals due to spikes.

The LIN driver in slave mode is configured for fixed Baud rate. Hence, it is mandatory that the master is also configured with the same data rate so that communication is possible between them. The ID 0x3C puts the driver into sleep mode. The driver can only be awakened by sending a dominant signal on the bus for a duration of 250 μs–5 ms. This causes the LIN module to generate a wake-up interrupt after which it resumes its normal operation.

The future enhancement will be to implement the application specific part and the network management part of the LIN 2.1 protocol which controls the precise functions assigned to each ID. Also, LIN can be implemented as a part of AUTomotive Open Systems Architecture (AUTOSAR) or NANOSAR architecture. This involves breaking down the code into controller specific part, driver specific part and also application specific part.

References

1. ReshimaBai and Phaneendra H. D.: Automotive Inter ECU Communication, International Journal of Scientific Research and Development (IJSRD), Vol. 3, Issue 3, pp. 3248–3249, May 2015.
2. Rajeshwari Hegde, Siddarth Kumar, and K. S. Gurumurthy: The Impact of Network Topologies on the Performance of the In - Vehicle Network, International Journal of Computer Theory and Engineering (IJCTE), Vol. 5, Issue 3, pp. 405–409, June 2013.
3. Last B., Houben H., and Rottner M.: Influence of Modern Diesel Cold Start Systems on the Cold Start, Warm-up and Emissions of Diesel Engines, 8th Stuttgart International Symposium, 2008.
4. Renesas RL78/F13, F14 Microcontroller User manual: Hardware, Rev. 2.00, May 2014.
5. Li G. X., Qin G.H., Liu W.J., and Zhang, J.D.: Design and Implementation of Vehicle Gateway Based on CAN Bus, Journal of Jilin University, Vol. 28, Issue 2, pp. 166–171, March 2010.
6. Schmidt E.G. et al.: Performance evaluation of FlexRay/CAN networks interconnected by a gateway, IEEE International Symposium on Industrial Embedded Systems (SIES), Trento, pp. 209–212, July 2010,.
7. Matthew Ruff, "Evolution of Local Interconnect Network (LIN) Solutions," in the *Proceedings of 58th IEEE Vehicular Technology Conference (VTC)*, vol. 5, ISSN: 1090 - 3038, pp. 3382–3389, Oct. 2003.
8. ST Microelectronics LIN 2.1 Driver suite User manual, Feb. 2014.

Storage Structure for Handling Schema Versions in Temporal Data Warehouses

Anjana Gosain and Kriti Saroha

Abstract The temporal data warehouse model, an extension of multidimensional model of data warehouses, has already been proposed in the literature to handle the updates of dimension data correctly. In order to maintain a complete history of data in temporal data warehouses (TDW), it is required to manage schema evolution with time, thereby maintaining the complete history of evolution of schema with schema versions as well as evolution of data defined under different versions of schema. This paper proposes an approach for managing schema versions in temporal data warehouses to allow for an effectual management of different versions of schema and their data.

Keywords Data warehouse · Temporal data warehouse · Schema versioning · Transaction time · Valid time · Bi-temporal

1 Introduction

Data warehouses (DWs) collect data from various heterogeneous and sometimes distributed source systems. They are designed to provide an efficient and fast access to data and enable the end users in making fast and better decisions [1, 2]. A DW schema and data are expected to get modified in several ways, to keep up with the application needs reflecting new user requirements and specifications. Different solutions have been proposed in the literature for handling revisions in DWs, namely schema and data evolution, schema versioning and temporal extensions. Schema and data evolution [3–10] approaches maintain one DW schema, update it and delete the previous state. This is a limited solution, as it may result in information loss because the

A. Gosain (✉)
USICT, GGSIPU, Dwarka, Gujarat, India
e-mail: anjana_gosain@hotmail.com

K. Saroha
SOIT, CDAC, Noida, India
e-mail: kritisaroha@gmail.com

© Springer Nature Singapore Pte Ltd. 2018 501
P.K. Sa et al. (eds.), *Progress in Intelligent Computing Techniques: Theory, Practice, and Applications*, Advances in Intelligent Systems and Computing 518, DOI 10.1007/978-981-10-3373-5_50

previous states are not preserved. This fact has led to schema versioning [11–17] approach that maintains all the history of the DW evolution as created by a series of schema revisions. But, not only the conceptual schema, its underlying data are also known to evolve continually with time and, thus, require attention and support for maintaining several different versions. Temporal data warehouses with support for schema versions [18–22] have been developed to fulfill this requirement. TDWs use the research achievements of temporal databases and manage the evolution of data by providing timestamps for dimension data to represent their valid time [23–26]. Two different temporal dimensions are generally used to represent the revisions; valid time (the time when a data are true), and transaction time (the time when a data were recorded in the database) [27].

But, despite all the proposals to manage schema evolution, none of them handles both schemata as well as their data with respect to multiple versions which is a highly required feature. Though profuse amount of efforts are spent in the area of TDWs regarding the temporal versioning of dimensional data, a thorough examination of versioning potentialities is yet to be done. Further, most of the works for TDWs concentrate mainly on dealing with updates in the structure of dimension instances [18–22] and, thus, only allow to track revisions for dimension data and not for multiple schema versions. Wrembel and Morzy [11] proposed the concept of Real and Alternate versions for multidimensional DWs but did not consider the possibilities of populating versions with instances from previous versions. Also, Golfarelli et al. [16] presented the notion of schema augmentation but did not focus on synchronous and non-synchronous mapping of data. Rasa et al. [28] proposed a temporal star schema by time-stamping dimension member and fact instances using valid time. Chamoni and Stock [18] presented an approach to store dimension structures together with their time stamps using valid times. Mendelzon and Vaisman [9] presented a temporal model for multidimensional OLAP and proposed a temporal query language (TOLAP). Their model stores information about the changes in the instances and structure of the dimensions, thereby providing support for schema evolution but did not record versions of data. Body et al. [14, 15] proposed a novel temporal multidimensional model to support the evolution of multidimensional structures. The approach proposed to time-stamp the level instances along with the hierarchical relations and fact data. However, only updates to the structure and instances of dimension have been discussed. Most of the approaches proposed in the area of TDWs mainly deal with the evolution of dimension instances using mainly the valid time. Moreover, schema versioning in TDWs has been dealt with only to a limited extent so far. Eder and Koncilia [19, 20, 29, 30] have presented COMET approach to represent the revisions applied to the transaction as well as structure data. This has been done by using the concept of time stamping the data elements with valid time. The model especially considers the evolution of dimension instances and neither considers schema evolution nor the evolution of cubes. They proposed mapping functions to allow conversions between structure versions but did not discuss the storage options for the versions.

In this paper, we present schema versioning with a broader perspective and discuss solutions for schema versions using valid time. The work uses the research achievements of temporal databases [23–26] and considers new design options along with a discussion on mapping instances from previous versions to new versions. The support for schema versioning involves actions at both, schema (intensional) as well as data (extensional) level. Also, it raises an important issue regarding the storage options for the several versions of schema and their data. The main objective of this work is to propose a strategy for handling multiple versions in temporal environment. The paper proposes an approach in the same direction. Two design solutions (central data storage and distributed data storage) are discussed to manage the extensional data when different versions of schema are created. Further, the concept of synchronous and non-synchronous mapping of data and schema is introduced in the context of temporal data warehouses.

The rest of this paper is structured as follows. Section 2 has an overview of the key concepts for the schema versioning and management of extensional data. Section 3 presents the proposed approach for schema versioning solutions including an example; Sect. 4 deals with some data modifications; and finally, Sect. 5 presents the conclusions and final considerations.

2 Schema Versions

In this section, we give the overview of valid-time, transaction-time and bi-temporal versioning of schema and discuss the methods that provide support for managing versions of schema in TDW. Schema versioning may be defined as transaction time, valid time or bi-temporal on the basis of temporal dimension used. When only one time dimension is used, the versioning is either transaction-time or valid-time schema versioning and when both time dimensions are used, it is bi-temporal schema versioning [29]. Since TDWs mainly use valid time for time-stamping data, so we would discuss design solutions with valid-time schema versioning. Valid-time schema versioning is vital for applications that require to handle retro- or proactive schema revisions. It is observed that retroactive revisions are routine in databases, and other decision-making applications these days, and can involve both extensional as well as intensional data [23]. Here, we consider valid-time schema versioning to manage present as well as old data and propose design solutions for managing multiple versions of schema in TDWs.

Valid-time Schema Versioning: In schema versioning along valid time, all schema versions are time-stamped with their related valid time. The new versions of schema become effective when their validity is realized. It provides support for retro- and proactive revisions. Using valid-time schema versioning, multiple schema versions are available to access and a single revision may impact multiple versions of schema, because each schema version, which has either complete or partial overlap with the valid-time interval of the revision gets affected.

2.1 Storage Choices for Extensional Data

Two different storage options are considered for managing extensional data, and the response of each one to the revisions applied to schema, or updates is discussed with the help of examples. The solutions are presented at the logical level [23]:

Central Data Storage: a single data repository is used to store the data corresponding to all schema versions according to an exhaustive schema containing all attributes specified in successive revisions applied to the schema.

Distributed Data Storage: different data repositories are used to store data for different versions of schema, where the format of each data repository is in synchronism with its associated schema version. A new storage is initialized by moving the records from the old repository into the new one following the corresponding schema revisions [23].

When both data and schema versioning are done using the same temporal dimensions, the mapping of data can be either synchronous or non-synchronous.

In *synchronous mapping,* the version of schema used for recording, extracting and updating data has the same validity of records with reference to the common temporal dimension [23].

In *non-synchronous mapping,* any version of schema may be used for extracting and updating data irrespective of the validity of the schema version. Here, the validity of schema is independent of the validity of data even with reference to the common temporal dimension(s) [23].

Central data storage is always synchronous. Distributed data storage may be synchronous or non-synchronous.

3 Proposed Approach

In valid-time schema versioning, a new version of schema is created by implementing the required revision to the schema version selected for modification that satisfies the validity condition. It may be possible that a single revision may get applied to more than one version of schema that qualifies the valid-time interval. Thus, any version of schema, which has complete or even partial overlap with the valid-time interval of the revisions, would be changed [23].

In valid-time schema versioning, the user is allowed to indicate the valid-time interval for revisions to be applied to the schema, thus providing the support for retro- and proactive revisions. Further, the revisions are not limited to the current schema version only as with transaction time but may be applied to any schema version selected for updation. The operations performed on the versions of schema and their associated data are represented with the help of figures and examples.

3.1 Management of Intensional Data

It is required to manage the multiple versions of schema created in respect to the desired revisions in the schema. Thus, the valid-time interval of the newly created version of schema is defined by the valid-time interval specified for the revision to be applied to the selected version of schema and the validity of any of the older versions of schema which have only partial overlap with the valid-time interval of the revision applied is confined accordingly. Whereas any of the older versions having complete overlap with the valid-time interval of the revision are archived and not deleted.

It is illustrated with the help of an example given in Fig. 1. Suppose we drop an attribute p4 from schema version (SV) with validity [t′, t″]. Figure 1a, b represents the states of schema versions before and after the revision, respectively. In the example, of all the schema versions (SV1 to SV4), only two (SV2 and SV4) are partially overlapped and one (SV3) is completely overlapped by valid-time interval [t′, t″].

The version related to [t1, t2] does not qualify the valid-time interval [t′, t″] and thus, it remains unaffected by the revision. The schema versions related to [t3, t4] and [t5, t6], partially overlaps the valid-time interval [t′, t″], are therefore affected by the revision and divided into two parts. The non-overlapping part remains unaffected by the revision and thus retains all of its attributes, whereas the over-lapping part would drop the attribute p4 and results in (p1..p3, p5) and (p1..p3, p5.. p8), respectively. The revision is also applied to the schema version relative to [t4, t5], which completely overlaps the valid-time interval [t′, t″]. Therefore, p4 is dropped from this schema version and results in (p1..p3, p5..p6), and the old schema version is deleted.

3.2 Management of Extensional Data

The management of extensional data can either be synchronous or non-synchronous with respect to the temporal dimension.

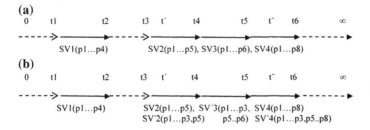

Fig. 1 Representation of states of schema versions before and after applying revisions

Central Data Storage Solution (Synchronous and Non-synchronous)

The central data storage is a single repository which stores all the extensional data using an exhaustive schema [23] that maintains all the attributes introduced by a series of revisions applied to a schema. If the revision to schema results in addition of an attribute, or a temporal dimension, the entire data repository is changed to the revised form. On the other hand, if the revision applied to schema is destructive for, e.g., results in the drop of an attribute, then the revision does not affect the schema and can only be recorded in the meta-schema. This is because no data can ever be discarded from the central storage. Data are therefore stored using the enlarged schema format defined so far, and the history of the revisions are recorded only in the meta-schema.

The original structure of data for each of the versions of schema can be easily restored with the help of data stored in meta-schema. This requires the knowledge about the current format of the central storage (including all the attributes described so far) as well as the format of each version of schema.

Distributed Data Storage Solution

The distributed data storage solution builds distinct data repositories for distinct versions of schema. Each version of schema accesses its own data repository. After applying a revision to schema, a new data repository is created as per the revisions. It is loaded with data records from the repository corresponding to the updated version of schema and modified according to the revisions applied. The data repositories associated with completely overlapped (deleted) versions of schema are also deleted. The data repositories connected to only partially overlapped versions of schema remain unaffected in the case of non-synchronous mapping, or their time stamps are confined in the case of synchronous mapping.

Distributed Data Storage Solution (Synchronous)

For synchronous mapping of the distributed data storage, the validity of data records in all the new repositories is confined to its intersection with the valid-time interval of the version of schema. The only difference with reference to the previous case is the management of data repositories associated with versions of schema that have partial overlap with the updates applied to the schema in case their data contain valid time. In that case, the valid-time interval of the record/tuple must be limited according to the valid-time interval of their corresponding schema version.

However, since the synchronous mapping is applied, if data records in data repository contain valid time, then the time stamps need to be confined accordingly.

Figure 2 shows the effect of a revision (addition of an attribute p3) which overlaps SV1 on [t2, t3], in case of synchronous mapping. A new schema version SV2 is created, which includes the attributes p1, p2 and p3. Figure 2a represents the original data repository SV1 with valid-time interval [t1, t3]. Figure 2c represents the temporal pertinence and the data contents of distributed data repositories. In the distributed data storage solution, creation of SV2 results in a new data repository. The original records are divided as per their valid-time intervals and are accordingly

Fig. 2 Synchronous
mapping of extensional data

P1	P2	From	To
A1	B1	t1	t3
A2	B2	t2	t3

SV1 (t1, t3)
(a) Initial data repository

P1	P2	P3	From	To
A1	B1	-	t2	t3
A2	B2	-	t2	t3

(b) Central data-storage
solution

P1	P2	From	To
A1	B1	t1	t3

SV1 (t1, t2)

P1	P2	P3	From	To
A1	B1	-	t2	t3
A2	B2	-	t2	t3

SV2 (t2, t3)

(c) Distributed data-storage solution

Fig. 3 Non-synchronous
mapping of extensional data

P1	P2	From	To
A1	B1	t1	t3
A2	B2	t2	t3

SV1 (t1, t3)
(a) Initial data repository

P1	P2	P3	From	To
A1	B1	-	t1	t3
A2	B2	-	t2	t3

(b) Central data-storage
solution

P1	P2	From	To
A1	B1	t1	t3
A2	B2	t2	t3

SV1 (t1, t2)

P1	P2	P3	From	To
A1	B1	-	t1	t3
A2	B2	-	t2	t3

SV2 (t2, t3)

(c) Distributed data-storage solution

split among the two repositories. Whereas, in the central data storage solution, the data in the repository are not divided as shown in Fig. 2b.

Distributed Data Storage Solution (Non-synchronous)

In non-synchronous distributed data storage solution, the format of each data repository is selected with respect to the related version of schema. Thus, a new data repository is created and populated by copying the records from the original affected repository as per the revisions suggested for the previous schema (update/drop of an attribute etc.). Figure 3 shows an example for the non-synchronous mapping of extensional data. If we consider the similar schema revision (as applied in Fig. 2) for the non-synchronous case, no divide is required on the valid-time interval for distributed data storage solution. For this example, the central data repository results coincide with the new data repository created for distributed data storage, and it may be noted that the data need not be duplicated in this case.

4   Data Revisions in Valid-Time Schema Versioning

In the following discussion, we illustrate the impact of distributed and central data storage on the modifications of data from different schema versions. Figure 4a, c shows the distributed and central repository of a valid-time relation, respectively. We assume non-synchronous versioning and apply the following changes.

Update the value of attribute B2 to B15 using SV1.
Update the value of attribute B2 to B25 using SV2.

Figure 4b shows the result for the distributed storage and Fig. 4d, e shows the results for the central storage. It may be noted that the sequence of applying the revisions does not affect the final outcome in case of distributed storage, since the data repositories are separated as shown in Fig. 4b. Whereas, the sequence of the revisions have an impact on the output in case of the central storage, as illustrated in Fig. 4d, e.

P1	P2	From	To
A1	B1	t0	t5
A2	B2	t2	t7

P1	P2	P3	From	To
A1	B1	-	t0	t5
A2	B2	-	t2	t7

(a) Distributed data-storage before modification

P1	P2	From	To
A1	B1	t0	t5
A2	B15	t2	t7

P1	P2	P3	From	To
A1	B1	-	t0	t5
A2	B25	-	t2	t7

(b) Distributed data-storage after modification

P1	P2	P3	From	To
A1	B1	-	t0	t5
A2	B2	-	t2	t7

(c) Central data-storage before modification

P1	P2	P3	From	To
A1	B1	-	t0	t5
A2	B15	-	t2	t7

B2 = 25 (1st operation)
B2 = 15 (2nd operation)

(d) Central data-storage after modification

P1	P2	P3	From	To
A1	B1	-	t0	t5
A2	B25	-	t2	t7

B2 = 15 (1st operation)
B2 = 25 (2nd operation)

(e) Central data-storage after modification

Fig. 4   Valid-time non-synchronous versioning

5 Conclusion

The proposed approach allows keeping track of data and schema history with valid time, which guarantees consistency of data and provides with alternatives that optimizes the space utilization for data storage. The valid-time versioning of schema and data as well as the interaction between them have been analyzed so as to explore the possible opportunities for the modeling of a temporal data warehouse that provides support for schema evolution. The selection for the kind of interaction depends on the degree of independence that is required. Various alternatives for the storage solutions of the extensional data have also been discussed. The selection of the type of the storage solution would depend on the available and required space for storage; for e.g., central data storage should be the choice if the available storage space is limited. The central storage solution does not duplicate data, even though it enlarges the format of the schema after applying the required revisions, thus may require more storage space. Whereas, the distributed storage requires to duplicate the data from the affected pool(s). The merits and demerits of each of the approaches are thoroughly examined in [23].

The distributed data storage solution does have an impact on the data model. Different data repositories would allow for independent evolutions of the same data as the revisions can be applied through different versions of the schema. In case of synchronous distributed storage, special applications that use both old and new schema versions need to be programmed again in order to reconstruct the entire history, because the data are partitioned in synchronicity with the different versions of schema. Whereas, non-synchronous mapping maintains a complete history of data for each version of schema. Thus, old applications may continue to operate correctly on old data as before, but new applications need be designed for new data that include the new attributes/data. Further, using synchronous distributed storage, as the data get updated only in the current data repository we cannot query data using an old definition of schema.

Moreover, if multi-schema queries are presented to the system, and if we are working with the central data storage solution, then the answer to the query would consist of only one table. However, if distributed data storage is applied, then a new schema need to be constructed including all the attributes required to provide the answer for the query. The handling of queries in distributed storage environment requires various data repositories and, thus, would involve some filtering. Future work would includes discussions on bi-temporal versioning in temporal data warehouses.

References

1. Inmon, W. H.: Building the Data Warehouse. Wiley and Sons. (1996).
2. Kimball, R., Ross, M.: The Data Warehouse Toolkit. John Wiley & Sons. (2002).
3. Benítez-Guerrero, E., Collet, C., Adiba, M.: The WHES Approach to Data Warehouse Evolution. Digital Journal e-Gnosis [online], http://www.e-gnosis.udg.mx, ISSN No. 1665–5745 (2003).
4. Blaschka, M., Sapia, C., Hofling, G.: On Schema Evolution in Multidimensional Databases. Proc. of Int. Conference on Data Warehousing and Knowledge Discovery (DaWaK). Lecture Notes in Computer Science, Vol. 1676. (1999) 153–164.
5. Blaschka, M.: FIESTA: A Framework for Schema Evolution in Multidimensional Information Systems. In 6th CAiSE Doctoral Consortium. Heidelberg (1999).
6. Hurtado, C. A., Mendelzon, A. O., Vaisman, A. A.: Maintaining Data Cubes under Dimension Updates. Proc. of Int. Conference on Data Engineering (ICDE). (1999) 346–355.
7. Hurtado, C. A., Mendelzon, A. O., Vaisman, A. A.: Updating OLAP Dimensions. Proc. Of ACM Int. Workshop on Data Warehousing and OLAP (DOLAP). (1999) 60–66.
8. Kaas, Ch. K., Pedersen, T. B., Rasmussen, B. D.: Schema Evolution for Stars and Snowflakes. Proc. of Int. Conference on Enterprise Information Systems (ICEIS). (2004) 425–433.
9. Mendelzon, A. O., Vaisman, A. A.: Temporal Queries in OLAP. Proc. of Int. Conference on Very Large Data Bases (VLDB). (2000) 242–253.
10. Vaisman, A., Mendelzon, A.: A Temporal Query Language for OLAP: Implementation and Case Study. Proc. of Workshop on Data Bases and Programming Languages (DBPL), Lecture Notes in Computer Science, Vol. 2397. Springer, (2001) 78–96.
11. Bębel, B., Eder, J., Konicilia, C., Morzy, T., Wrembel, R.: Creation and Management of Versions in Multiversion Data Warehouse. Proc. of ACM Symposium on Applied Computing (SAC), (2004) 717–723.
12. Bębel, B., Królikowski, Z., Wrembel, R.: Managing Multiple Real and Simulation Business Scenarios by Means of a Multiversion Data Warehouse. Proc. of Int. Conference on Business Information Systems (BIS), Lecture Notes in Informatics (2006) 102–113.
13. Bębel, B., Wrembel, R., Czejdo, B.: Storage Structures for Sharing Data in Multi-version Data Warehouse. Proc. of Baltic Conference on Databases and Information Systems, (2004) 218–231.
14. Body, M., Miquel, M., Bédard, Y., Tchounikine, A.: A Multidimensional and Multiversion Structure for OLAP Applications. Proc. of ACM Int. Workshop on Data Warehousing and OLAP (DOLAP). (2002) 1–6.
15. Body, M., Miquel, M., Bédard, Y., Tchounikine, A.: Handling Evolutions in Multidimensional Structures. Proc. of Int. Conference on Data Engineering (ICDE). (2003) 581.
16. Golfarelli, M., Lechtenbörger, J., Rizzi, S., Vossen, G.: Schema Versioning in Data Warehouses. Proc. of ER Workshops. Lecture Notes in Computer Science, Vol. 3289. (2004) 415–428.
17. Malinowski, E., Zimanyi, E.: A Conceptual Solution for Representing Time in Data Warehouse Dimensions. In 3rd Asia-Pacific Conf. on Conceptual Modelling, Hobart Australia (2006) 45–54.
18. Chamoni, P., Stock, S.: Temporal Structures in Data Warehousing. Proc. of Int. Conference on Data Warehousing and Knowledge Discovery (DaWaK). Lecture Notes in Computer Science, Vol. 1676. (1997) 353–358.
19. Eder, J., Koncilia, C.: Changes of Dimension Data in Temporal Data Warehouses. Proc. of Int. Conference on Data Warehousing and Knowledge Discovery (DaWaK). Lecture Notes in Computer Science, Vol. 2114. (2001) 284–293.
20. Eder, J., Koncilia, C., Morzy, T.: The COMET Metamodel for Temporal Data Warehouses. Proc. of Conference on Advanced Information Systems Engineering (CAiSE). Lecture Notes in Computer Science, Vol. 2348. (2002) 83–99.

21. Letz, C., Henn, E. T., Vossen, G.: Consistency in Data Warehouse Dimensions. Proc. of Int. Database Engineering and Applications Symposium (IDEAS). (2002) 224–232.

22. Schlesinger, L., Bauer, A., Lehner, W., Ediberidze, G., Gutzman, M.: Efficiently Synchronizing Multidimensional Schema Data. Proc. of ACM Int. Workshop on Data Warehousing and OLAP (DOLAP). (2001) 69–76.

23. De Castro, C., Grandi, F., Scalas, M., R.: On Schema Versioning in Temporal Databases. In: Clifford, S., Tuzhilin, A. (eds.): Recent Advances in Temporal Databases. Springer-Verlag, Zurich Switzerland (1995) 272–294.

24. Grandi, F., Mandreoli, F., Scalas, M.: A Generalized Modeling Framework for Schema Versioning Support. In Australasian Database Conference (2000) 33–40.

25. Grandi, F., Mandreoli, F., Scalas, M.: A Formal Model for Temporal Schema Versioning in object oriented Databases. Technical report CSITE-014-98, CSITE - CNR (1998).

26. Serna-Encinas, M.-T., Adiba, M.: Exploiting Bitemporal Schema Versions for Managing an Historical Medical Data Warehouse: A Case Study. In Proc. of the 6th Mexican Int. Conf. on Computer Science (ENC'05), IEEE Computer Society. (2005) 88–95.

27. SOO, M.: Bibliography on Temporal Databases, ACM SIGMOD Record, March (1991).

28. Agrawal, R., Gupta, A., Sarawagi, S.: Modeling Multidimensional Databases. IBM Research Report, IBM Almaden Research Center (1995).

29. Eder, J.: Evolution of Dimension Data in Temporal Data Warehouses. Technical Report 11, Univ. of Klagenfurt, Dep. of Informatics-Systems (2000).

30. Eder, J., Koncilia, C., Morzy, T.: A Model for a Temporal Data Warehouse. In Proc. of the Int. OESSEO Conference. Rome Italy (2001).

LVCMOS-Based Frequency-Specific Processor Design on 40-nm FPGA

Aditi Moudgil and Jaiteg Singh

Abstract This paper shows a power efficient processor design using LVCMOS and HSTL I/O standards. We have compared the performance of our processor through different frequencies. To increase the performance, code of the processor has been remodified by using I/O standards. On remodification of code at different operating frequencies, a significant reduction in logic power, clock power, IO power, signal power and thus total power dissipation is observed. By doing power analysis, the best results are obtained by using LVCMOS12. This paper provides a novel FPGA-based design for power efficient processor.

Keywords Processor · FPGA · HSTL · LVCMOS · Thermal aware design · Energy efficient design

1 Introduction

Computing device frameworks are turning out to be progressively universal and a part of the worldwide base, bringing about expensive establishments of PC frameworks. With the increase in computing applications and need of IT among individuals, the productive advancements are being created. PCs and other computing applications positively have a major effect universally; however, the other part of the innovation utilization is disturbing. For whatever time that PCs are running, oblige energy is being utilized, resulting in CO_2 discharge [1], and also other gases, which are effortlessly infused into our biological system [2].

A. Moudgil (✉) · J. Singh
Department of Computer Science, Chitkara University, Chandigarh, Haryana, India
e-mail: aditi.modgil@chitkara.edu.in

J. Singh
e-mail: jaiteg.singh@chitkara.edu.in

© Springer Nature Singapore Pte Ltd. 2018   513
P.K. Sa et al. (eds.), *Progress in Intelligent Computing Techniques: Theory,*
Practice, and Applications, Advances in Intelligent Systems and Computing 518,
DOI 10.1007/978-981-10-3373-5_51

Table 1 Factors versus actions leading to green computing

Factors	Solutions
High company energy costs	Using high energy rated stars for electronic products
Degraded life of hardware	Switching off the display when not in use
High IT maintenance activities	Power supply management
Occupied space on the data centre floor	Using storage efficient techniques

Fig. 1 Techniques to optimize processor

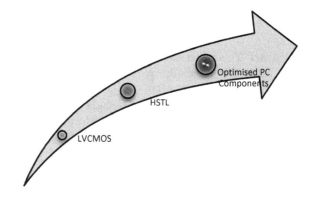

Xilinx ISE [3] (Integrated Synthesis Environment) is a product apparatus created by Xilinx for union and examination of HDL outlines, empowering the engineer to orchestrate ("accumulate") their plans, perform timing investigation, inspect RTL charts, reproduce a configuration's response to various boosts, and arrange the objective gadget.

Green computing: [4, 5] green computing is the study of making IT resources energy efficient. It includes manufacturing, designing, and efficiently using computers and associated subsystems. To promote the idea of green computing, energy star rating [6] was introduced for electronic devices; energy star rating of the electronic devices effects imperishable computing. The more is the star rating, the less is the power consumption. The factors escalating carbon emissions and probable solutions are briefed in Table 1.

HSTL: [7] high-speed transceiver logic (HSTL) standard deals with the test conditions of device designs that operate in logic range of 0 V to 1.5 V and signals either single ended or differential ended.

LVCMOS: low-voltage complementary metal oxide semiconductor [8] operates on the lowest voltage, and it is most productive and is known for lowest power dissipation.

We have tested our processor at I/O standard (LVCMOS and HSTL) with different frequencies on it. And, energy efficiency was analysed in respect to different LVCMOS and HSTL IO standard used in processor (Figs. 1 and 2).

The processor shown in Fig. 3 accepts 8 bit data input signal and outputs an 8 bit data output.

Fig. 2 Components of
processor

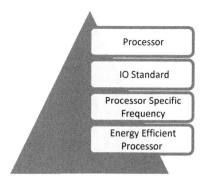

Fig. 3 RTL of processor

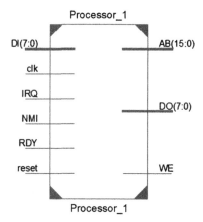

2 Thermal and Power Analysis

We designed a processor. Then we analysed the power dissipated at different processor frequencies namely (I3, I5, I7, and Motorola) as shown in Table 2.

On 1.2 GHz, there is a reduction of 3.22% in the clock power, 30% reduction in the logic power, 28% reduction in the signal power, 49.68% reduction in the IOs power, 0.92% reduction in the leakage power when we use LVCMOS12 instead of HSTL_I as shown in Fig. 4 and Table 3.

Finally, we get 23.79% reduction in the total power when LVCMOS12 is used.

After applying an I/O standard LVCMOS12 on 1.7 GHz processor, reduction of 17.65% is observed in clock power and reduction of 23.07% is observed in logic power and 30.30% in signal power as shown in Fig. 5 and Table 4. Also IO power is depreciated by 44.33%, leakage power by 0.92%, and thus total power is depreciated by 23.79% when LVCMOS12 is used instead of HSTL_II as shown in Fig. 5 and Table 2.

Table 2 Operating frequency of different processor

Processor	Frequency (GHz)
I7	3.0
I5	3.6
I3	2.5
Moto-E	1.2
Moto-X	1.7

Fig. 4 Power dissipation by a processor on 1.2 GHz

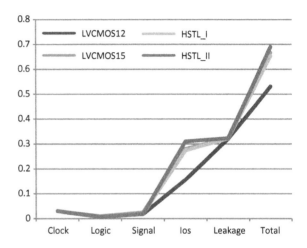

Table 3 Power dispersion by a processor at 1.2 GHz

	Clock	Logic	Signal	IOs	Leakage	Total
LVCMOS12	0.03	0.007	0.018	0.156	0.32	0.531
LVCMOS15	0.030	0.010	0.025	0.280	0.322	0.667
HSTL_I	0.031	0.007	0.019	0.274	0.322	0.653
HSTL_II	0.031	0.007	0.019	0.31	0.323	0.690

Fig. 5 Power dissipation by a processor o1.7 GHz

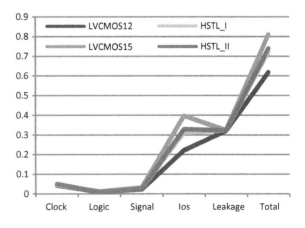

Table 4 Power dispersion by a processor at 1.7 GHz

	Clock	Logic	Signal	IOs	Leakage	Total
LVCMOS12	0.042	0.01	0.023	0.221	0.321	0.618
LVCMOS15	0.044	0.013	0.033	0.397	0.324	0.811
HSTL_I	0.051	0.01	0.027	0.312	0.323	0.723
HSTL_II	0.051	0.01	0.027	0.329	0.323	0.740

Fig. 6 Power dissipation by a processor on 2.5 GHz

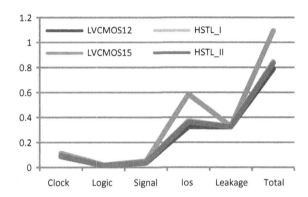

Table 5 Power dispersion by a processor at 2.5 GHz

	Clock	Logic	Signal	IOs	Leakage	Total
LVCMOS12	0.088	0.015	0.037	0.325	0.323	0.788
LVCMOS15	0.113	0.020	0.051	0.584	0.327	1.094
HSTL_I	0.093	0.015	0.041	0.373	0.324	0.846
HSTL_II	0.093	0.015	0.041	0.361	0.324	0.834

On 2.5 GHz, frequency reduction in clock power, logic power, signal power by applying I/O standard LVCMOS12 is observed as 22.12, 25, 27.45%. Minimization in IOs power and leakage power is 44.35 and 1.22%. As a result reduction of 27.97% in the total power is observed when LVCMOS12 is used instead of different HSTL IO standards as shown in Fig. 6 and Table 5.

On 3.0 GHz, frequency decrement in clock power (18.03%), logic power (21.74%), signal power (28.33%) is analysed as shown in Fig. 7 and Table 6.

Also 45.57% reduction in IOs power and 1.51% reduction in leakage power are observed as shown in Fig. 7 and Table 6.

As a result, 28.88% reduction in the total power is observed when LVCMOS12 is used instead of HSTL_II.

On 3.6 GHz, there is a reduction of 20.57% in the clock power, 22.22% reduction in the logic power, 21.21% reduction in the signal power, 51.90% reduction in the IOs power, 1.50% reduction in the leakage power as shown in Fig. 8 and Table 7.

Fig. 7 Power dissipation by
a processor on 3 GHz

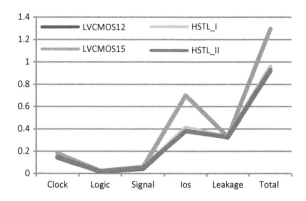

Table 6 Power dispersion by a processor at 3.0 GHz

	Clock	Logic	Signal	IOs	Leakage	Total
LVCMOS12	0.145	0.018	0.043	0.39	0.324	0.921
LVCMOS15	0.183	0.023	0.060	0.700	0.329	1.295
HSTL_I	0.15	0.018	0.051	0.411	0.325	0.956
HSTL_II	0.15	0.018	0.051	0.381	0.325	0.926

Fig. 8 Power dissipation by
a processor on 3.6 GHz

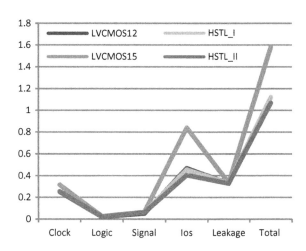

Table 7 Power Dispersion by a processor at 3.6 GHZ

	Clock	Logic	Signal	IOs	Leakage	Total
LVCMOS12	0.251	0.021	0.052	0.468	0.327	1.12
LVCMOS15	0.316	0.027	0.066	0.840	0.332	1.582
HSTL_I	0.255	0.022	0.061	0.456	0.327	1.121
HSTL_II	0.255	0.022	0.061	0.404	0.327	1.068

Thus, 32.49% reduction in the total power is observed when we use LVCMOS12 instead of HSTL_I as shown in Fig. 8 and Table 7.

3 Conclusion

When remodification of the code of processor was done and energy utilization was calculated at 3.6 GHz operating frequency, we observed a reduction of 20.57% in the clock power, 22.22% reduction in the logic, 21.21% reduction in the signal power, 51.90% reduction in the IOs power, 1.50% reduction in the leakage power and thus 32.49% reduction in the total power when LVCMOS12 was used instead of other standards. So, the paper concludes that LVCMOS12 I/O standard is the best suited IO standard for efficiency as compared to LVCMOS15, HSTL_I and HSTL_II when we are designing an energy efficient processor on FPGA.

4 Future Scope

We used LVCMOS and HSTL I/O standard, and we can possibly remodify the code using other IO standards such as GTL, PCIX, PCI33, PCI66 and many other I/O standards for making an energy proficient processor. We can use FPGA Virtex 5, instead of Virtex 6. Other frequencies can be used to test these IO standards.

References

1. Murugesan, San. "Harnessing green IT: Principles and practices." IT professional 10.1 (2008): 24–33.
2. Harmon, Robert, et al. "From green computing to sustainable IT: Developing a sustainable service orientation." System Sciences (HICSS), 2010 43rd Hawaii International Conference on. IEEE, 2010.
3. Book, Xilinx Data. "Xilinx." Inc., San Jose, CA (1998).
4. Sharmila Shinde, S., S. Simantini Nalawade, and A. Ajay Nalawade. "Green computing: go green and save energy." Int. J. Adv. Res. Comput. Sci. Softw. Eng 3.7 (2013): 1033–1037.
5. Garg, Saurabh Kumar, Chee Shin Yeo, and Rajkumar Buyya. "Green cloud framework for improving carbon efficiency of clouds." Euro-Par 2011 Parallel Processing. Springer Berlin Heidelberg, 2011. 491–502.
6. http://www.energyrating.gov.au/.
7. Virtex-6 FPGA Select IO Resources User Guide http://www.xilinx.com/support/documentation/userguides/ug361.pdf.
8. Saini, Rishita, et al. Low Power High Performance ROM Design on FPGA Using LVDCI I/O.

Improvised Bat Algorithm for Load Balancing-Based Task Scheduling

**Bibhav Raj, Pratyush Ranjan, Naela Rizvi, Prashant Pranav
and Sanchita Paul**

Abstract The development of computing system has always focused on performance improvements driven by the demand of applications by customers, scientific and business domain. Cloud computing has emanated as a new trend as well as required domain for the efficient usage of computing systems. As the applications operating in cloud environments are becoming popular, the load is also rising on the servers and the traffic is increasing rapidly. In this paper, a new metaheuristic algorithm has been discussed known as improvised Bat algorithm and the case study of it is explained with proper example. The improvised Bat algorithm works on Min-Min, Max-Min and Alpha-Beta pruning algorithm for population generation and then uses the Bat algorithm for determining the sequence of execution of tasks to keep it minimum.

Keywords Algorithm · Cloud · Computing · Task · Load

B. Raj (✉) · P. Ranjan · N. Rizvi · P. Pranav · S. Paul
Department of Computer Science and Engineering, Birla Institute of Technology,
Mesra, Ranchi 835215, Jharkhand, India
e-mail: raj.bibhav@gmail.com

P. Ranjan
e-mail: pratyushranjan@live.com

N. Rizvi
e-mail: naelarizvi92@gmail.com

P. Pranav
e-mail: prashantpranav19@gmail.com

S. Paul
e-mail: sanchita07@gmail.com

© Springer Nature Singapore Pte Ltd. 2018   521
P.K. Sa et al. (eds.), *Progress in Intelligent Computing Techniques: Theory,*
Practice, and Applications, Advances in Intelligent Systems and Computing 518,
DOI 10.1007/978-981-10-3373-5_52

1 Introduction

Cloud computing focuses on maximizing the capabilities of the shared resources. Cloud resources are not only shared among multiple users but are also dynamically reallocated as per demand. Load balancing distributes the workload among all the computing resources. In cloud computing, task scheduling determines proper sequencing of tasks to virtual machines for maximum utilization of the resources and increasing throughput. It focuses on keeping the available resources equally busy and avoids overloading any one of the machine or resource with many tasks. Load balancing is done so that every virtual machine in cloud system does the same amount of work, increase the throughput and minimize the response time. The algorithm schedules the task in such a way that the load is balanced effectively. The improvised Bat algorithm works using the Min-Min and Max-Min optimization technique to allocate and reduce the execution time of the tasks requested to the VMs. After the population is optimized, the Bat algorithm executes and determines the sequence of execution of the allotted tasks keeping the execution time minimum, thus reducing the load. Section 2.1 explains the proposed algorithm, and Sect. 3 shows the case study of the algorithm with Sect. 4 detailing the result obtained through case study. Section 5 gives the conclusion and limitations of the proposed work and also suggests some future work.

2 Proposed Methodology

2.1 Improvised Bat Algorithm

a. The Proposed system is working on the new metaheuristic method called Bat Algorithm.
b. The improvisation is also done with the method by utilizing the Min-Min and Max-Min algorithm for generating an enhanced and optimized population of the virtual bats.
c. Min-Min algorithm: It enumerates the minimum finishing time of the tasks and then selects the minimum from them. It then schedules the resource taking minimum execution time, and the available time is added to all other tasks.
d. Max-Min algorithm: It computes the maximum finishing time of the tasks and then selects the minimum from them. It then schedules the resource having minimum execution time, and the available time is added to all.
e. Population generation: It generates the population on the basis of scheduling
f. List keeping the processing time and availability time of each node.
g. By applying the Min-Min and Max-Min, a more preferable population is generated for a better optimized result. Now, pulse frequency (f) for each node (Xi) is initialized; f lies in [0, fmax].

h. Also, pulse rates (r) and loudness (Ai) are also initialized;

   I. 'r' lies in [0,1]: 0—no pulse and 1—max pulse.
   II. 'Ai' lies in A0—1 and Amin—0.

i. Now, the Bat algorithm is applied till it reaches an optimum result. The Bat algorithm works on the property of echolocation and the frequency of the ultrasonic waves which they produce.

j. It is the nature of bat that they compute their target as well as any obstacles in the flying path with the rate of emission and waves produced by them for guidance. They decrease the loudness or frequency of sound produced if they reach their prey and stops temporarily the emission of the waves.

2.2 Pseudocode

Objective function:-f(x)= [x1,x2,……, xd]t
Initialized population Xi; I = 1,2,……,n and velocity Vi
Apply Min-Min and Max-Min algorithm
Reconsider the population for further process
The pulse rate ri and loudness Ai is initialized
Initialize pulse frequency fi for each node Xi
While(t<maximum no. of generation)
      Generate the solution by making adjustments in frequency
      Update the velocity, location/solution [fromequ–2&4]
      if(rand>ri)
          opt for a solution from the best solution
          muster a local solution from the selected solution
      end if
      if(rad.<Ai & f(xi)<f(x*)) Acquire
          the new solution raiser
          and reduce Ai.
      End if
      Order the bats and acquire the current best x*
End while
Post process result ,visualization
Virtual Bats movement:-
      $F_i = f_{min} + (f_{max} - f_{min}) * b$ -------2]
      $V_i(t) = V_i(t-1) + (X_i(t) - X^*) * fi$-------[3]
      $X_i(t) = X_i(t-1) + V_i(t)$--------[4]
      b= [0,1]
New solution:-
      $X_{new} = X_{old} + eA(t)$
      e = [-1,1]
      A(t) = <A(t)>-----------avg. loudness at time 't'

3 Case Study

Phase I: The case study is drawn taking '10' virtual machines and '5' tasks having different makespan on different VMs. The initial population for the algorithm was taken from this matrix (ref. Table 1.), and then, the Min-Min, Max-Min and Alpha-Beta pruning algorithm was applied which gave different result for different conditions of allotment to the VMs.

Phase II: Now, for allotment the optimization algorithms are used. The results thus we get for the allotment of the tasks for Min-Min and Max-Min were different and are shown below:

In Min-Min and Max-Min scheduling technique, makespan means the largest operating time of any machine on which the tasks are requested. In Fig. 1 and Fig. 2, Y-axis denotes the operating times of the machines and X-axis denotes the different machines. For Min-Min—'240' and Max-Min—'230', for '5' tasks/cloudlets allotted to '10' VMs. One technique may outperform the other, and assignments of

Table 1 Execution times

	T1	T2	T3	T4	T5
VM1	160	110	200	110	230
VM2	180	130	210	190	290
VM3	200	150	190	120	240
VM4	210	170	220	180	280
VM5	250	190	180	130	250
VM6	280	210	230	170	270
VM7	170	230	170	140	260
VM8	150	250	240	160	210
VM9	220	270	160	150	200
VM10	240	290	250	100	190

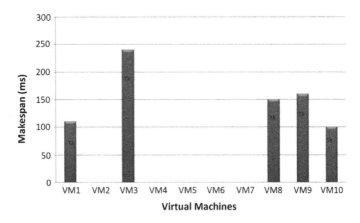

Fig. 1 Min-Min algorithm assigning tasks

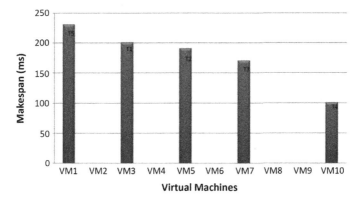

Fig. 2 Max-Min algorithm assigning tasks

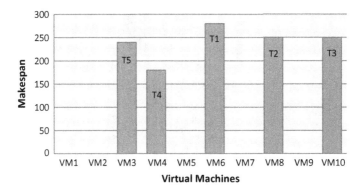

Fig. 3 Tasks assignment using Alpha-Beta pruning

tasks on VMs may change. Population generated for each method is compared keeping the VMs fixed and varying the tasks.

Alternatively, we can do the combined assignment in a more computation reducing method known as Alpha-Beta pruning. In this method, the search is done in a DFS mode and the current best of the node traversed is taken either as alpha (if encountered a maximizing node) and beta (if encountered a minimizing node). This method generates an optimal maximized result. This method is used to do the assignment of the tasks at each stage. In alpha-beta pruning, the maximum makespan calculated was '250' ms (Fig. 3).

Phase III: **Execution of Bat Algorithm for each Allotment**:

In phase III, the Bat algorithm is executed for each of the allotments done by the Min-Min, Max-Min and Alpha-Beta to the VMs. In this phase, the execution times or the processing time of each of the tasks in each allotment is compared. The following tables discuss the processing times, and the comparative studies of the tasks are displayed (Tables 2, 3 and 4).

Table 2 Bat algorithm with Min-Min allotment

Iteration 1		Iteration 2		Iteration 3		Iteration 4		Iteration 5	
Velocity	Execution time	Velocity	Execution time	Velocity	Execution time	Velocity	Execution time	Velocity	Execution time
V1 = 45	T1 = 45	V1 = 56.25	T1 = 101.25	V1 = 83.43	T1 = 185.18	V1 = 139.36	T1 = 324.05	V1 = 251.08	T1 = 575.13
V2 = 33	T2 = 33	V2 = 35.25	T2 = 68.25	V2 = 41.80	T2 = 110.05	V2 = 55.47	T2 = 165.52	V2 = 82.97	T2 = 248.49
V3 = 48	T3 = 48	V3 = 61.50	T3 = 109.50	V3 = 93.83	T3 = 203.33	V3 = 159.97	T3 = 363.30	V3 = 292.99	T3 = 655.84
V4 = 30	T4 = 30	V4 = 27.75	T4 = 57.75	V4 = 27.99	T4 = 85.74	V4 = 27.99	T4 = 113.73	V4 = 27.99	T4 = 141.72
V5 = 72	T5 = 72	V5 = 103.50	T5 = 175.50	V5 = 177.33	T5 = 352.83	V5 = 327.56	T5 = 680.39	V5 = 628.55	T5 = 1308.94
Best	30	Best	57.75	Best	85.74	Best	113.73	Best	141.72

Table 3 Bat algorithm with Max-Min allotment

Iteration 1		Iteration 2		Iteration 3		Iteration 4		Iteration 5	
Velocity	Execution time	Velocity	Execution time	Velocity	Execution time	Velocity	Execution time	Velocity	Execution time
V1 = 40	T1 = 40	V1 = 49.86	T1 = 89.86	V1 = 79.73	T1 = 175.58	V1 = 119.66	T1 = 298.75	V1 = 229.84	T1 = 545.12
V2 = 34	T2 = 34	V2 = 36.05	T2 = 70.05	V2 = 43.47	T2 = 117.15	V2 = 59.77	T2 = 171.31	V2 = 89.35	T2 = 261.70
V3 = 46	T3 = 46	V3 = 60.05	T3 = 106.05	V3 = 90.56	T3 = 200.03	V3 = 148.37	T3 = 352.40	V3 = 278.68	T3 = 636.80
V4 = 27	T4 = 27	V4 = 24.75	T4 = 51.75	V4 = 25.80	T4 = 77.55	V4 = 25.80	T4 = 103.35	V4 = 25.80	T4 = 129.15
V5 = 68	T5 = 68	V5 = 98.6	T5 = 166.6	V5 = 166.66	T5 = 329.33	V5 = 302.49	T5 = 651.47	V5 = 582.95	T5 = 1098.90
Best	27	Best	51.75	Best	77.55	Best	103.35	Best	129.15

Table 4 Bat algorithm with Alpha-Beta allotment

Iteration 1		Iteration 2		Iteration 3		Iteration 4		Iteration 5	
Velocity	Execution time	Velocity	Execution time	Velocity	Execution time	Velocity	Execution time	Velocity	Execution time
V1 = 84	T1 = 84	V1 = 106.5	T1 = 190.5	V1 = 159.78	T1 = 301.59	V1 = 235.54	T1 = 498.32	V1 = 315.68	T1 = 602.68
V2 = 75	T2 = 75	V2 = 90.75	T2 = 165.75	V2 = 175.78	T2 = 316.54	V2 = 255.74	T2 = 418.64	V2 = 335.32	T2 = 536.75
V3 = 75	T3 = 75	V3 = 90.75	T3 = 165.75	V3 = 175.78	T3 = 316.54	V3 = 255.74	T3 = 418.64	V3 = 335.32	T3 = 536.75
V4 = 54	T4 = 54	V4 = 51.25	T4 = 105.25	V4 = 51.25	T4 = 156.50	V4 = 51.25	T4 = 207.75	V4 = 51.25	T4 = 248.68
V5 = 72	T5 = 72	V5 = 74.25	T5 = 146.25	V5 = 82.78	T5 = 389.53	V5 = 107.65	T5 = 502.33	V5 = 149.36	T5 = 723.52
Best	54	Best	105.25	Best	156.50	Best	207.75	Best	248.68

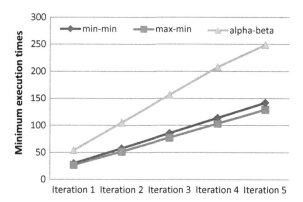

Fig. 4 Comparative chart of the processing times

The above shows all the processing times of the tasks which are allotted to the VMs with different methods of optimization and with the application of Bat algorithm the variation in their makespans.

4 Result Comparison and Discussion

The Bat algorithm is also a heuristic method of optimizing the problems, and it also functions on multiple solution paths. It is a method which is used for generation of an optimal solution.

Figure 4 shows that the red graph line which is Max-Min allotted tasks process is the better one for the execution by the algorithm. The graph chart infers that for the consideration of all the tasks allotment we have to consider the minimum makespan algorithm to be executed by the Bat algorithm which can be seen in the figure.

5 Conclusion, Limitations and Future Direction

In this paper, we discussed the allotment of different tasks on the virtual machines, and their execution times are compared. In the result, we observed that the allotment of tasks and minimum execution time is gained from Min-Min and Max-Min algorithm, while in Alpha-Beta pruning method, the tasks execution is optimistically distributed, but the overall execution/processing time is greater than any other. As part of future work, we would also like to explore some benefits from the optimization methods used and other algorithms such as ACO, PSO.

References

1. Peenaz Pathak, Er. Kamna Mahajan – "A Review on Load Balancing in Cloud Computing"- International Journal Of Engineering And Computer Science ISSN:2319-7242 Volume 4 Issue6 June 2015, PageNo.12333–12339
2. Akshat Dhingra, SanchitaPaul - "A Survey of Energy Efficient Data Centres in a Cloud Computing Environment"-International Journal of Advanced Research in Computer and Communication Engineering Vol.2, Issue10, October 2013
3. Kalyani Ghuge, Prof. Minaxi Doorwar – "A Survey of Various Load Balancing Techniques and Enhanced Load Balancing Approach in Cloud Computing" - International Journal of Emerging Technology and Advanced Engineering (ISSN2250-2459, ISO9001:2008 Certified Journal, Volume 4, Issue 10, October 2014)
4. Kun Li, Gaochao Xu, Guangyu Zhao, Yushuang Dong, Dan Wang– "Cloud Task scheduling based on Load Balancing Ant Colony Optimization"- 2011Sixth Annual China Grid Conference
5. Mayanka Katyal, Atul Mishra – "A Comparative Study of Load Balancing Algorithms in Cloud Computing Environment"- http://www.publishingindia.com
6. Akshat Dhingra and Sanchita Paul– "Green Cloud:Heuristic based BFO Technique to Optimize Resource Allocation"- Indian Journal of Science and Technology, Vol 7(5), 685–691, May 2014
7. Pardeep Kumar, Amandeep Verma – "Independent Task Scheduling in Cloud Computing by Improved Genetic Algorithm"- International Journal of Advanced Research in Computer Science and Software Engineering Volume 2, Issue 5, May 2012, ISSN:2277128X
8. Nada M. Al Sallami – "Load Balancing in Green Cloud Computation" - Proceedings of the World Congress on Engineering 2013 Vol II, WCE2013, July 3–5, 2013, London, U.K.
9. Velagapudi Sreenivas, Prathap. M, Mohammed Kemae – "Load BalancingTechniques: Major Challenge In Cloud Computing– A Systematic Review"
10. Paulin Florence and V. Shanthi – "A Load Balancing Model Using Firefly Algorithm In Cloud Computing"-Journal of Computer Science 10(7):1156–1165, 2014 ISSN:1549-3636
11. Raja Manish Singh Abhishek Kumar Priyanka Karn, Dr. Sanchita Paul– "Task Scheduling in Cloud Computing using ATM Approach"- International Journal of Engineering Research & Technology (IJERT), ISSN:2278-0181, Vol. 4 Issue 04, April-2015
12. Xin-She Yang– "A New Metaheuristic Bat-Inspired Algorithm"
13. Devipriya, S., and C. Ramesh. "Improved Max-min heuristic model for task scheduling in cloud", 2013 International Conference on Green Computing Communication and Conservation of Energy (ICGCE), 2013
14. Zhan, Zhi-Hui, Xiao-Fang Liu, Yue-Jiao Gong, Jun Zhang, Henry Shu-Hung Chung, 8 and Yun Li. "Cloud Computing Resource Scheduling and a Survey of Its Evolutionary Approaches", ACM Computing Surveys, 2015

dMDS: Uncover the Hidden Issues of Metadata Server Design

Ripon Patgiri, Dipayan Dev and Arif Ahmed

Abstract The Big Data is a huge unmanageable set of data that call for store, process, and analyze these data. The Big Data is going colossal per day as well as metadata size. It is the time to bring in the distributed Metadata Server (dMDS) to appease the hunger of data scientist to conquer the dilemma of storing, processing, and analyzing the hysterical data. The dMDS makes stride to carry out the intricate research problems of the file system in Big Data to make impeccable technology. In this study, we investigate properly to get the bottom of the things of Metadata Server (MDS).

Keywords Metadata server · Metadata management · MDS · Large-scale computing · Big Data · Clustered file system · Distributed file system

1 Introduction

The Big Data is emerging with flying color, has become the 'hot potato' in the IT industries as well as academy, and is now the most popular buzzword in the world. The adopting Big Data technology is right as rain in the industries to deal with large-scale computation. The Big Data is cash cow for most of the IT industries, because the Big Data deals with very complex data set. The Big Data comprises volume, velocity, variety [1], veracity, value [2], visualization, virtual, validity,

R. Patgiri (✉)
Department of Computer Science & Engineering, National Institute
of Technology Silchar, Room No. CS30, Silchar 788010, Assam, India
e-mail: ripon@cse.nits.ac.in; ripon.patgiri@gmail.com

D. Dev
PromptCloud Technology Pvt Ltd., Bengaluru, Karnataka, India
e-mail: dev.dipayan16@gmail.com

A. Ahmed
National Institute of Technology Silchar, Silchar, Assam, India
e-mail: arif@cse.nits.ac.in

© Springer Nature Singapore Pte Ltd. 2018   531
P.K. Sa et al. (eds.), *Progress in Intelligent Computing Techniques: Theory,*
Practice, and Applications, Advances in Intelligent Systems and Computing 518,
DOI 10.1007/978-981-10-3373-5_53

variability [3–9], and complexity [10]. The Big Data industries face these 9Vs daily and innovates the way to handle these 9Vs. Thus, the Metadata Server (MDS) is a very small cardinal part of Big Data to enhance the Big Data technology. The rise of Big Data has empowered MDS as exigent and momentous. This rise gives exponential growth of metadata as well as data size. In this case, stand-alone MDS is not enough to accommodate large-sized metadata. Besides, the stand-alone MDS cannot support the millions of metadata request per second. Therefore, the scalability issue demands a clustering of MDS which results distributed MDS (dMDS). The dMDS is a set of hardware dedicated to serve only metadata request which is comprised of low-cost commodity hardware or high-end servers. Therefore, the dMDS assuages the condition of scalability. The purpose of the dMDS must converge with the same price of stand-alone MDS with low-sized metadata. No doubt, stand-alone MDS is far better than the dMDS in low-sized metadata. We have found that the most of successful file system implements only stand-alone MDS. However, some file system uses clustered MDS. In the case of Big Data, the stand-alone MDS cannot scale exabytes of data to be stored in the file system, and the clear disadvantage is its bottleneck in accessing the metadata. The trillions of user request to a single MDS clearly slow down the entire system. This is a well-known issue, and we must obviate this issue through dMDS. Albeit, the dMDS can serve the metadata in a scale of infinite, Dr.Hadoop [11], CephFS [12] for instance, but fails in throughput, and stuck in synchronization. Therefore, this challenge gives us the opportunity to work on dMDS to build a perfect technology and deploy Metadata as a Service (MaaS).

The most dMDS is dealing with fault-tolerant system, namely Dr. Hadoop [11], DROP [13], CalvinFS [14], IndexFs [15], and ShardFS [16]. The fault tolerance can be attained by replication, journaling, erasure code, RAID, and de-duplication. But, the most of the dMDS design outlines vaguely about fault tolerance. Moreover, most systems depend on stand-alone MDS, which is the most prone to failure. We have found that the most of the parameters of dMDS are overlooked in designing dMDS. For example, what are the constructive measurement of the MDS, if the MDS crashes? Does the dMDS design support hot-standby or any failover mechanism? What is the countermeasurement of the faulty MDS? The fault tolerance is the most important factor to take care of before going to design dMDS, since most of the metadata is kept in RAM. Furthermore, the users are cash cow of the company. The user put their faith in the company. The name and fame of the companies involves with faith of users. Once a metadata is lost, the everything will be lost. In addition, the heterogeneity is one of the elements to be weighed. The heterogeneity is also another prominent challenge in distributed system. We cannot make simple assumption that all MDS will process with equally likely same speed. One or two MDS can also be a straggler candidate. The straggler MDS cannot serve more amount of users, while others are serving huge amount of metadata requests. Meantime, the users are waiting for response in the straggler MDS. In this case, the dMDS can scale very high, but fails to deliver the service of client requests.

In this investigation, we have found that the most of the designers have not considered most significant characteristics of MDS, which may harm the file system and performance of MDS very seriously. Due to this overlook, the designs are not deplorable in the real world. The metadata is kept in RAM to retrieve the information as fast as light, but not durable. In this paper, we present the current state-of-the-art MDS parameters in designing.

The article is structured as follows: Sect. 3 describes the taxonomy of Metadata Server. Section 2 exposes MDS operations. Section 4 disposes the various methods used to construct dMDS. Section 5 discloses the various issues in designing dMDS. Section 6 exposes the future possibilities of MDS design. And finally, the article concluded with Sect. 7.

2 Prime Metadata Operations

The number of metadata operations varies depending on the underlying MDS architecture and design. The operations of $O(1)$ may also vary; for example, querying metadata request may take $O(1)$ time complexity, but querying to one MDS produces query to other MDS. Albeit it takes $O(1)$ time complexity, performance differs a lot—IndexFS [15], for example. The create metadata operation is used to create directory, and directories are created in hierarchical order. The nature of the hierarchical directory structure automatically forms tree-based MDS without adapting another strategy to serve metadata queries. Now, turn into tree-based dMDS environment, where create metadata operation performs exactly equal amount of directory with the number of MDS required to store in the RAMs. For example, if dMDS keeps a directory in three MDSs, then it need to perform three create metadata operation and communicate with the three MDSs. On the other hand, the deletion operation deletes directory or file and this operation is rare in most of the distributed or clustered file system. Single deletion operation takes at least $O(log_t n)$ time complexity in tree-based MDS, where t is the maximum children and n is the total number of node. The hash-based deletion takes $O(1)$ time complexity. Another metadata operation, the readdir, is the most frequent metadata operation performed in a file system. The readdir should perform in $O(1)$ time complexity, but it takes $O(log_t n)$ in tree-based metadata. Conversely, the rename metadata operation is very rare in case. Rename metadata operation must reflect to all other MDS containing the same information. Rename metadata operation causes distributed lock in the dMDS to ensure that all other MDS must have same information reflected with newer version of the name of the metadata. Similarly, the move metadata operation is also very rare. The movement of metadata must reflect all the MDS those having the same information.

3 Types of Metadata Servers

The metadata is the heart and soul of a file system which contains information about other data for frequent use. Metadata is key to ensuring that resources will survive and continue to be accessible into the future [17]. The metadata is the most crucial part of a file system to enhance the performance. The metadata can define scalability and performance of a file systems.

3.1 Stand-Alone MDS

There is no doubt that the stand-alone MDS is the best, provided the metadata size fits in main memory or the metadata size is very low. This is obvious due to latency issue in MDS cluster. The primary challenge is to avoid dMDS, but it cannot support ultra-large file system, for example, HDFS NameNode [18], GFS, MooseFS, and Quantcast File System. The fact is that the few size of metadata can accommodate terabytes of data to be stored in the storage system. On the contrary, a few GB of data can eat up the intact storage of an MDS, as addressed in [19, 20]. The total size of small-sized files is very less to store in the storage media, but nearly equivalent with the metadata size. Hence, the very small-sized files problem is the burning issue of the stand-alone MDS. Instead, the stand-alone MDS does not have distributed lock inside the MDS, synchronization among the MDS, latency issue, and re-adjustment cost of load balancing. But, it has dozens of disadvantages too, namely hot-standby issue, data recovery issue, bottleneck issue, hotspot issue, and scalability issue. The stand-alone MDS must have some failover mechanism; otherwise, failure or faulty MDS leads entire to cluster down. In this case, the synchronization issue also arises.

3.2 Distributed MDS (dMDS)

The designing dMDS cluster is an open challenge. Many parameters are to be taken care of, and many of these parameters can not be obviated. The parameters of designing dMDS cluster are keeping the lowest communication overhead, infinite scalable, disaster recoverable, no hotspots, the lowest latency, hot-standby (i.e., no downtime), transparency of metadata, capability of fine-grained fault tolerance, very good locality of references, very good load balancing, and very high throughput [11, 13]. The dMDS is cluster of MDS, only for high scalability. This turns into many parameters to be considered, for instance, load balancing, latency issues, hotspot issues, etc. Further, the dMDS often faces the problem of throughput, deployed many resources with low utilization rate becomes barrier in revenue. The dMDS has benefited of scalability, load balancing, and bottleneck. "**At least, the dMDS confronts the problem of synchronization and network access on update. It is impossible to obviate these problems.**" It is impossible to obviate these problems. Moreover, the performance of all MDS and loads is not same. The heterogeneity can be a barrier in dMDS too. Apart from those parameters, the usability of the dMDS is most an important issue. The most system can obviate dMDS, but people have to engage dMDS with their file system. The dMDS is very useful in where the metadata-intensive computing is performed. The industries have enough experience in slowing down with *BTree* or B^+Tree file system in large scale.

4 Methods Used to Design dMDS

The hash-based MDS hashes the file paths and returns to the location of the files. The hash-based MDS has several advantages, such as very good load balancing, fast computation of the location of the data, and easy to design dMDS. The hash-based MDS removes the hotspot problems, but it delivers a very poor locality of references, because the files are scattered throughout the MDS cluster. "**The hash-based MDS has the best load-balancing capability with worst locality of references.**" The main problem of hash-based MDS is renaming and moving a file, and few such operations can cause to slow down the entire cluster, for instance, Dev et al. [11], Xu et al. [13], two-level hash/table-based MDS [21], and Xiao et al. [16]. Nevertheless, the renaming and moving a file is rare in the case. Examples of hash-based MDS are CephFS, Vesta, AbFS, and LustreFS. There are two most useful hash-based MDS methods, and these are the consistent hashing [22] and locality-preserving hashing (Lph) [23]. Another approach, the replica-based dMDS, is good solution, but it gives high latency due to disk read or SSD read, for instance. The dMDS must store its data exclusively in RAM, contrasting to CalvinFS [14] and Wnag et al. [24]. The MRFS [25] deals with the popular metadata entries, which are replicated to other MDSs that query them frequently during a predefined period. Other than replicated, the subtree portioning exhibits [15, 16] very good locality of references, but it shows hotspot problem. The particular MDS can receive a huge metadata request, while others receive none. This phenomenon may cause serious damages to the file systems, and it is not once in a blue moon. Moreover, the most popular file has millions of users to access its metadata. These users request to a single Metadata Server for metadata, results slowdown the server. That particular server becomes bottleneck. Consequently, the other MDSs have none, only like funneling oil in reverse direction. "**The tree partitioning has the best locality of references with least load-balancing capability.**" The static subtree partitioning never balanced the load of a particular server, once partitioned remain same, and fixed MDS server will serve the Metadata. The outstanding characteristic of static subtree partitioning is the excellent cache performance, for instance, AFS, NFS, Coda, and Sprite. The metadata can be foisted somewhere, but, once allocated an MDS, never changed its entire lifetime. On the contrary, to eschew the bottleneck of dMDS, dynamic subtree partitioning balanced the load dynamically. The salient feature of dynamic subtree partitioning also shows excellent cache performance as like static subtree partitioning—CephFS [26], for example.

The Bloom filter was introduced by Burton H. Bloom in 1970 [27], which has got wider application area. The Bloom filter data structure is used to test whether a given item belongs to a set or not. The Bloom filter is used in metadata management and introduces HBA [28, 29]. Further, Group-based Hierarchical Bloom filter Array (G-HBA), CBF [30], and multidimensional Bloom filters [31] are extended works of Bloom filter.

5 Issues in dMDS Design

The MDS plays a very big role in the field of Big Data in solving problem. The MDS can enhance the performance of Big Data, and it has significant corelation with Big Data. The heart, MDS, without that all function of file system stops, because information is kept in the MDS or dMDS. Therefore, the design must take care of every bit of the issues pertained to it. "**A dMDS must comprises of at least high scalability, durability, fine-grained fault tolerance, high throughput, and high availability.**" Moreover, read-efficient dMDS performs better than write-efficient dMDS, while reading metadata and vice versa. The design decision is on the court of designer to design the read-efficient of write-efficient dMDS. "**Read-efficient dMDS and write-efficient dMDS are head and tails of MDS that never comes with single side.**"

The stand-alone MDS clearly shows bottleneck of file system in millions of metadata request. The metadata resides inside RAM. Stand-alone MDS never has the problem of consistency, since no synchronization takes place. Another advantage of stand-alone MDS is maximum throughput. However, the weather never remains same and so, times have changed. The data increases exponentially and fulfilling the users' requirement become nearly impossible by a standalone MDS, and therefore, call for dMDS. Further, the Hadoop NameNode has experienced with excessive metadata generation with very small-sized files. The size of the files is less than 64 MB, and this turns into a huge size of metadata without having a huge size of data to store in storage media. The actual data to be stored and processed is very less, but the metadata size is really high due its vast set of small-sized file. The solution of the small-sized file problem has been introduced by many researchers, for instance, HAR [20] and HAR+ [19]. The small file problem seriously affects the performance of the file system [32], and the article [33] shows how the small-sized file affects in execution time of job.

The fault tolerance is the big issue in MDS designing. For instance, one MDS fails, or faulty, then what will be the corrective steps to be taken care of. What is an administrative cost? Is there the least amount of hiccup time in endurance? Does the design measured single-point failure (such as Apache Hadoop that has a single point of failure)? To overcome this problem, the ultimate solution is *hot-standby* MDS, but this design increases network traffic in the cluster. The rename/modify/write operation causes synchronization issues in dMDS. This is solving one problem by inviting another problems. The replication is one most suitable techniques to design fault-tolerant dMDS, but the replication strategy may slow down the process. Each MDS in dMDS has to synchronize on every update to ensure consistency. The synchronization makes the process to slow down because one MDS has to communicate with other MDS. This process involves network access where a latency and network traffic may be another issue for the system. The replication system incurs lot of network access and may cause network traffic. Therefore, replication is done periodically to reduce the communication cost. Another solution to dMDS is the hot-standby, which provides seamless service when an MDS fails. When an MDS fails, user can get the metadata from hot-standby MDS, but incur communication cost. The MDS has to synchronize on every updates with the master and hot-standby MDS. Furthermore, the journaling system can be used, and it is the most useful techniques in ensuring the data lost and data recovery. The metadata can be journaled on update periodically to ensure fault tolerance. Finally, the de-duplication removes the unnecessary duplicate data, and reduces space requirement to store.

The hotspot is the most frequent access by users in a certain time period. The frequency of file access rate increases exponentially, and after some period (one week, a month, or a year), the frequency of accessing the file decreases. Such kind of problem arises in Google and YouTube very frequently. The hotspot leads an MDS to overload by billions of user request, and other MDSs are sitting idle. Merely, the overloaded MDS cannot serve all requests in time and cause to slow down the entire system. The dMDS is designed apropos to balance the load of all user requests, but results can contrast. The hash-based solution gives an idealistic resolution to this hotspot issues. On the other side, the hash-based places the metadata scatteringly which causes poor locality of reference in cache memory. This is the trade-off designing dMDS. Moreover, the designing an MDS is not an issue at all, but the designing infinite scalability is the greatest challenge. An MDS can hold much information about terabytes of data, and thus, two or more MDSs increase the scalability of the file system. And, therefore, augmenting new MDS can scale many metadata of thousands of petabytes. Summing up, thousands of MDS can serve nearly unimaginable data size. In dMDS, metadata has to be distributed among the MDS and this distribution process needs time to become stable. In the mean time, the progress of a task may be halted due to this process. The metadata distribution must ensure minimal transfer of metadata and minimal communication and become stable as fast as possible. "**At least, some time is required to detect a faulty MDS, since there is no silver bullet**." Furthermore, failure of an MDS causes hiccup in the system. The hot-standby can take over the charge of failure at the drop of a hat.

The communication cost is also a big issue in designing MDS. The designer must ensure minimal communication. The number of RPC should not be more than one or two, because the communication consumes lots of times, even it may takes up to seconds. The communication cost includes network traffic, data lost in network, link failure, bandwidth, and network rotational latency. On the other hand, the detection of liveliness of an MDS is done using heartbeat. If the MDS is unable to respond the heartbeat, then after some time period, the MDS is marked as dead or faulty. It happens due to the network traffic generated by the cluster. Moreover, the problem arises on when the network partitioned among the MDS. In addition, the metadata must consistent, but inconsistency may be tolerated. To maintain the consistency, network communication must perform, which incurs sometime. In-memory MDS does not have durability; if the entire system shuts down, then data will be lost. The load balancing is another issue in MDS design. The hash based has good load balancing, but poor cache performance. The subtree partition is not good at load balancing. The MDS should not be affected by hotspot problem. The MDS must have a capability of balancing the load dynamically. Further, an MDS can crash at any time for any unforeseen reason. An MDS stores its data in RAM, and RAM is volatile in nature.

Therefore, there is a requirement journaling of metadata, or flushing metadata to permanent storage (disk or SSD) to ensure durability of whatever the update is. The disaster recovery of any system is a challenge for past [34] and future. The disaster recovery system depends on design decision of the system. Generally, MDSs are run in a farmhouse where all the other servers are running. What happens, if all MDSs are damaged and irrecoverable? The MDS server performance may be affected by disaster recovery system. The challenge is disaster-recoverable MDS without affecting the MDS performance.

6 Future Vision

The future can be anything, but the data are growing exponentially. The growth of data certainly increase the metadata. In any large-scale system, there must be metadata to enhance the process of retrieving data. One aspect of MDS is that the machine capabilities are increasing, namely processing power, RAM size, cache size, network speed, and many more. Therefore, the future MDS is dMDS, which will outperform the stand-alone MDS. Another aspect of MDS is that the stand-alone MDS outperforms dMDS in the present scenario in the case of limited metadata size. The dMDS should outperform stand-alone MDS in the same conditions. The dMDS is emerging, and the future MDS is dMDS. Further, IDC reports the digital universe data size as 44ZB, while giant companies are approaching to EB, namely Google, Facebook, NSA, and Amazon. Therefore, MDS will be obsolete very soon, and the dMDS is the hot potato of research. Moreover, Metadata as a Service (MaaS) is one of the most prominent upcoming research areas in the field of Big Data. The MaaS can serve many terabytes of metadata, because of dMDS. The dMDS is intrinsic in-memory database where durability also plays a vital role in research. We have found that the current approaches of MDS are only for few GB metadata, but the future is beyond terabytes. The metadata-based computation needs huge RAM size for supporting huge sized metadata and supports millions of request per seconds.

7 Conclusion

In this 360° investigation, we have found that there are dozens of parameters to be considered till now which has been continuously omitting or overlooking. We have shown all weakness of the dMDS design parameters which will be very helpful to the research community in designing of ultra-modern dMDS design. We hope that this investigation serves as a major milestone in the designing of upcoming dMDS design. The dMDS designs are missing lots of consideration, namely durability, consistency, latency, network traffic, heterogeneity, and many more.

References

1. Doug Laney, "3D Data Management: Controlling Data Volume, Velocity, and Variety", Gartner, file No. 949, 6 February 2001, URL: http://blogs.gartner.com/doug-laney/files/2012/01/ad949-3D-Data-Management-Controlling-Data-Volume-Velocity-and-Variety.pdf
2. M Ali-ud-din Khan, M F Uddin, and N Gupta, "Seven Vs of Big Data: Understanding Big Data to extract Value", 2014 Zone 1 Conference of the American Society for Engineering Education (ASEE Zone 1), 3–5 April 2014, DOI:10.1109/ASEEZone1.2014.6820689
3. Gema Bello-Orgaz, Jason J. Jung, and David Camacho, "Social big data: Recent achievements and new challenges", Information Fusion, Volume 28, March 2016, Pages 4559, DOI:10.1016/j.inffus.2015.08.005
4. Jianzheng Liu, Jie Li, Weifeng Li, and Jiansheng Wu, "Rethinking big data: A review on the data quality and usage issues", ISPRS Journal of Photogrammetry and Remote Sensing, In Press, DOI:10.1016/j.isprsjprs.2015.11.006
5. Cheikh Kacfah Emani, Nadine Cullot, and Christophe Nicolle, "Understandable Big Data: A survey", Computer Science Review, volume 17, 2015, pages 7081, DOI:10.1016/j.cosrev.2015.05.002
6. Xiaolong Jin, Benjamin W. Wah, Xueqi Cheng and Yuanzhuo Wang, "Significance and Challenges of Big Data Research", Big Data Research, volume 2, number 2, 2015, pages 5964, DOI:10.1016/j.bdr.2015.01.006
7. Yuri Demchenko, Cees de Laat, and Peter Membrey, "Defining Architecture Components of the Big Data Ecosystem", 2014 International Conference on Collaboration Technologies and Systems (CTS)", 2014, pages 104–112, DOI:10.1109/CTS.2014.6867550
8. landmark.solutions, "The 7 pillars of Big Data", A White Paper of Landmark Solutions, Retrieved 17, December, 2015 from URL: https://www.landmark.solutions/Portals/0/LMSDocs/Whitepapers/The_7_pillars_of_Big_Data_Whitepaper.pdf
9. Eileen McNulty, "Understanding Big Data: The Seven Vs", Retrieved 17, December, 2015 from URL: http://dataconomy.com/seven-vs-big-data/
10. Monica Bulger, Greg Taylor and Ralph Schroeder, "Engaging Complexity: Challenges and Opportunities of Big Data", London: NEMODE, 2014.
11. Dipayan Dev and Ripon Patgiri, "Dr.Hadoop: an infinite scalable metadata management for HadoopHow the baby elephant becomes immortal", Frontiers of Information Technology & Electronic Engineering, volume-17, number-1, pages 15–31, 2016, DOI:10.1631/FITEE.1500015.
12. Sage A. Weil, Kristal T. Pollack, Scott A. Brandt, Ethan L. Miller, "Dynamic Metadata Management for Petabyte-Scale File Systems", Proceedings of the 2004 ACM/IEEE Conference on Supercomputing (SC'04).
13. Quanqing Xu, R V Arumugam, K L Yong, and S Mahadevan, "DROP: Facilitating Distributed Metadata Management in EB-scale Storage Systems", 2013 IEEE 29th Symposium on Mass Storage Systems and Technologies (MSST), 6–10 May 2013, Long Beach, CA, 2013,
14. Alexander Thomson, and Daniel J. Abadi, "CalvinFS: Consistent WAN Replication and Scalable Metadata Management for Distributed File Systems", FAST'15: Proceedings of the 13th USENIX Conference on File and Storage Technologies, February 1619, 2015, Santa Clara, CA, USA., 2015.
15. Kai Ren, Qing Zheng, Swapnil Patil, and Garth Gibson, "IndexFS: Scaling File System Metadata Performance with Stateless Caching and Bulk Insertion", SC14: International Conference for High Performance Computing, Networking, Storage and Analysis, 16–21 Nov. 2014, New Orleans, LA., pages 237–248, 2014.
16. Lin Xiao, Kai Ren, Qing Zheng, and Garth A. Gibson, "ShardFS vs. IndexFS: Replication vs. Caching Strategies for Distributed Metadata Management in Cloud Storage Systems", SoCC '15 Proceedings of the Sixth ACM Symposium on Cloud Computing, August 27–29, 2015, Kohala Coast, HI, USA., pages 236–249, 2015,
17. NISO, "Understanding metadata", NISO Press, 2004, address- Bethesda, MD, ISBN: 1-880124-62-9.

18. HDFS, "HDFS Architecture Guide", [Online], Retrieved on 10 January, 2016, URL: https://hadoop.apache.org/docs/r1.2.1/hdfs_design.html.

19. Dipayan Dev, and Ripon Patgiri, "HAR+: Archive and Metadata Distribution! Why Not Both?", The proceedings of 2014 International Conference on Computer Communications and Informatics, pages 1–6, Coimbatore, India, 8–10 January 2015, DOI:10.1109/ICCCI.2015.7218119.

20. HAR, "Hadoop Archive", [online], retrieved on 20 January, 2016 from, URL: https://developer.yahoo.com/blogs/hadoop/hadoop-archive-file-compaction-hdfs-461.html.

21. Antonio F. Díaz, Mancia Anguita, Hugo E. Camacho, Erik Nieto, and Julio Ortega, "Two-level Hash/Table approach for metadata management in distributed file systems", Journal of Supercomputing, volume 64, issue 1, 2013, pages 144–155, DOI:10.1007/s11227-012-0801-y.

22. David Karger, Eric Lehman, Tom Leighton, Matthew Levine, Daniel Lewin, and Rina Panigrahy, "Consistent hashing and random trees: distributed caching protocols for relieving hot spots on the World Wide Web", The twenty-ninth annual ACM symposium on Theory of computing, May 04–06, 1997, El Paso, TX, USA.

23. A. Chin, "Locality-Preserving Hash Functions for General Purpose Parallel Computation", Algorithmica, Volume-12, number-2, pages 170–181, 1994, DOI:10.1007/BF01185209.

24. Feng Wang, Jie Qiu, Jie Yang, Bo Dong, Xinhui Li, and ying Li, "Hadoop High Availability through Metadata Replication", CloudDB'09, November 2, 2009, Hong Kong, China.

25. Jiongyu Yu, Weigang Wu, Di Yang, and Ning Huang, "MRFS: A Distributed Files System with Geo-replicated Metadata", Algorithms and Architectures for Parallel Processing, volume 8631, pages 273–285, 2014, DOI:10.1007/978-3-319-11194-0_21.

26. Sage A. Weil, Kristal T. Pollack, Scott A. Brandt, and Ethan L. Miller, "Dynamic Metadata Management for Petabyte-Scale File Systems", In the proceedings of the 2004 ACM/IEEE conference on Supercomputing, 2004.

27. Burton Howard Bloom, "Space/time trade-off in hash coding with allowable errors", Communications of the ACM, volume-13, number-7, pages 422–426, 1970, DOI:10.1145/362686.362692.

28. Yifeng Zhu, Hong Jiang, and Jun Wang, "Hierarchical Bloom Filter Arrays (HBA): A Novel, Scalable Metadata Management System for Large Cluster-based Storage", CLUSTER '04, Proceedings of the 2004 IEEE International Conference on Cluster Computing, pages 165–174, 2004.

29. Yifeng Zhu, Hong Jiang, Jun Wang, and Feng Xian, "HBA: Distributed Metadata Management for Large Cluster-Based Storage Systems", IEEE transactions on parallel and distributed systems, volume 19, number 6, pages 750–763, 2008, DOI:10.1109/TPDS.2007.70788.

30. R. Anitha, and Saswati Mukherjee, "CBF: Metadata management in cloud computing", Third International Conference on Computational Intelligence and Information Technology, 2013. CIIT 2013., volume-40, number-8, pages 272–278, DOI:10.1049/cp.2013.2602.

31. Zhisheng Huo, Limin Xiao, Qiaoling Zhong, Shupan Li, Ang Li, Li Ruan, Shouxin Wang, and Lihong Fu, "MBFS: a parallel metadata search method based on Bloomfilters using MapReduce for large-scale file systems", The Journal of Supercomputing, pages 1–27, 2015, DOI:10.1007/s11227-015-1464-2.

32. Dipayan Dev, and Ripon Patgiri, "A Survey of Different Technologies and Recent Challenges of Big Data", Proceedings of 3rd International Conference on Advanced Computing, Networking and Informatics-ICACNI 2015, DOI:10.1007/978-81-322-2529-4_56

33. Dipayan Dev and Ripon Patgiri, "A Deep Dive into the Hadoop World to Explore Its Various Performances", in the book- Techniques and Environments for Big Data Analysis: Parallel, Cloud, and Grid Computing, pp 31–51, DOI:10.1007/978-3-319-27520-8_3

34. Yoichiro Ueno, Noriharu Miyaho, and Shuichi Suzuki, "Disaster recovery mechanism using widely distributed networking and secure metadata handling technology", UPGRADE-CN '09 Proceedings of the 4th edition of the UPGRADE-CN workshop on Use of P2P, GRID and agents for the development of content networks, pages 45–48, 2009, DOI:10.1145/1552486.1552514.

35. Qing Zheng, Kai Ren, and Garth Gibson, "BatchFS: Scaling the File System Control Plane with Client-Funded Metadata Servers", PDSW'14 Proceedings of the 9th Parallel Data Storage Workshop, 2014, DOI:10.1109/PDSW.2014.7.

An Unsupervised Method for Attribute Identification from a Natural Language Query

Rohith Bhaskaran and B.R. Chandavarkar

Abstract Identifying which attributes the user querying is an important step in providing a Natural language (NL) search interface to a relational database. In this paper, we discuss an unsupervised approach for identifying the target database attributes from natural language (NL) query. This is a knowledge base-driven method which can be adopted into any domain with very little effort. Initially, we created a knowledge base using background information about the database domain. Then used a probabilistic algorithm to calculate the semantic dependency between different nodes in the knowledge base. When processing the query, this dependency score will be used to resolve the target attribute.

Keywords Database natural language interface · Database and query processing

1 Introduction

Finding relevant data quickly and efficiently from a relational database requires the end user to be familiar with query languages like SQL and he/she should also be aware of the structure of the underlying database. Unfortunately, in most cases, they are not. So they require additional support from IT experts. For businesses, it is extremely important to get required information in time.

Software Engineers spend 30% of their time creating custom reports for their clients [1]. This can be avoided up to an extend by providing an efficient and approachable search interface to the database. Assisting or automating the search process by providing a natural language interface will be very helpful. This will reduce the response time and involvement of IT experts for this trivial task.

R. Bhaskaran (✉) · B.R. Chandavarkar
National Institute of Technology, Mangalore, Karnataka, India
e-mail: vrarohith@gmail.com
URL: http://cse.nitk.ac.in

B.R. Chandavarkar
e-mail: brcnitk@gmail.com

© Springer Nature Singapore Pte Ltd. 2018   543
P.K. Sa et al. (eds.), *Progress in Intelligent Computing Techniques: Theory,*
Practice, and Applications, Advances in Intelligent Systems and Computing 518,
DOI 10.1007/978-981-10-3373-5_54

According to usability studies of Kaufmann et al. [2] regarding query language interfaces, Natural language interfaces is most acceptable and most preferable over other methods such as GUI-based and menu-driven methods. Even though Natural language search interfaces are not common because of the high costs of their development and domain customizations, which often requires linguistic engineers and domain experts.

In this paper, we discuss an knowledge base-driven method to identify the attributes of the underlying database from the natural language query. This is often the first step in processing a NL query. This proposed method can be customized in to any domain with small cost. Firstly, we present our approach for attribute identification, and in the later part, we evaluate its coverage, portability, and scalability.

2 Related Works

There are several publications discussing the use of Natural language processing to provide a usable interface to relational databases. Majority of them are ontology-based methods.

QuestIO [3] is a question answering interface to query structured data. This works by converting NL queries into formal queries in SeRQL. For this conversion, relation between phrases are gathered from the ontology and ranked with a similarity score. This similarity score is calculated based on position and inverse of the distance in the ontology. This system does not require training.

FREyA [4] is an advancement over QuestIO. FREyA uses syntactic parsing along with ontology-based method described in QuestIO. When FREyA fails to understand a query, it asks the user some questions regarding the query and domain. This interactive feedback and clarification dialogues helps FREyA to improve the accuracy and precision.

In [5] S. Kara et al. discusses about an ontology-based information retrieval system. This system uses a well-prepared ontology on specific domain to answer the queries. Using this ontology and the data, multiple levels of indices are created and stored. Then, these indices will be used to search over the data using Apache Lucene.

3 Approach

Since this is a knowledge base (KB)-driven approach, the first and important step is the preparation of the KB. The first step is creation of KB with the background information from the database domain. Then, this KB will be used by query pre-processing module to identify the target attributes.

3.1 Knowledge Base Creation

Knowledge Base is the critical component in this approach. The quality of knowledge base is a key factor determining the effectiveness and coverage of the entire system. So the KB should be created based on a trusted data source. For our implementation, we used Wikipedia as a trusted data source for the KB. Also, we used a graph-based approach for KB implementation.

First step of KB creation is fetching relevant Wikipedia page according to the database attribute name. In the knowledge base, a ROOT node is created for each of the attribute in the database. Next step is cleaning the data by removing unnecessary HTML tags and annotations. After that the entire content is sentence tokenized and tagged with part of speech (POS) information. This POS information is required for the chunker. In the implementation, we used OpenNLP for POS tagging and chunking. In the next step, we will shallow parse the POS tagged sentences and extract all the noun phrases (NP).

3.1.1 Calculating Dependency Score

We use 'Dependency score' to calculate the dependency or relatedness of noun phrases with each other in the knowledge base. In other words, dependency score of two noun phrases is the probability for those noun phrases to be semantically dependent. For calculating this dependency, we used Pearson's χ^2 test [6]. It is an efficient method to check whether two random variables X and Y are independent or not. So we can use {1 − Probability for X and Y to be independent} as the dependency score [7].

Advantage of using Pearson's Chi square test over frequency count is that this statistic test will normalize the dependencies as probability values. If we are just considering the occurrence frequency count, the dependency score will highly depend on knowledge base contents. That is, if two pairs of words occurs same amount of time together does not mean that they are equally related.

Let X and Y be two random variables. Our null hypothesis, H_0 is, variables X and Y are independent. i.e., $P(X|Y) = P(X)$. We can calculate the χ^2 value using contingency table shown in Table 1 with Eq. 1. According to Pearson, χ'^2 will follow an approximate χ^2 distribution.

Table 1 Contingency table for occurrence of X and Y over a sample size N

$Y|X$	X	$\neg X$	Total
Y	n_{11}	n_{12}	$n_{1.}$
$\neg Y$	n_{21}	n_{22}	$n_{2.}$
Total	$n_{.1}$	$n_{.2}$	N

Table 2 Contingency table for Asia and Continent

X|Y	Continent	¬Continent	Total
Asia	36	14	50
¬Asia	30	25	55
Total	66	39	105

$$\chi'^2 = \frac{N(n_{11}n_{22} - n_{12}n_{21})^2}{n_{1.}n_{2.}n_{.1}n_{.2}} \sim \chi^2 \tag{1}$$

In our case, X and Y will be noun phrases and our sample space will be all the documents collected to KB creation. In KB, separate nodes will be created for each database attribute. These attribute nodes will also be considered for calculating dependency score. For example consider 'Asia' and 'Continent' be 2 words we want to calculate the dependency. Let Table 2 be the contingency table for these 2 words.

$$\chi^2 = \frac{105(36 \times 25 - 14 \times 30)^2}{50 \times 55 \times 66 \times 39} = 3.418 \tag{2}$$

We need to calculate the probability level, α such that $\chi^2 = \chi_c^2$. Where χ_c^2 is the critical chi square value. So α will be the maximum probability value such that our null hypothesis will stay true. In this example $\alpha = 0.0645$. So dependency score for 'Asia' and 'Continent' will be $(1 - \alpha) = 0.9355$. This score will be stored as edge weight for the edge between these nodes in the knowledge base.

If the value of $(1 - \alpha)$ is below a specified threshold, the dependency score for that particular pair will be taken as 0. This is to avoid unreliable dependency scores in the KB.

3.2 Query Preprocessing

This phase uses natural language query, the pre-prepared knowledge base and the meta-data about the database to identify the target attributes in the underlying database. As the first step the algorithm tries to identify the attribute name by only using the database meta-data. If it fails, then it uses the KB for this purpose.

When the database meta-data is not enough to resolve the attribute name, the algorithm will calculate the weights of shortest path between query terms and each ROOT node in the KB (The pseudo code for `shortestPathWeight` is given in Algorithm 1) and returns the ROOT node with largest value. The database attribute corresponding to this ROOT node will be taken as target attribute

```
 1  Function shortestPathWeight()
        input  : The knowledge base G, query and root node rootNode
        output: Total dependency score for the given rootNode and query
 2      weight = 0
 3      foreach term in query.splitAndClean() do
 4          termNodes = G.getNodesContainingTerm('NP', term)
 5          if termNode then
 6              foreach node in termNodes do
 7                  relationships = G.getShortestPath(rootNode, node)
 8                  temp_wt = 0
 9                  foreach rel in relationships do
10                      temp_wt = temp_wt + rel.depScore
11                  end
12                  weight = weight + temp_wt
13              end
14          end
15      end
16      return weight
```

Algorithm 1: Calculate dependency between query terms and root node

The method `shortestPathWeight` calculates the a dependency score for a query with a given root node. This method finds the shortest route (using `getShortestPath()`) from root node to each relevant token in the query (which is returned by `splitAndClean()`). This algorithm will iterate through each edge and calculate the total dependency score for that path. Here, we are considering only the shortest path. For more better results, we can consider all the paths from root to that particular node. The overall weight for a query for a given root node can be defined as Eq. 3.

The weight W, for a query Q and a rootNode R is given as:

$$W = \forall_t \left(W + \sum depScore(x_i, x_{i+1}) \right) \tag{3}$$

where t is the relevant tokens in query Q, and x_i and x_{i+1} are the adjacent nodes in the path from root node R to t. i.e., It will calculate shortest path from each root node to each relevant term in the query. Then calculate sum of dependency scores of each path. The total dependency score of a root node is the sum of dependency scores calculated for all path from that root node.

4   Evaluation

For the evaluation purpose, we used the world database provided by Oracle. To evaluate how well our system works with natural language queries, we created a set of 25 queries, and we divided them into two different groups, queries which selects a

Table 3 Evaluation results

	Group 1		Group 2	
	Proposed system	Correctness (%)	Proposed system	Correctness (%)
Correctly identified	8	80	7	46.7
Partially identified	0	0	5	16.7
Wrongly identified	1	0	2	0
Failed to identify	1	0	1	0
Total	10	80	15	63.4

single database attribute and queries which selects multiple attribute. Also the results are categorized into four as follows.

- Correctly identified: Identified all the attributes correctly.
- Partially identified: Identified some of the attributes correctly but not all. (For this category 50% is considered as correct for calculating correctness of the algorithm.)
- Wrongly identified: Identified attribute is not the target attribute of the query.
- Failed to identify: System returns nothing, since enough information is not available on the KB.

In the first group of 10 questions, 8 of them are correctly identified, 1 of them wrongly identified and for one query the system failed to identify.

For the second group, out of 15 only for 7 queries the attributes are identified correctly. For 5 queries, the system partially identified the attributes. In the remaining 3, two were wrong and for 1 query, the system failed. These results are shown in Table 3.

This system gives an overall correctness of 80% for queries that select a single attribute as output, while it gives correct answer for 63.4% for queries with multiple target attributes.

The accuracy of this system is heavily depends on the quality and coverage the knowledge base. There is no way to make the coverage 100%, but adding data to the KB adaptively will enhance the coverage.

This system can be ported to any domain with very little efforts. The only change required is to change the data sources used for knowledge base creation.

5 Conclusion and Future Work

In this paper, we presented an approach to identify target attributes from a natural language query against an underlying relational database. It utilizes a knowledge base prepared by incorporating domain information. This is our first step in creating a complete NL search interface to relational databases. This method can be further improved by improving the quality and coverage of the knowledge base. In this paper, we used shortest path between root node and the target node to calculate the dependency. Instead, we can find out all paths between these node and select the path with greater dependency score. This will improve the efficiency of the algorithm, but it will take much more time to calculate.

This method along with an efficient algorithm for constrain extraction can be used to develop a full-fledged NL search interface.

References

1. R. Crompton, "Class and stratification", 3rd edition. Cambridge. Journal of Social Policy, 2008
2. Kaufmann, E., Bernstein, A.: "How useful are natural language interfaces to the semantic web for casual end-users?", Proceedings of the Forth European Semantic Web Conference, Innsbruck, Austria, June 2007
3. V. Tablan, D. Damljanovic, and K. Bontcheva, "A Natural Language Query Interface to Structured Information", Proceedings of the 5th European semantic web conference on The semantic web: research and applications, 2008
4. D Damljanovic, M. Agatonovic, and H Cunningham, "FREyA: an Interactive Way of Querying Linked Data Using Natural Language", Proceedings of the 8th international conference on The Semantic Web, 2012
5. S. Kara, O. Alan, O. Sabuncu, S. Akpnar, N.K. Cicekli, F.N. Alpaslan, "An ontology-based retrieval system using semantic indexing", Information Systems, Volume 37, Issue 4, June 2012, Pages 294–305
6. Erich L. Lehmann, Joseph P. Romano, "Testing Statistical Hypotheses", 3rd Edition, Pages 590–598
7. Ping Chen, Chris Bowes, Wei Ding and Max Choly, "Word Sense Disambiguation with Automatically Acquired Knowledge", IEEE Intelligent Systems, Issue 04, Aug 2012, Pages 46–55

Author Index

© Springer Nature Singapore Pte Ltd. 2018
P.K. Sa et al. (eds.), *Progress in Intelligent Computing Techniques: Theory,*
Practice, and Applications, Advances in Intelligent Systems and Computing 518,
DOI 10.1007/978-981-10-3373-5

Printed in the United States
By Bookmasters